Validation of Communications Systems with SDL

Validation of Communications Systems with SDL

The Art of SDL Simulation and Reachability Analysis

Laurent Doldi
TransMeth Sud-Ouest, France

WILEY

Telephone (+44) 1243 779777

Email (for orders and customer service enquiries): cs-books@wiley.co.uk
Visit our Home Page on www.wileyeurope.com or www.wiley.com

Other Wiley Editorial Offices

John Wiley & Sons Inc., 111 River Street, Hoboken, NJ 07030, USA

Jossey-Bass, 989 Market Street, San Francisco, CA 94103-1741, USA

Wiley-VCH Verlag GmbH, Boschstr. 12, D-69469 Weinheim, Germany

John Wiley & Sons Australia Ltd, 33 Park Road, Milton, Queensland 4064, Australia

John Wiley & Sons (Asia) Pte Ltd, 2 Clementi Loop #02-01, Jin Xing Distripark, Singapore 129809

John Wiley & Sons Canada Ltd, 22 Worcester Road, Etobicoke, Ontario, Canada M9W 1L1

Wiley also publishes its books in a variety of electronic formats. Some content that appears in print may not be available in electronic books.

British Library Cataloguing in Publication Data

A catalogue record for this book is available from the British Library

ISBN 0-470-85286-0

Typeset in 10/12pt Times by Laserwords Private Limited, Chennai, India
Printed and bound in Great Britain by Antony Rowe Ltd, Chippenham, Wiltshire
This book is printed on acid-free paper responsibly manufactured from sustainable forestry in which at least two trees are planted for each one used for paper production.

To my parents

To Martine
To Elsa

Contents

Preface

This book is the second in a series; after a first book explaining SDL (Specification and Description Language) and about how to build an SDL model [Doldi01], this book explains how to validate the model by simulation. The two books share the same SDL case study, a simplified version of V.76, a protocol standardized by the ITU (International Telecommunications Union).

I have written this book because, after using ObjectGeode™ and Tau SDL Suite™ (both from Telelogic) for more than ten years, I felt that the numerous powerful features available in such SDL tools needed to be clearly explained to a wide audience, instead of being confined to severe technoweenies.

Readers who want to practise the exercises described in the book must contact the SDL tool vendors (see their updated list on www.sdl-forum.org), who generally provide free licenses for evaluation or cheap licenses for universities.

The first versions of ObjectGeode and Tau SDL Suite including a simulator, named Geode and SDT at that time, have been released around 1989. However, to my knowledge, 13 years after that release, this is the first book published on validation of SDL systems by simulation.

Very few other commercial tools or languages provide such a range of features for the validation and development of communications systems and software.

Some may question the need for this book, as the SDL tools have their own documentation. The answer is that the documentation of each tool, assuming one of your colleagues has not taken it away, contains thousands of pages, which is not always organized to present first the basic simulation notions and then to introduce progressively more advanced features. In the book, every notion presented is illustrated by a hands-on systematic example, which has been actually executed on the two simulation tools, with direct explanations.

Although this book describes how to validate telecommunication systems, it can be used to validate the behavior of other kinds of real-time systems that can be modeled by communicating state machines.

I hope that this book will reveal to students or managers the power of SDL simulation, and will help designers and developers in the validation of their SDL models.

Laurent Doldi

Foreword

'Better', 'faster' and 'cheaper' are the master words nowadays: how to build the best product, spend less and finish in time. Every project manager knows this triptych: every time she or he starts a new project, it could turn into a nightmare. CMM-I, Six Sigma, COCOMO II, MDA, MDD, XML and others are answers that have resulted from different industries involving the development of complex systems.

Modeling techniques have had significant quality and productivity impact in domains ranging from business processes to embedded real-time applications. The Unified Modeling Language (UML), Model Driven Architecture (MDA), Component Based Development (CBD), Use Case Maps, Message Sequence Charts (MSCs), and Specification and Description Language (SDL) all support modeling concepts that help reduce the impedance mismatch between models of the problem domain and designs in the solution domain.

Programming languages are no longer the necessary and sufficient condition to success. All effort is put in product development to better manage the process. Requirement management, system and software architecture, and model development are part of the artifacts of a good system or software development process.

Simulation, one of the most acclaimed requirements to UML 2.0, helps software engineers who simulate the software architecture as well as its design for a better verification and validation of what is being built. Simulation can only be based on a formal language with a clearly defined syntax and semantics. Formal abstract languages such as SDL and the upcoming UML 2.0, are the answers for modelers with concerns such as verifying architecture models and design models.

Why simulate? What to simulate? How to simulate Answers to these questions are found in this book, which is a result of the vast experience Laurent Doldi has acquired in the more than two decades during which he was involved as a consultant engineer, in complex system engineering, in tool development and in teaching classes. Whether your concern is verification or validation, you will find in this book a systematic and practical approach aimed at engineers. It will guide you through the use of tools to perform simulation, architecture or design debugging by getting coverage of your requirements expressed as test cases using MSCs, meet your quality expectations and get a faster return on investment not only for the tools but also during product development.

Jamel Marzouki
Distinguished Member of the Technical Staff, Motorola Labs, Schaumburg, IL

1

Introduction

1.1 VALIDATION OF COMMUNICATIONS SYSTEMS

Communications systems and software are more and more difficult to develop: they include complex features such as wireless and mobile access, under strong constraints such as low size, weight or power consumption, interworking, total interoperability, security, short time-to-market and low cost.

Respecting such an array of constraints requires a high quality of the specifications or standards used for their development. This is why the specifications of many communications systems are based on SDL (Specification and Description Language) or at least contain SDL parts describing complex behaviors. Examples of such systems are the GSM second-generation mobile telephony system, the UMTS third-generation mobile telephony system, the ETSI HiperLAN 2 Broadband Radio Access Network or the IEEE 802.11 wireless Ethernet local area network.

Validation of such systems by simulation of SDL models is useful, for example, at the following stages:

- when standards are created by the organizations, to check that the behavior of the system is correct, to generate Message Sequence Charts (MSCs) (sequence diagrams) illustrating typical use cases, or to generate TTCN (Tree and Tabular Combined Notation) test cases to test the conformance of future implementations;
- before the implementation of a standard by a company, because standards rarely contain a finished SDL model ready to be translated into the application code;
- to provide nonambiguous low-error specifications to a contractor, enabling a quicker and less expensive implementation;
- after changes in the specifications, to check that the system has not regressed.

During all these stages, the simulation allows the detection of specification or design-level anomalies, preventing them to be embedded in the implementation. Once the code is loaded into a target device embedded into a complex test environment, each error detected is more difficult and expensive to analyze than during an SDL model simulation: the error can come not only from the specification but also from the coding, from the testing environment, from the hardware and so on.

Also, SDL simulation enables the execution of the specification before the target hardware and software platform is available: board, board support package, compiler and so on.

Validation of Communications Systems with SDL: The Art of SDL Simulation and Reachability Analysis.
Laurent Doldi ISBN: 0-470-85286-0

The SDL Simulators, especially in exhaustive mode, quickly find error scenarios far beyond human imagination: bugs that could appear after millions of devices have been sold, revealed by a modification of their environment, can be detected and fixed during the specification or design phase.

Concerning safety-critical systems such as fault-tolerant aircraft systems architectures, medical or car devices, the validation by simulation can prove formally that their specified behavior is correct, according to a set of criteria.

1.2 SDL, LANGUAGE TO MASTER COMPLEX SYSTEMS DEVELOPMENT

1.2.1 Overview of SDL

SDL stands for Specification and Description Language. It is standardized by the ITU (International Telecommunication Union) in the Z.100 Recommendation [SDL00, SDL99].

In SDL-92[1], the architecture is modeled as a system containing blocks, as depicted in Figure 1.1. Each block may contain either blocks or processes. Each process contains an extended finite state machine. State machines communicate by exchanging signals through channels (or signal routes).

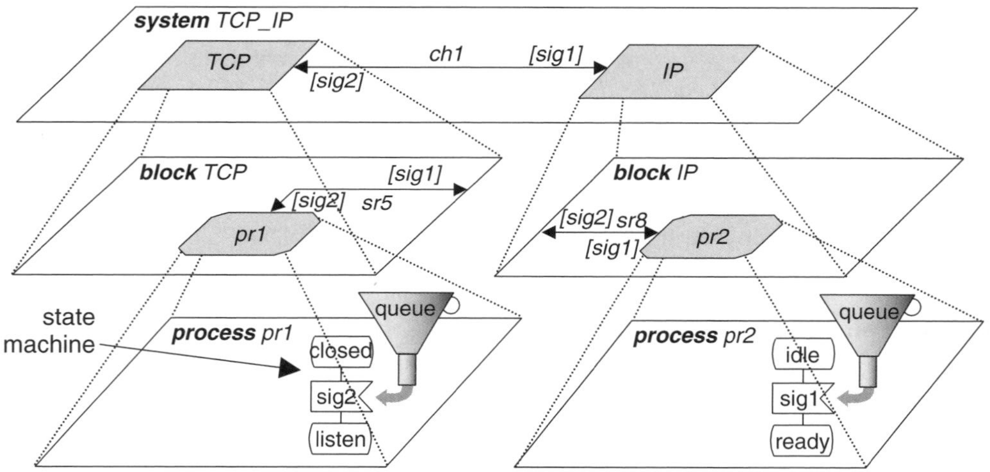

Figure 1.1 Schematic view of an SDL-92 description

Signals (*sig1* or *sig2* in Figure 1.1) arriving on a state machine are queued. By consuming a signal from its queue, a state machine executes a transition from one state to another state. During the execution of a transition, a wide range of actions can be performed by a state machine: signal transmission to another state machine, assignment, procedure or operator call, loop, process instance creation and so on. The execution semantics of SDL is accurately described and includes the semantics of actions in state machine transitions.

Data types are described using predefined types or constructs such as Integer, Boolean, Character, struct, Array, String, Charstring, Powerset. ASN.1 can be used in an SDL model

[1] SDL-92 means the 1992 version of SDL plus the corrections introduced in Addendum 1 to Recommendation Z.100 of 1996, sometimes called SDL-96.

to describe more complex data types, using constructs such as choice (similar to union in C), optional fields, Bitstring or Octetstring and providing standardized encoding rules.

SDL is object-oriented: it provides the notions of classes, inheritance, polymorphism (in SDL-2000) and so on found in object-oriented programming languages.

SDL is frequently used with MSCs, similar to UML (Unified Modeling Language) Sequence Diagrams.

1.2.2 Benefits provided by SDL

SDL being a graphical language enables you to visually design models, instead of using only a textual notation. SDL provides graphical structuring features (blocks, etc.), state machines and communication through signals that are not available in programming languages such as C++ or Java.

During the modeling process and after its completion, and because SDL has a complete semantics, the SDL description can be rapidly checked and debugged using the powerful tools available today (see Figure 1.2), namely the compilers and simulators, enabling very fast model correction.

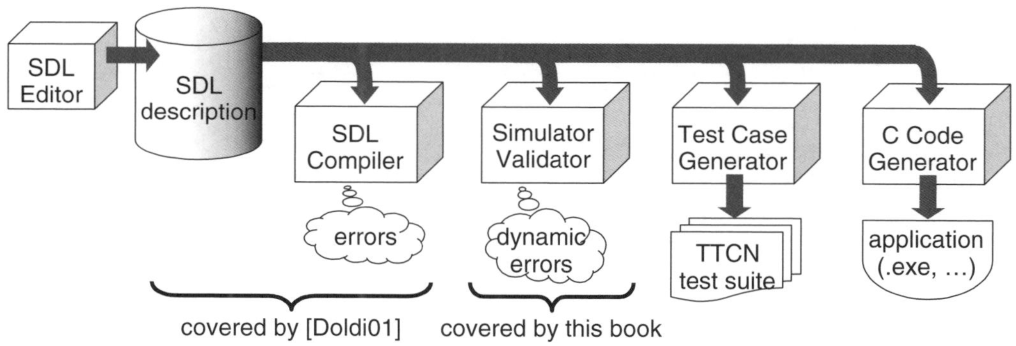

Figure 1.2 Life of an SDL description

Bugs are found and corrected before the implementation begins. SDL simulators provide high caliber debugging features, from symbol by symbol stepping to automatic simulations using various strategies (random, exhaustive, bit-state, supertrace etc.) coupled to automatic error detection by observers. Simulation scripts allow automatic nonregression testing of the SDL description in a few seconds by automatic replay of scenarios, with observers on-line checking the SDL behavior. Simulators generate MSCs representing a visual trace of the simulation.

After testing the SDL description, code generators can autocode it: it is not necessary to write a single line of code to get the application running, except when communicating with non-SDL parts or optimizing performance if severe constraints exist. Just by pressing a button, a code generator produces, without manual coding errors, one or several binaries running on one or several computers or boards, without executive or under Unix, win32, Posix™, VxWorks™, VRTX™, Chorus™, PSOS™ and so on. Execution on the target system produces a visual trace in the form of MSC sequence diagrams.

Also, test cases in TTCN or in another test language can be generated by very sophisticated tools, ranging from transformation of an interactive simulation scenario into TTCN to automatic generation of TTCN test cases covering all SDL symbols or automatic generation of TTCN test cases corresponding to user-defined test purposes.

1.3 SIMULATION LIFE CYCLE

The life cycle of simulation can be split into three steps, as illustrated in Figure 1.3:

1. Production of an SDL model ready for simulation, starting either from a textual specification, as in the V.76 case study presented in the book, or from an SDL model, for example coming from a protocol standard, or from legacy code that must be reverse-engineered because no useful specification documents exist;
2. Interactive simulation, offering a good level of validation and automatic nonregression testing;
3. Exhaustive simulation, the top level in validation, reserved for safety-critical or cost-critical systems. Cost-critical means that leaving errors in the system would be very expensive, because its code will be embedded into millions of devices.

Simulation produces a low-default executable specification, plus reference MSCs (sequence diagrams) that provide an excellent documentation and which can be used as test cases to test the system implementation: as they have been generated by simulation, they are consistent with the validated SDL model.

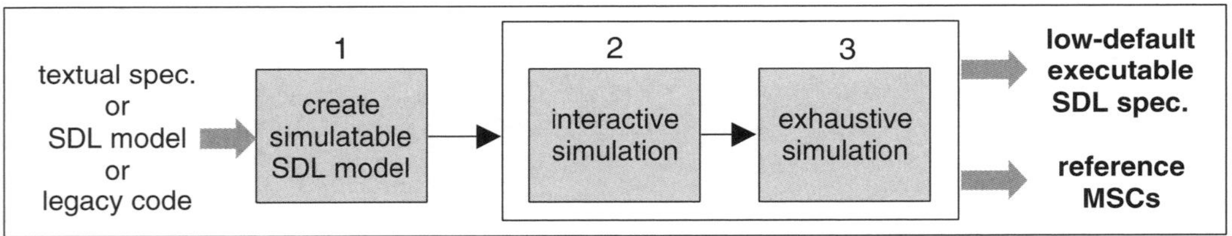

Figure 1.3 The three simulation steps

Step 1 is shown in Figure 1.4: if an SDL model exists for the system, it must be compiled. If there are errors, the SDL model must be corrected. Some SDL models may need to be completed, for example to add missing data type declarations. If no SDL model exists, a decision must be taken to use or not to use SDL: if the system is not complex or not safety-critical, such an investment is not necessary. Beware of systems that seem to be simple but are not, in terms of behavior. Then, if the system (or a part of it) cannot be modeled using extended finite state machines communicating with signals through queues (a kind of mailbox), use another language, such as a synchronous language. Otherwise, continue with Step 2.

Step 2 is shown in Figure 1.5: first, the main scenarios (the use cases) are simulated step by step. Each scenario is stored (files *.scn*, *.com* or *.cui*) to be replayed automatically later. Each MSC trace is also stored. When an error is found, the SDL model (or an specification if it is wrong) is corrected.

To detect automatically when the SDL model is wrong or when it is correct, observers can be created. Then the SDL model is simulated together with its observers, replaying the scenarios stored previously, to check that the observers work correctly. Again, when an error is found, the SDL model is corrected.

To replay scenarios automatically, scripts can be written: after each model correction or evolution, all the test scenarios are automatically replayed in a few seconds and the simulator reports any error or any violation detected by the observers.

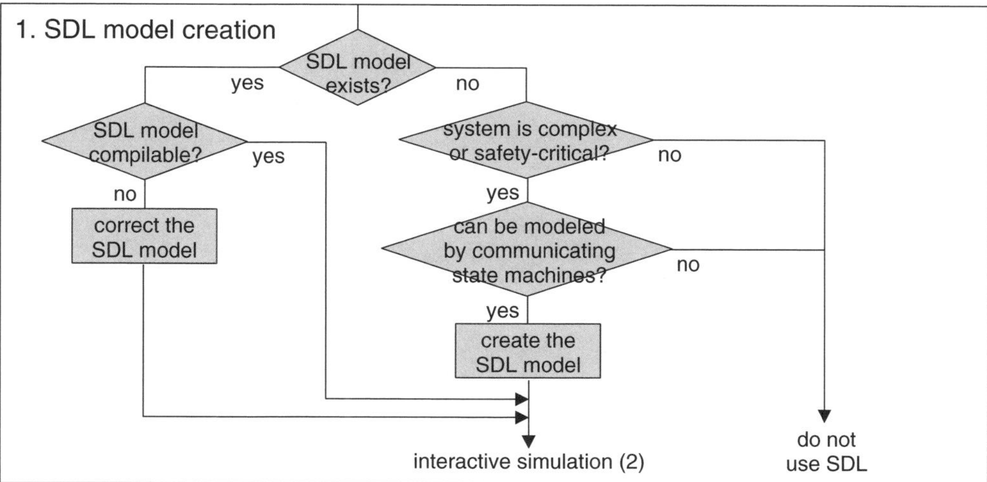

Figure 1.4 Step 1: SDL model creation

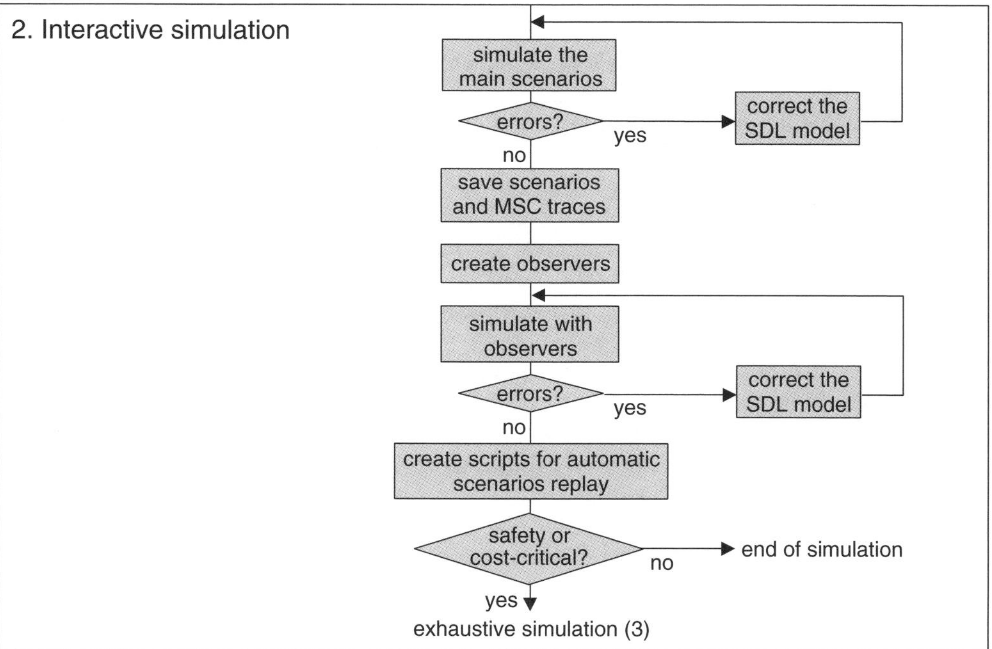

Figure 1.5 Step 2: interactive simulation

Simulation detects SDL symbols never executed: more scenarios must be created to try to cover them.

If the system is not safety- or cost-critical, this level of simulation is sufficient. Otherwise, continue with Step 3.

Step 3 is shown in Figure 1.6: exhaustive (or bit-state) simulation is run. When an error or an observer violation is found, the SDL model (or an observer) is corrected. The scenarios

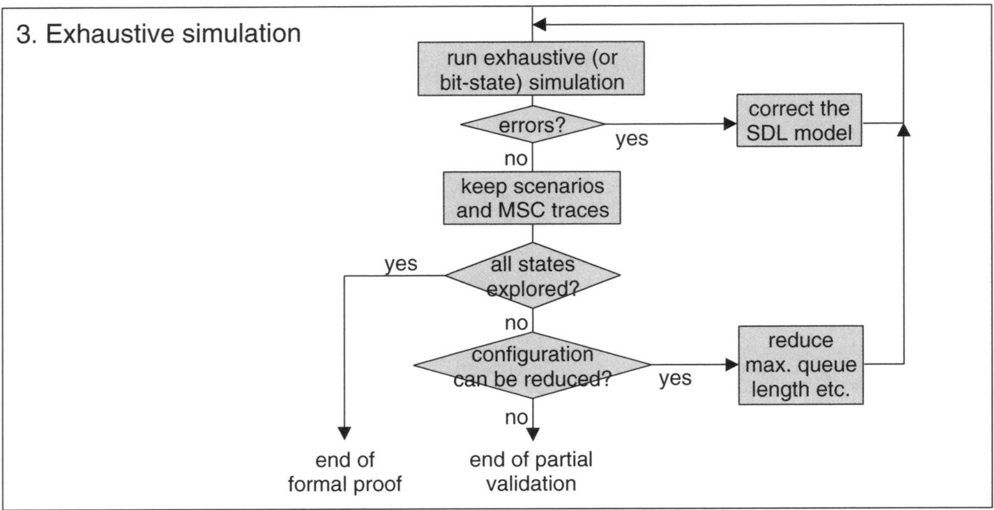

Figure 1.6 Step 3: exhaustive simulation

generated by the simulation must be kept, together with the MSC traces of the scenarios that satisfy the observers.

If all the reachable states of the SDL model have been explored, the simulation is finished: the SDL model correctness has been formally proved (according to the configuration used).

Otherwise, because the large number of global states prevents exploring them, the model configuration must be reduced, for example, by allowing one or two signals only per process input queue, limiting the exploration depth and so on. If the configuration cannot be reduced, the simulation is finished, and the validation is partial because some global states remain unexplored.

The simulator also detects SDL symbols never simulated: they indicate SDL transitions or branches that could be removed, or reveal missing test signals or test values to be transmitted to the SDL model by the simulator.

All these steps are detailed in the book, illustrated by numerous hands-on exercises.

1.4 CONTENTS OF THE BOOK

This book is divided into eight chapters. Chapter 1 is the present introduction. Chapter 2 is a quick tutorial on SDL-92, the language used for the exercises in the rest of the book. Chapter 3 contains the simplified version of the V.76 protocol specification, some analysis MSCs and the corresponding SDL model used during the simulation exercises. Chapter 4 explains how to validate the V.76 SDL model using interactive simulation. Chapter 5 introduces observers, what they can detect, how to build them and how to use them during interactive simulation. Chapter 6 describes random simulation. Chapter 7 presents exhaustive simulation, how to use it with observers and other simulation algorithms such as bit-state or liveness. Chapter 8 illustrates other simulator features such as calling external C code or adding buttons to the simulators.

Each chapter contains hands-on exercises with solutions for the two main SDL tools commercially available: ObjectGeode and Tau SDL Suite, both from Telelogic.

1.5 TOOLS AND PLATFORMS USED

The exercises of the book have been developed using the following commercial off-the-shelf tools, both developed by Telelogic[2]:

- ObjectGeode Version 4.2 for Windows
- Tau SDL Suite Version 4.0 for Windows

The Unix versions of these tools are very similar to their Windows version. The main difference is the way they are launched: it is generally performed by double clicking on an SDL file in Windows and by typing a command in Unix.

The V.76 SDL model used in the book and its associated files can be downloaded in ObjectGeode and Tau SDL Suite formats on *ftp://ftp.wiley.co.uk/pub/books/ldoldi/*.

[2] ObjectGeode has been developed by Verilog, with France Telecom R&D (former CNET) know-how. Then, in 1999, Telelogic has acquired Verilog.

2

Quick Tutorial on SDL

For readers not familiar with SDL, this chapter presents the most frequently used constructs of SDL-92. A detailed step-by-step tutorial on SDL-92 [SDL92] and a presentation of SDL-2000 are provided in [Doldi01].

2.1 STRUCTURE OF AN SDL MODEL

2.1.1 System, block and process

SDL provides the following entities to structure a description:

- system: top level, outermost construct;
- block: must be contained in the system or in a block;
- process: must be contained in a block;
- service (optional): must be contained in a process;
- procedure (optional): can be placed anywhere.

Figure 2.1 shows an example of a system containing three blocks.

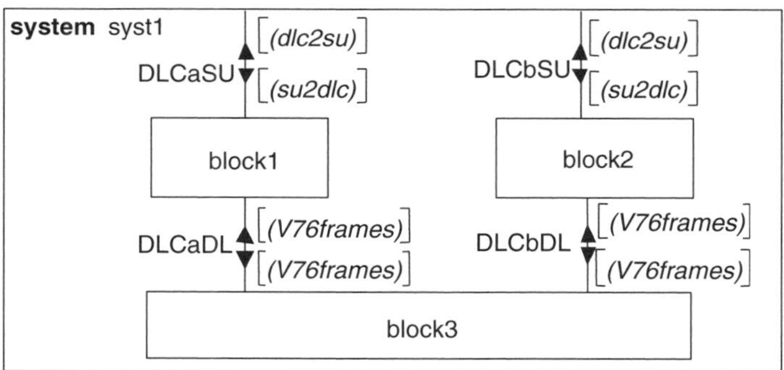

Figure 2.1 System *syst1*

Figure 2.2 shows that *block1* contains two blocks.

Figure 2.3 represents the contents of *block1_1* and *block3*. *Block1_1* contains one process, and *block3* contains two processes. The state machines contained in processes (or services) instances communicate together or with the environment by transmitting and receiving signals (or remote variables or procedures) through channels and signal routes.

Validation of Communications Systems with SDL: The Art of SDL Simulation and Reachability Analysis.
Laurent Doldi © 2003 John Wiley & Sons, Ltd ISBN: 0-470-85286-0

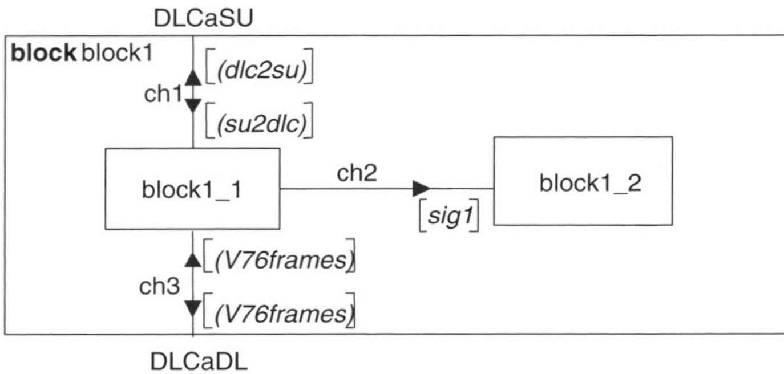

Figure 2.2 Block *block1*

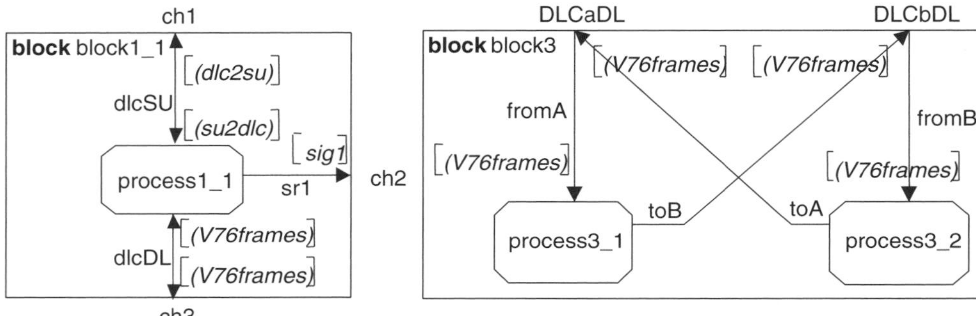

Figure 2.3 Blocks *block1_1* and *block3*

2.1.2 Scope of declarations

A declaration is used to define signals, data types, variables and so on. In SDL, a declaration is visible in the current entity and its children. For example, a signal declared in the system is visible in the whole system. Variables cannot be declared in a block or in a system; therefore global variables do not exist in SDL.

2.1.3 Process

A process, if it does not contain any service, contains a state machine. Figure 2.4 shows a process in which the variable *V76par* has been declared. A variable is local to a process and thus is not visible, for example, from another process. Our process contains only one state, *ready*.

Each process has one or more instances, running in parallel, and independent. Two numbers are used to specify the number of instances for a process:

- the first number indicates the initial number of instances (when the system is started),
- the second number indicates the maximum number of instances running at a certain moment.

Figure 2.5 depicts the two main possibilities for the number of instances.

Each process instance contains four implicit variables:

- *self*: contains the Pid of the current instance,

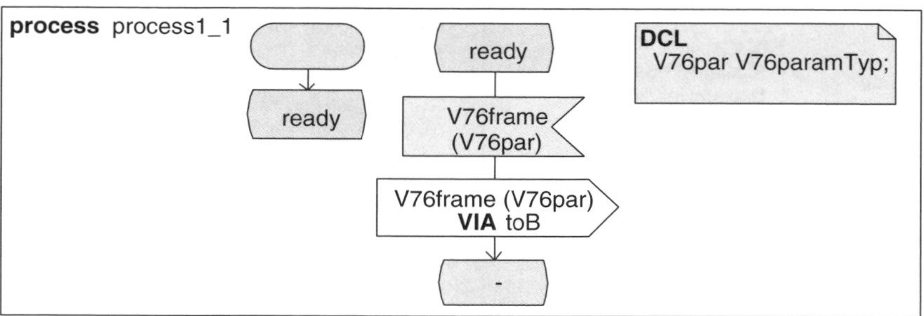

Figure 2.4 A process

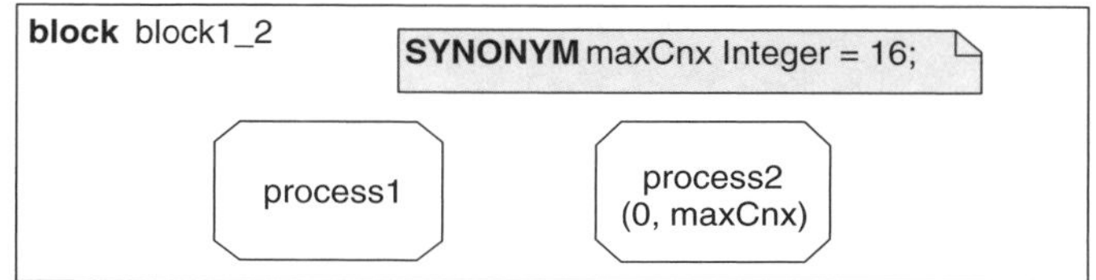

Process name	Number of instances at system startup	Maximum number of instances
process1	1	no limit
process2	0	16

Figure 2.5 Examples of number of instances

- *sender*: contains the Pid of the instance that sent the last signal consumed,
- *parent*: contains the Pid of the instance that created the current instance,
- *offspring*: contains the Pid of the last instance created.

Those four variables identify each process instance; this is necessary, for example, to output a signal to a particular process instance.

2.1.4 Procedure

Figure 2.6 shows a process containing the definition of procedure *sendFrame*, and a call to this procedure. The procedure *sendFrame* is called in a procedure call symbol, passing the three parameters corresponding to the formal parameters declared in the procedure header.

2.2 COMMUNICATION

In SDL, the state machines communicate together or with the environment by transmitting and receiving signals (or remote variables or procedures) through channels and signal routes.

2.2.1 Signals

First, a signal must be declared. Figure 2.7 provides examples of signal declarations. A signal may carry one or more parameters. In our example, only *L_ReleaseInd* carries a parameter, of type *releaseCause*.

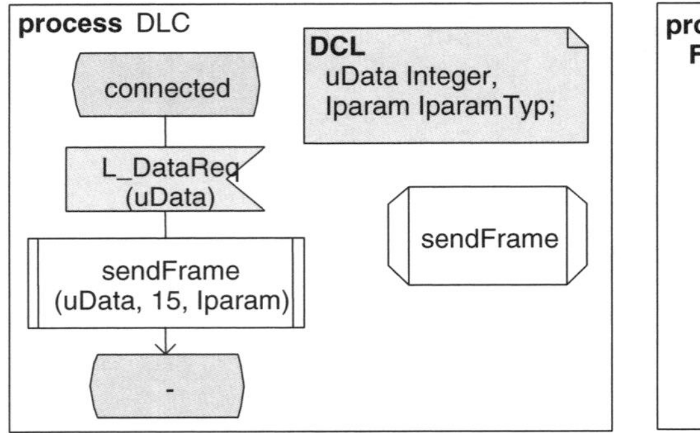

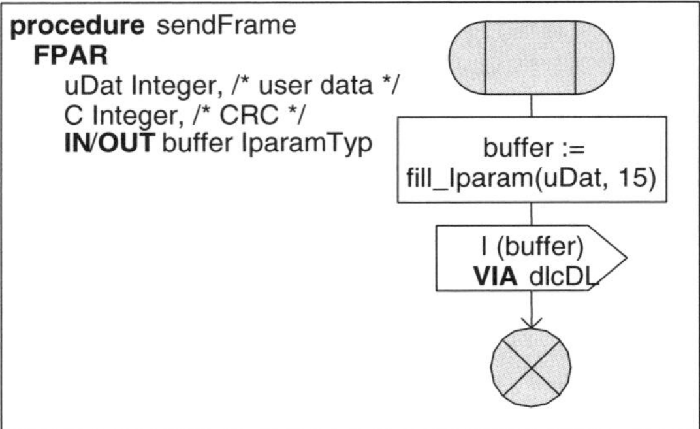

Figure 2.6 Example of procedure

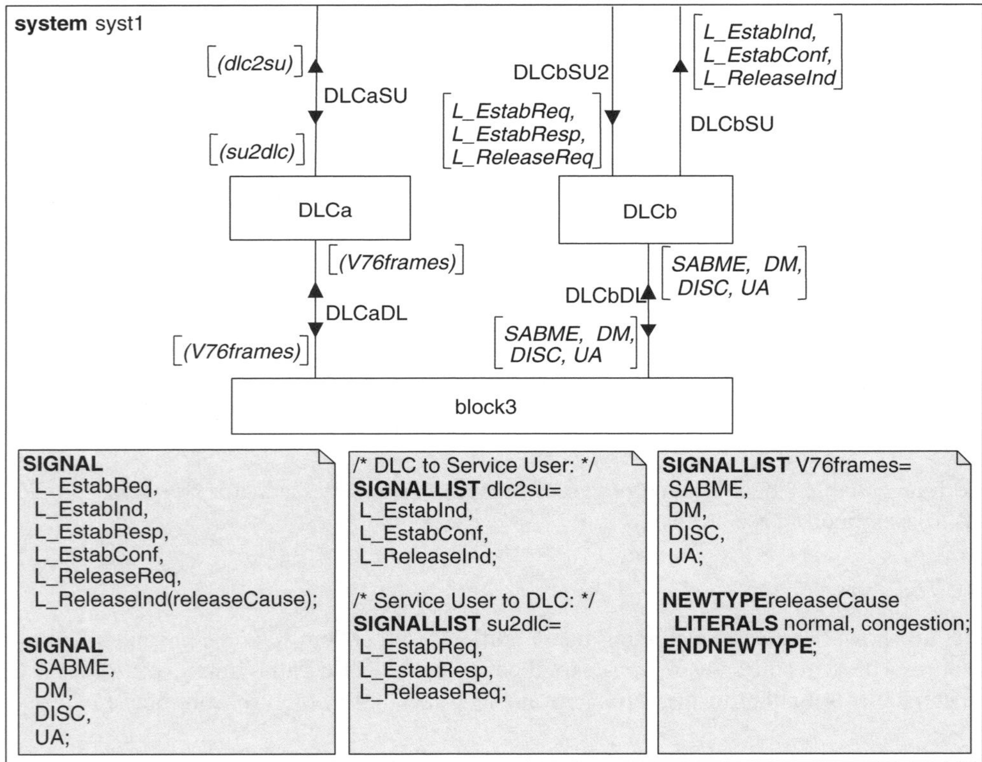

Figure 2.7 Communication at system level

Once declared, a signal can be inserted into the "wires" (channels and routes). The signal name can be directly inserted into the [ ] symbols of the channel or signal route, or the signal name can be inserted into a signal list (like in the channel *DLCaSU* on the left part of Figure 2.7).

The "wires" that you see in SDL descriptions are either channels or signal routes: if a "wire" is connected to a process, it is a signal route, otherwise it is a channel.

2.2.2 Channel

A channel can carry signals either in one or in both directions. Figure 2.8 shows channels *ch1* and *ch2* carrying signals to *block_b_1*, while channel *ch3* carries signals from *block_b_1* to outside. Channel *ch4* carries signals both in and out of *block_b_1*.

By default, a channel contains a FIFO (First In First Out) queue used to delay the signals[1].

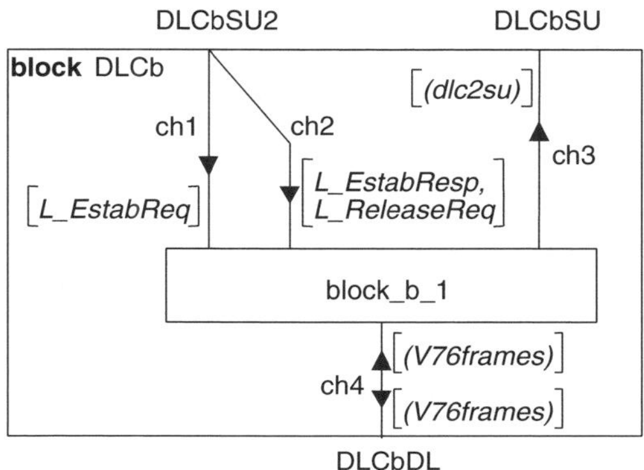

Figure 2.8 Channels in block *DLCb*

2.2.3 Signal route

The difference between channels and signal routes is that signal routes do not delay signals. Figure 2.9 shows an example of four signal routes.

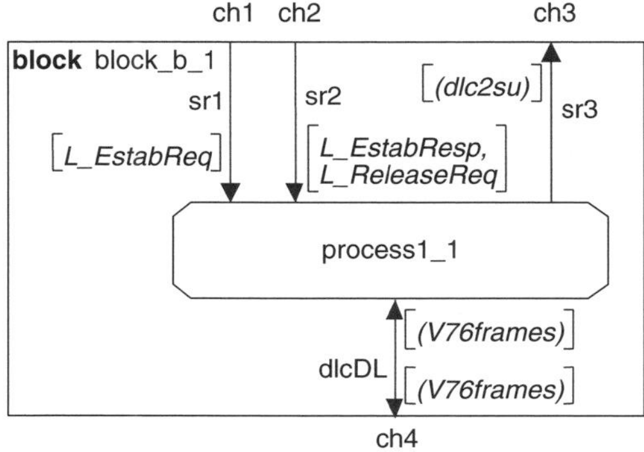

Figure 2.9 Signal routes in *block_b_1*

2.3 BEHAVIOR

2.3.1 Structure of a transition

As depicted in Figure 2.10, a transition in a process begins with a state, immediately followed by an input or a spontaneous transition, or a priority input, or a continuous signal or a save. No other symbol is allowed after a state.

[1] When using an SDL editor, an option allows you to specify if you want the channel to delay signals or not. In ObjectGeode and Tau SDL Suite, by default, channels do not delay signals.

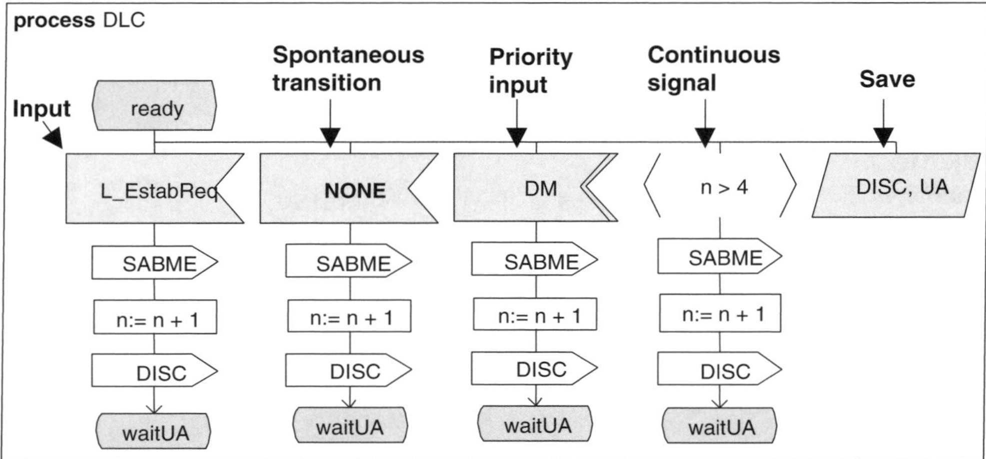

Figure 2.10 Symbols legal after a state

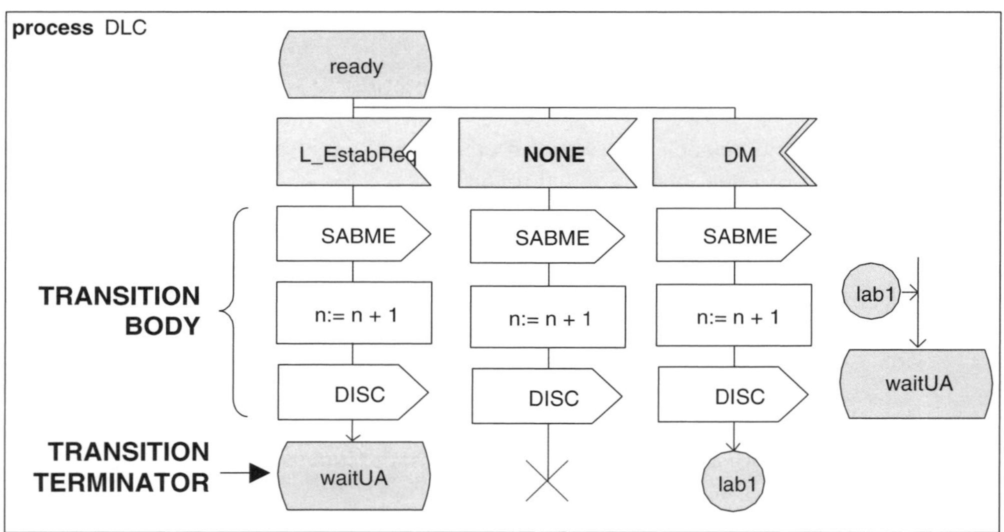

Figure 2.11 Transition structure

Figure 2.11 shows the structure of a transition: the symbols after an input, or a spontaneous transition, or a priority input or a continuous signal are called the transition body. The transition body must not contain any input or continuous signal. Following the transition body we find the transition terminator, which can be a nextstate, a stop, a join (*lab1* in the figure) or a return (for procedures only).

2.3.2 Start

Every process must contain exactly one start symbol. When a process instance is created, the first transition ready to be executed is the transition beginning from the start symbol.

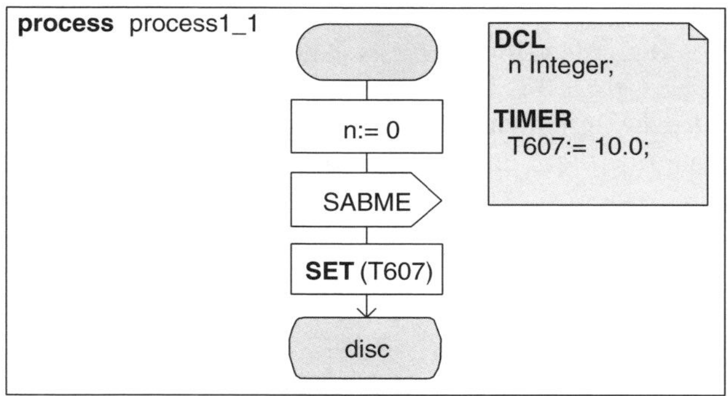

Figure 2.12 Start transition

Figure 2.12 shows an example of start transition: *n* is set to 0, signal *SABME* is transmitted and timer *T607* is started before going to state *disc*.

2.3.3 States

States must be defined using a state symbol. A common mistake is to confuse the notions of state and nextstate: Figure 2.12 is incorrect, because state *disc* is not defined; Figure 2.13 is correct, because state *disc* is defined.

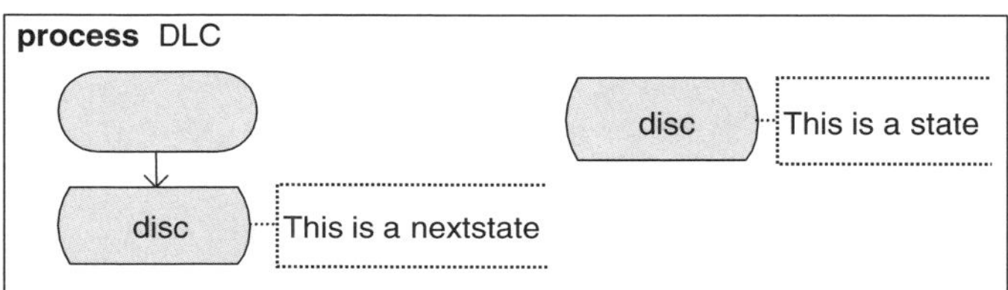

Figure 2.13 State *disc* defined

When the character "-" is entered in a nextstate, as in Figure 2.14, it means that after executing the transition, the state will remain unchanged.

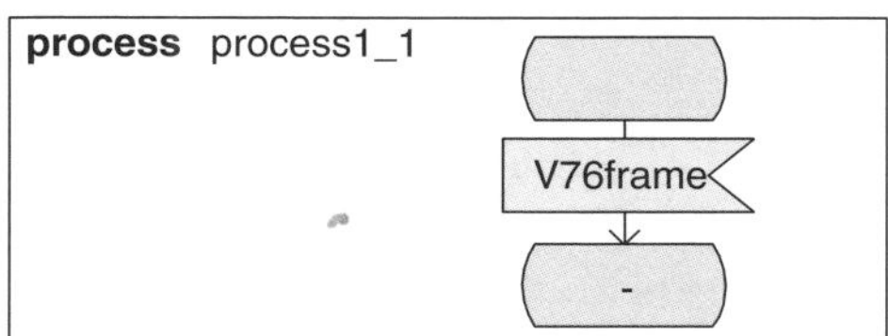

Figure 2.14 Dash nextstate

2.3.4 Input

Signals reaching a process instance are stored into its FIFO queue. When performing an input, the first signal in the queue (the oldest one) is removed from the queue, and the values of the signal parameters, if any, are assigned to the variables specified in the input symbol.

Note, that if the first signal in the queue is not present in any input below the current process state, the signal will be discarded (lost). This is called an implicit transition.

In Figure 2.15, left part, the FIFO queue of process *display* contains the signals *blue*, *green* and *red*. Process *display* being in state *idle*, the signal *blue* is discarded (lost), and then *green* is input, leading to state *resizing*. Signal *red* is now first in the queue.

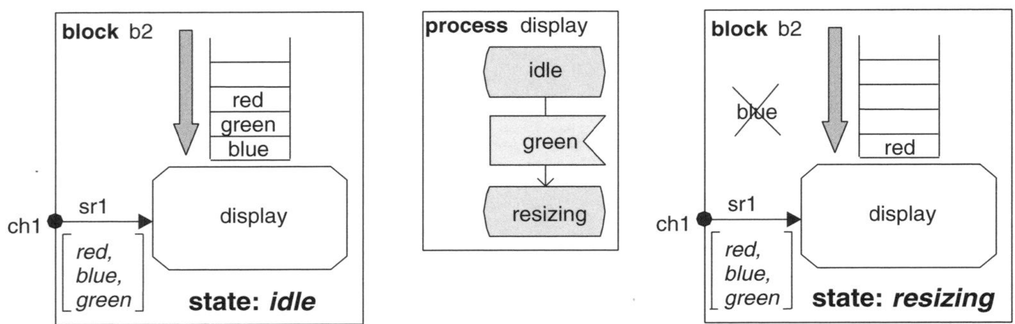

Figure 2.15 The FIFO queue of one instance of process *display*

2.3.5 Save

To avoid losing signal *blue* as in Figure 2.15, we add a save symbol below state *idle*, as depicted in Figure 2.16. When a signal is saved, it stays in the input queue at the same position, and the next signals in the queue are examined to see if they can be input, saved or discarded.

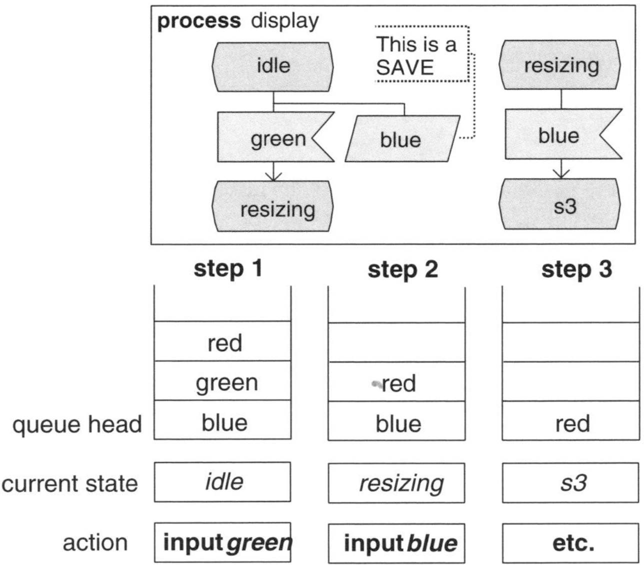

Figure 2.16 Saving signal *blue*

Reading the table in Figure 2.16 helps you understand how the save works:

1. From state *idle*, *blue* is first in the queue: it remains here because it is saved, and the next signal in the queue, *green*, can be input, leading to state *resizing*.
2. From state *resizing*, *blue* is input, and we go to state *s3*.

2.3.6 Variables

Variables are used to store data in process instances. Variables cannot be declared in systems or blocks: global variables do not exist in SDL.

Figure 2.17 shows an example of variable declaration and usage: the variable *n* of type Integer is declared, set to 0 upon process instance start, and then incremented by 1 each time a *disc* signal is input. We remind you that if, for example, two instances of process *DLC* are created, each instance has its own variable *n* in its context.

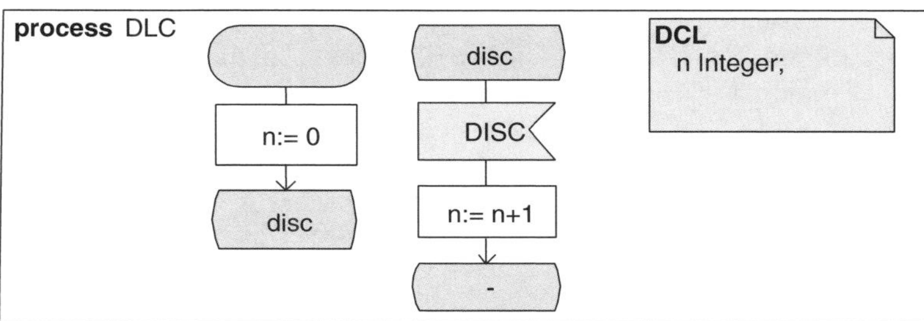

Figure 2.17 Example of variable

2.3.7 Stop

After executing a stop symbol, the process instance and its associated input queue and the signals it contains are immediately destroyed. Figure 2.18 shows an example of stop.

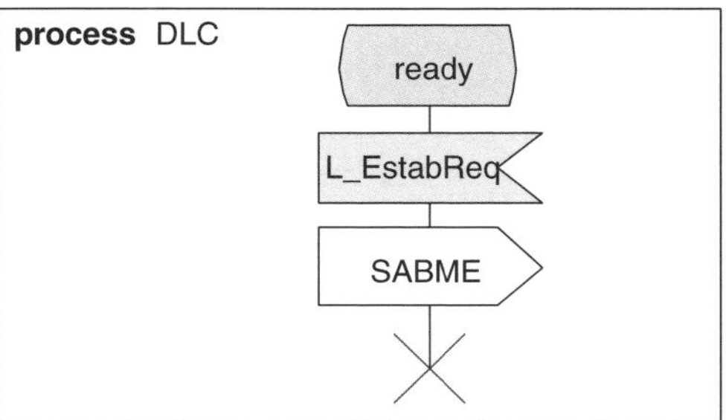

Figure 2.18 Example of stop

2.3.8 Task

Figure 2.19 shows two task examples. The first one simply performs $n := n + 1$ and the second one contains informal text (sometimes called informal task).

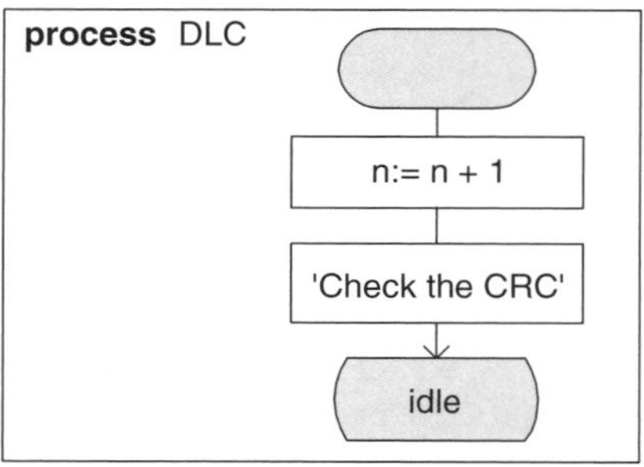

Figure 2.19 Example of tasks

2.3.9 Create

SDL allows dynamic creation of process instances. Figure 2.20 shows a process *DLCmaster* creating instances of process *DLC*. The dashed line between the two processes shows who creates who, but is optional. The process instance creation is actually performed by the create request symbol seen inside *DLCmaster* on the right.

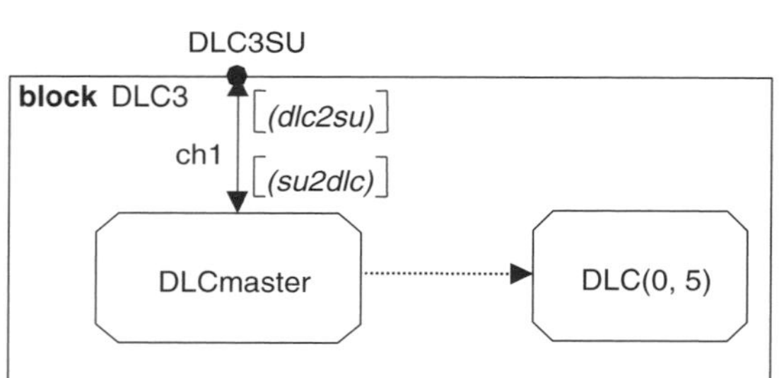

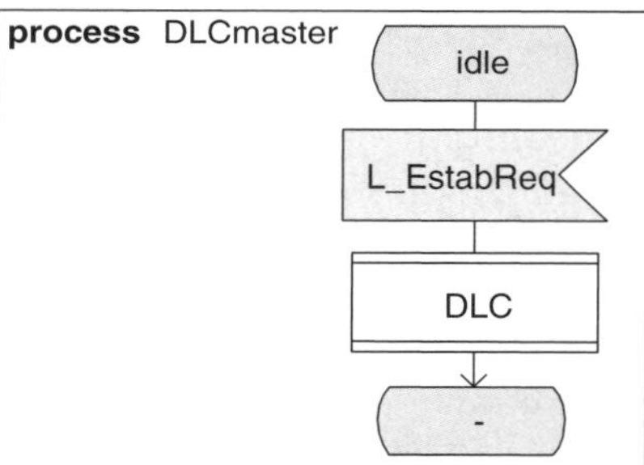

Figure 2.20 Example of process creation

Every process instance contains an implicit variable called offspring. After a create, offspring contains the Pid (Process identification) of the created instance, or Null if the create failed.

2.3.10 Output

Output is used to transmit a signal to another process instance. In Figure 2.21, signal *s1* is transmitted, with parameter values *True* and *15*.

If more than one process instance can be reached by the signal, as in Figures 2.22 and 2.23, it is safer to use VIA or TO to specify which instance must receive the signal.

Figure 2.22 shows how to use VIA to send signal *red* through *sr1*, where process *screen* cannot have more than one instance.

Figure 2.23 shows the use of TO to send signal *red* to a certain instance of process *screen*, which has a maximum of three instances. Before the output of *red*, variable *p* must be filled with the Pid of the target instance of *screen*: this is generally done using an array.

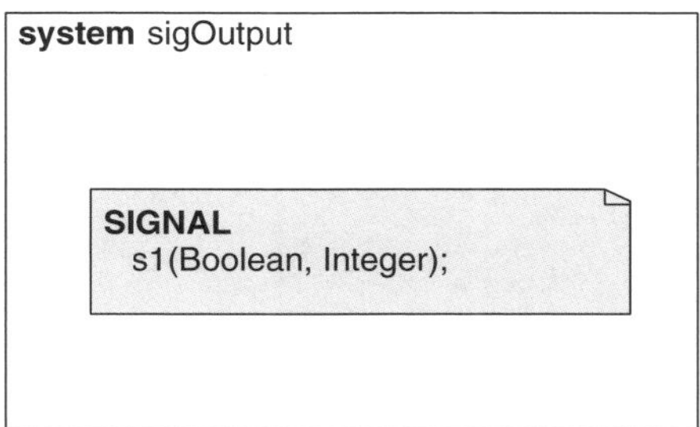

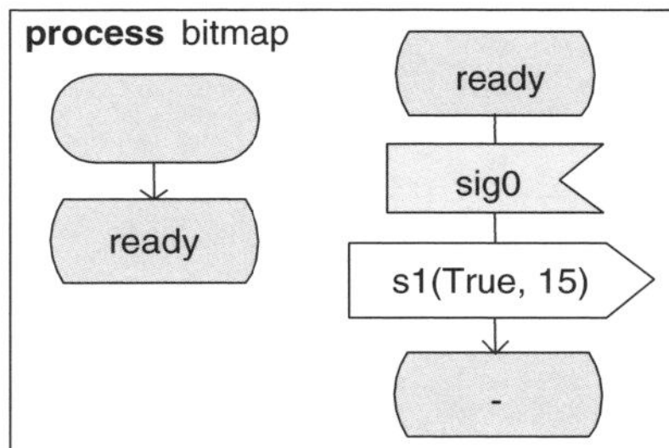

Figure 2.21 Signal parameters in output

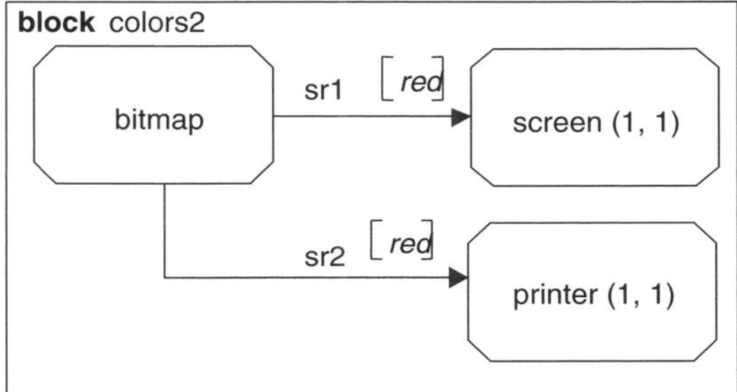

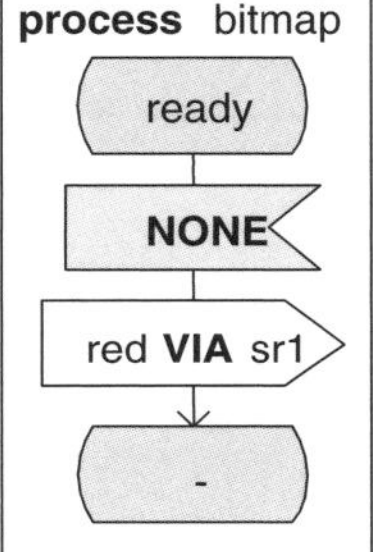

Figure 2.22 Using VIA in output

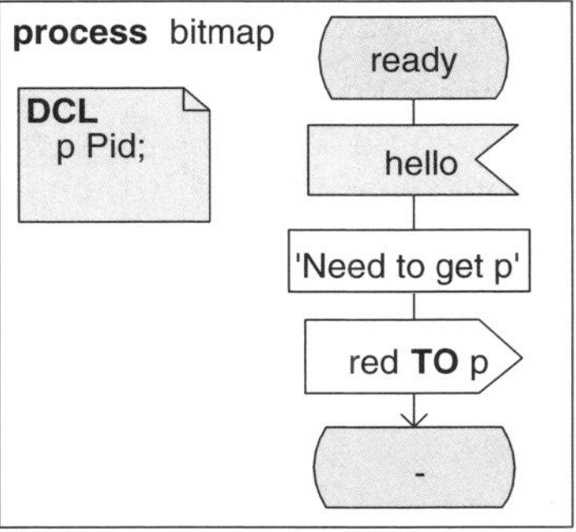

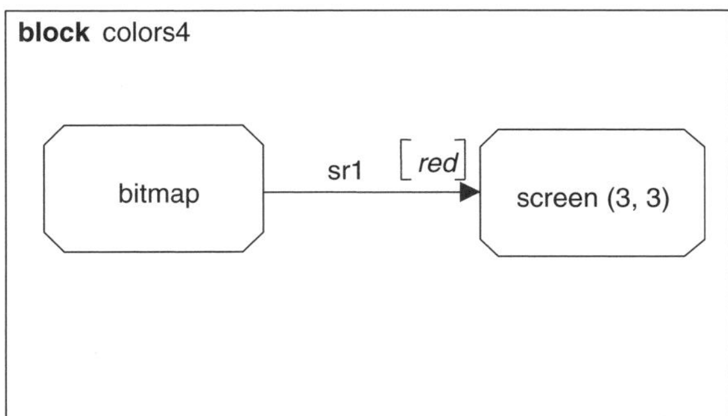

Figure 2.23 Output to a Pid

2.3.11 Decision

A decision is used to branch according to the value of an expression. In Figure 2.24, the decision tests the value of n: if $n > 3$, then the left branch is executed, else the right branch.

2.3.12 Timers

In SDL, the expression NOW contains the current value of the global time. Figure 2.25 shows an example of a timer used to monitor a response. First, the timer *T201* is declared (right part). Then after output *IAM*, *T201* is started using SET: timer *T201* will time out at NOW+15.0.

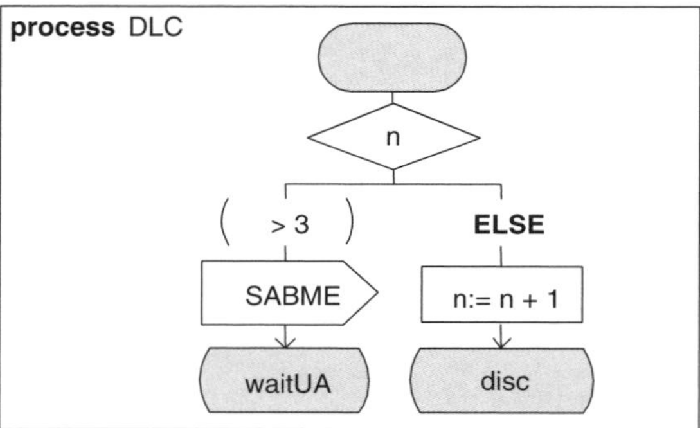

Figure 2.24 Decision example

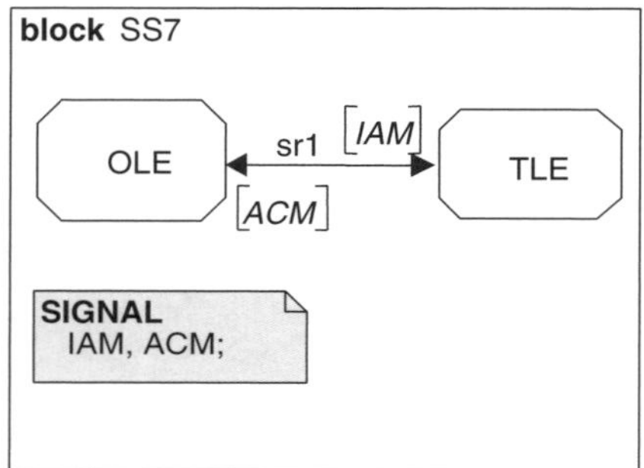

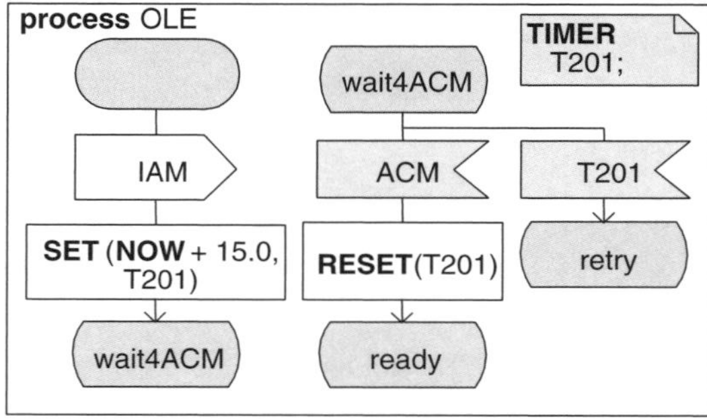

Figure 2.25 Timer example

From state *wait4ACM*, we either input the response to *IAM, ACM*, and we stop the timer using RESET, or we input the timer signal *T201* because *ACM* arrived more than *15.0* time units after SET.

2.4 DATA TYPES

2.4.1 Predefined data

Predefined data types in SDL are

- Boolean: True, False
- Character: 'A', '8', etc.
- String: generic string (not only a string of characters)
- Charstring: 'Example of charstring'
- Integer: −45, 0, 36700, etc.
- Natural: null or positive Integer
- Real: 23.5 etc.
- Array: generic array
- Powerset: generic set
- Pid: to identify process instances
- Duration, Time: used in timers.

2.4.2 Array

The Array generator is used to create arrays containing any element type and indexed by any type[2].
Example:

```
/* An array of 5 Integers: */
NEWTYPE intTable
 Array(itIndex, Integer)
ENDNEWTYPE;
SYNTYPE itIndex = Integer CONSTANTS 0:4 ENDSYNTYPE;
DCL t intTable, x Integer;
TASK t:= (. 13 .); /* puts 13 into the 5 array elements. */
TASK t(2):= 127;
TASK x:= t(4);
```

2.4.3 Synonym and syntype

Synonyms are used to define constants.
Example:

```
SYNONYM maxCount Natural = 127;
SYNONYM Yes Boolean = True;
SYNONYM No Boolean = False;
```

Syntypes are often used to define intervals, for example, to index an array. A syntype may or may not contain a range condition (e.g. 0:4).
Example:

```
/* Integers 0, 1, 2, 3 and 4: */
SYNTYPE itIndex = Integer CONSTANTS 0:4 ENDSYNTYPE;
SYNTYPE logical = Boolean ENDSYNTYPE;
```

2.4.4 Newtype

Using NEWTYPE allows building your own types based on the predefined SDL types. A NEWTYPE may also contain operator signatures or definitions.

2.4.4.1 Literals

Literals are used to define enumerated values, as shown in Figure 2.26.

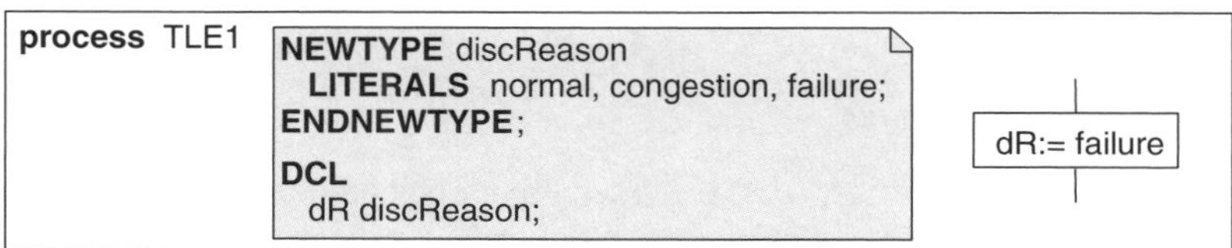

Figure 2.26 Newtype with literals

[2] As opposed to C, SDL arrays indexes may not start at 0.

2.4.4.2 Struct

Struct is used inside a NEWTYPE to define a data structure, as illustrated in Figure 2.27.

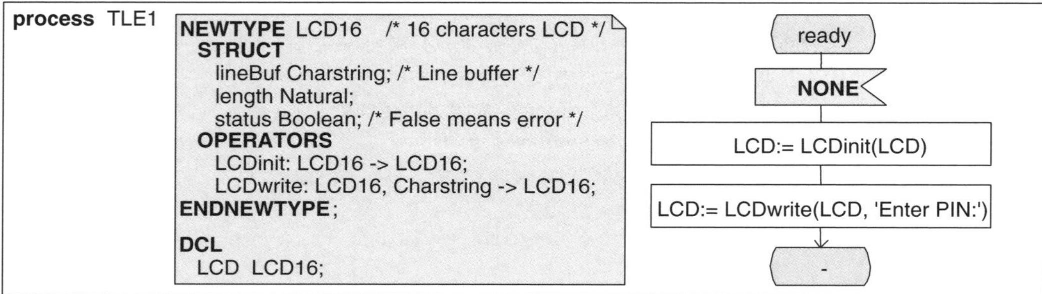

Figure 2.27 Newtype containing a struct and operators

2.4.4.3 Operator signature

You have three possibilities to describe an operator: using algorithmic notation (textual or graphical), using the C language[3] or using axioms.

2.5 CONSTRUCTS FOR BETTER MODULARITY AND GENERICITY

2.5.1 Package

To reuse type definitions easily in several systems, they can be moved into a package, as shown in Figure 2.28. The package can then be imported in a system as shown in Figure 2.29.

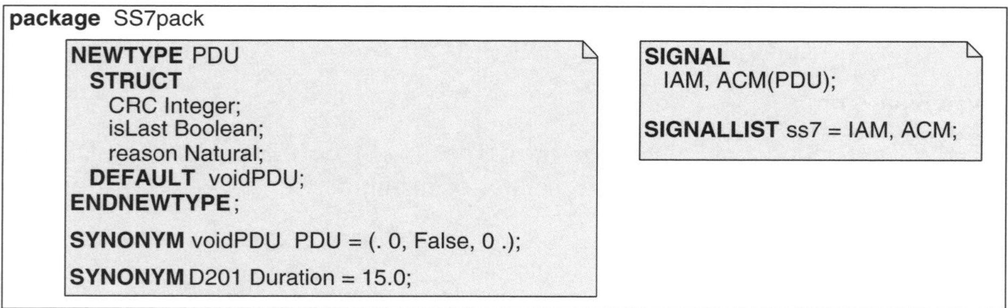

Figure 2.28 The package *SS7pack*

2.5.2 Types, instances and gates

To allow reuse of structural entities, SDL provides object-oriented features: any structural entity may be a type; thus we can use system types, block types, process types and service types.

[3] This is not part of SDL. SDL tools generally provide extensions (placed in comments) to allow implementing an operator in C, with various options for every need. This is useful to interface the C generated from an SDL description with existing codes such as device drivers.

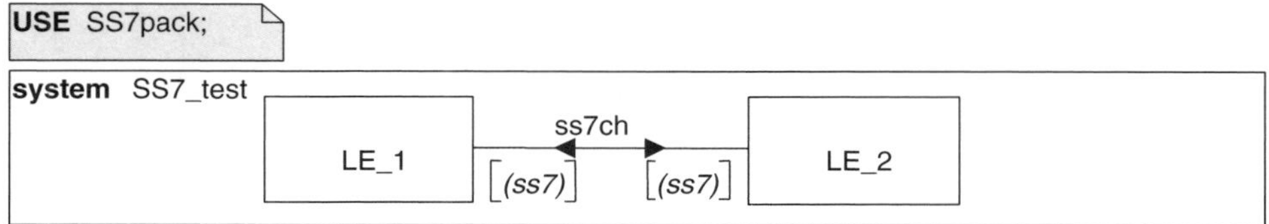

Figure 2.29 The system *SS7_test* using the package *SS7pack*

2.5.2.1 Block type

A block type allows you to create a kind of reusable component. The two blocks in Figure 2.29 have been replaced, as illustrated in Figure 2.30, by

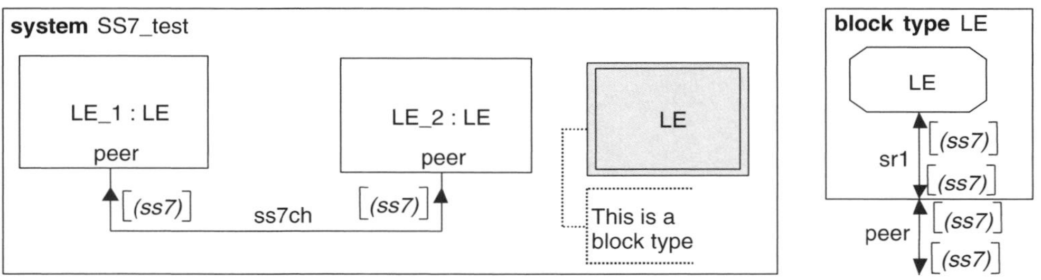

Figure 2.30 The blocks *LE_1* and *LE_2* based on block type *LE*

- the block type *LE*, containing the previous block contents, and
- two block definitions *LE_1* and *LE_2* based on *LE*.

You can see one gate *peer* in Figure 2.30: this gate is used to connect the signal route *sr1* in *LE* to the channel *ss7ch* in *SS7_test*.

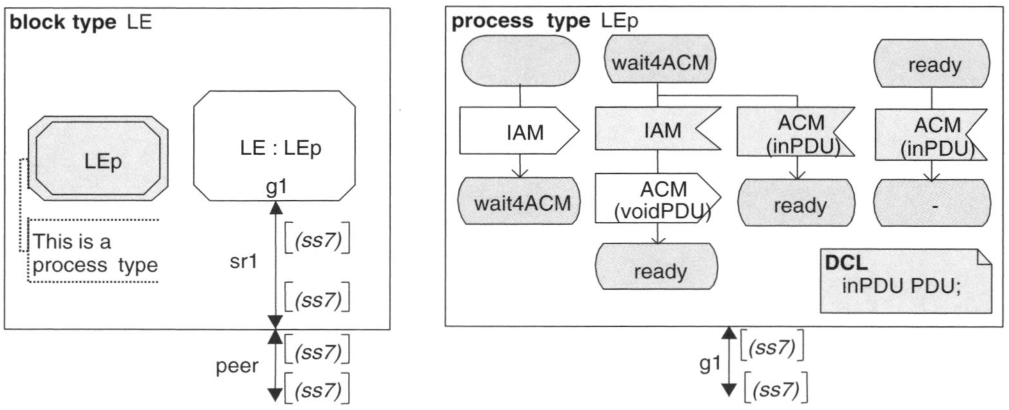

Figure 2.31 The process *LE* based on process type *LEp*

2.5.2.2 Process type

The process *LE* in Figure 2.29 has been replaced, as illustrated in Figure 2.31, by

- the process type *LEp*, containing the previous process contents,
- a process definition *LE* based on *LEp*.

2.5.3 Specialization

In SDL, a type can inherit from another type: for example, a block type *SS7* could inherit from a block type *protocol*, or a signal *sig1* could inherit from a signal *sig2*. In addition, the structure of a type or the transitions it contains can be redefined in the subtype.

3

The V.76 Protocol Case Study

3.1 PRESENTATION

The system used for the case study is a simplified version of the protocol described in the ITU-T V.76 Recommendation based on the Link Access Procedure for Modems (LAPM). This recommendation describes a protocol to establish Data Link Connections (DLCs) between two modems and to transfer data over those connections.

For a detailed step-by-step tutorial on SDL and how to create the simplified V.76 model used here, see [Doldi01].

The V.76 SDL model and associated files can be downloaded in ObjectGeode and Tau SDL Suite formats on *ftp://ftp.wiley.co.uk/pub/books/ldoldi/*.

To illustrate the protocol and the terms used in the ITU-T V.76 Recommendation, we have depicted in Figure 3.1 two Service Users (SU), A and B, communicating through the V.76 protocol layer.

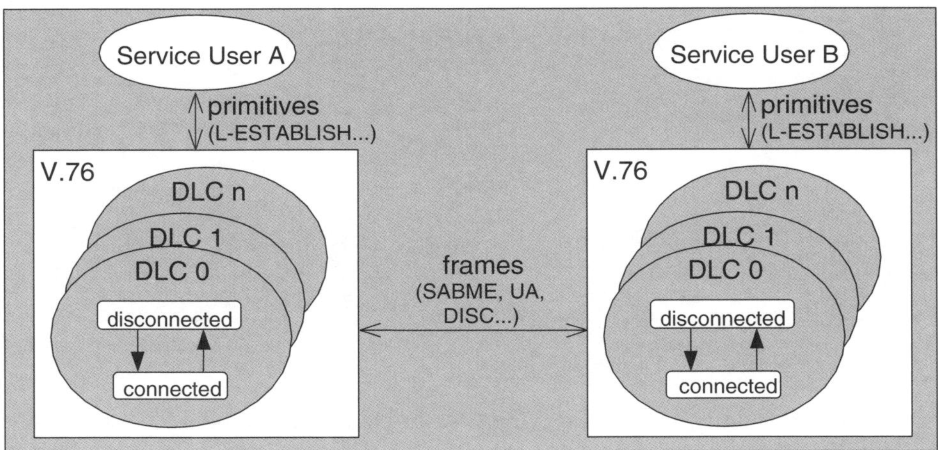

Figure 3.1 Communication between A and B through the simplified V.76

Several connections can exist in parallel: SU A may establish DLC number 0 to transmit voice to or from SU B; while DLC 0 is running, SUA may establish DLC number 1 to transmit data to or from SU B.

Validation of Communications Systems with SDL: The Art of SDL Simulation and Reachability Analysis.
Laurent Doldi ISBN: 0-470-85286-0

An example of this scenario is provided in Figure 3.2: first A and B perform an eXchange IDentification (XID); then A establishes DLC 0; (DLC 0 on sides A and B are in state connected); data is transferred through DLC 0 (between SU A and B); another XID occurs; finally the DLC 0 is released.

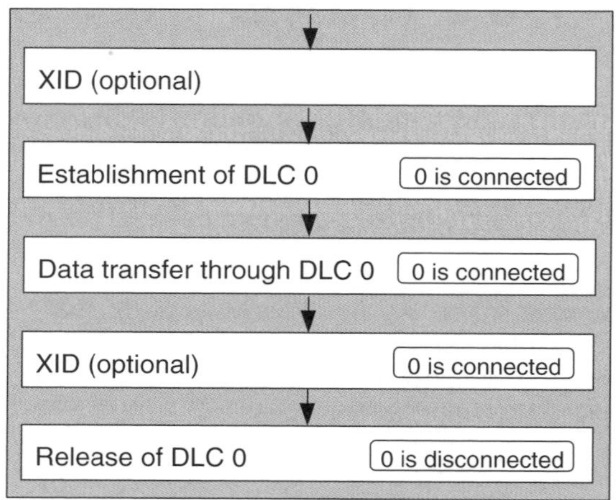

Figure 3.2 Example of V.76 scenario

We remind you in Figure 3.3 of the usual conventions for signal naming in protocols; the right part also shows those conventions mapped on the architecture depicted in Figure 3.1.

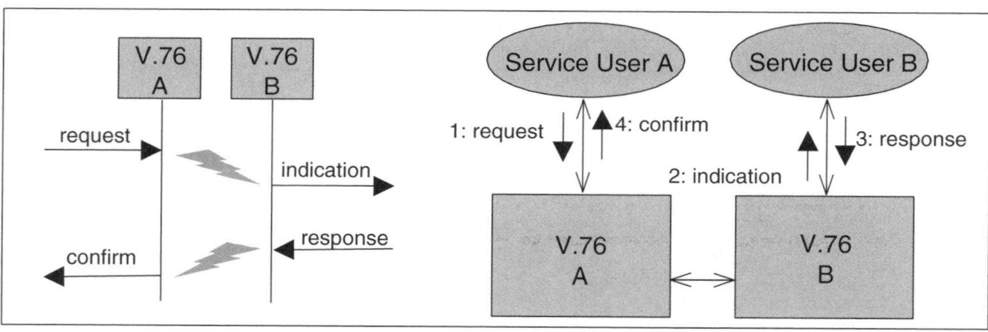

Figure 3.3 Conventions used for signal naming

A request on one side is generally followed by an indication on the other side of the connection; then, if the layer above accepts, it transmits a response, translated into a confirm on the side originator of the request.

3.2 SPECIFICATION OF THE V.76 PROTOCOL

3.2.1 Abbreviations used

DISC	DISConnect
DLC	Data Link Connection entity
DM	Disconnect Mode

I	Information
SABME	SET ASYNCHRONOUS BALANCED MODE EXTENDED
SU	Service User
XID	eXchange IDentification

3.2.2 Exchange identification procedures (XID)

Upon receipt[1] of an L-SETPARM request primitive from its SU (the layer on top of V.76), the DLC entity shall transmit an XID command frame.

On receipt of an XID command frame, the DLC shall issue an L-SETPARM indication primitive to its SU.

Upon receipt of an L-SETPARM response primitive from its SU, the DLC shall transmit an XID response frame.

On receipt of an XID response frame, the DLC shall inform its SU by an L-SETPARM confirm primitive.

3.2.3 Establishment of a data link connection

Establishing a DLC[2] means going from a disconnected to a connected state to allow the transfer of user data.

On receipt of an L-ESTABLISH request primitive from its SU, the V.76 shall attempt to establish the DLC. The DLC entity transmits a Set Asynchronous Balanced Mode Extended (SABME) frame, the retransmission counter shall be reset and timer T320 shall then be started.

A DLC entity receiving an SABME command, if it is able to establish the DLC (as indicated by receipt of an L-ESTABLISH response primitive from the SU in response to an L-ESTABLISH indication primitive), shall

- respond with an Unnumbered Acknowledge (UA) response;
- consider the DLC as established and enter the connected state.

If the SU is unable to accept establishment of the DLC (as indicated by an L-RELEASE request primitive from the SU in response to an L-ESTABLISH indication primitive), the DLC entity shall respond to the SABME command with a Disconnect Mode (DM) response.

Upon reception of the UA, the originator of the SABME command shall stop timer T320 and consider the DLC as established (i.e. enter the connected state) and inform the SU by using the L-ESTABLISH confirm primitive.

Upon reception of a DM response, the originator of the SABME command shall inform its SU of a failure to establish the DLC (by issuing an L-RELEASE indication primitive).

If timer T320 expires before the UA or DM response is received, the DLC entity shall retransmit the SABME command as above, restart timer T320 and increment the retransmission counter.

After retransmission of the SABME command N320 times and failure to receive a response, the DLC entity shall indicate this to the SU by means of the L-RELEASE indication primitive. The value of N320 is 3.

---

[1] A data link connection can be established without being preceded by an XID procedure.

[2] More than one DLC can run in parallel, numbered 0, 1 and so on. This number is indicated in the L-ESTABLISH request.

3.2.4 Information transfer modes

Once in the connected state, information transfer may begin.

3.2.4.1 Transmitting I (Information) frames

Data received by the DLC entity from the SU by means of an L-DATA request primitive shall be transmitted in an I frame[3].

3.2.4.2 Receiving I frames

When a DLC entity receives an I frame, it shall pass the information field of this frame to the SU using the L-DATA indication primitive.

3.2.5 Release of a DLC

The SU requests release of a DLC[4] by use of the L-RELEASE request primitive, then the DLC entity shall initiate a request for release of the connection by transmitting the disconnect (DISC) command.

All outstanding L-DATA request primitives and all associated frames in queue shall be discarded.

A DLC entity receiving a DISC command while in the connected state shall transmit a UA response. An L-RELEASE indication primitive shall be passed to the SU and the disconnected state shall be entered.

If the originator of the DISC command receives either a UA response or a DM response, indicating that the peer DLC entity is already in the disconnected state, it shall enter the disconnected state.

The DLC entity that issued the DISC command is now in the disconnected state and will notify its SU.

3.3 ANALYSIS MSCs FOR THE V.76 PROTOCOL

In order to better understand the protocol, we have created five Message Sequence Charts (MSCs) (similar to UML Sequence Diagrams) illustrating the main behaviors. Such an approach is recommended, especially if the system is complex, before starting the SDL model.

The MSC in Figure 3.4 is named *xid1* and shows two DLC entities *A* and *B* performing an XID transmission, as described in Section 3.2, initiated by DLC *A*.

The MSC in Figure 3.5 shows two DLC entities *A* and *B* performing a DLC establishment (a connection), as described in Section 3.2.3, initiated by DLC *A*.

The MSC in Figure 3.6 shows two DLC entities *A* and *B* performing data (I frames) transfer from *A* to *B*, as described in Section 3.2.4 – to simplify, we consider that the information to

[3] A number in the L-DATA request indicates through which DLC the data must be transmitted.

[4] More than one data link connection can run in parallel, numbered 0, 1 etc. The number identifying the connection to release is indicated in the L-RELEASE request.

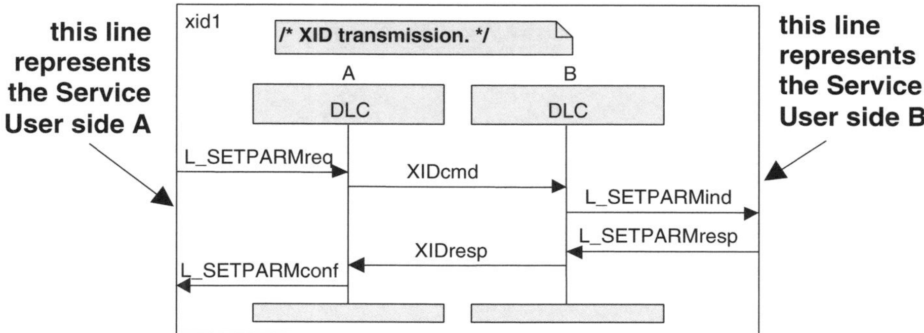

Figure 3.4 The MSC *xid*1: XID (eXchange IDentification

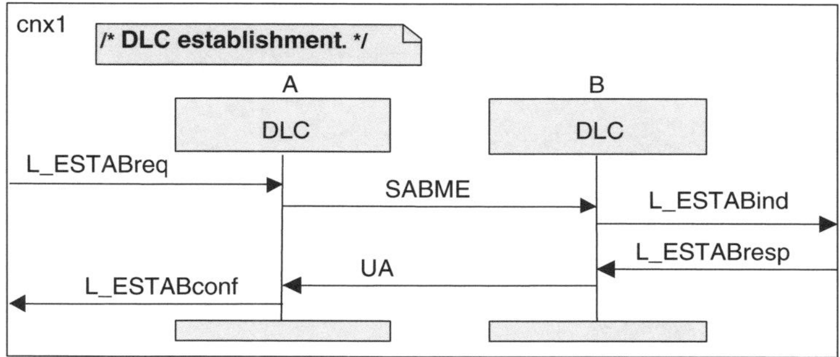

Figure 3.5 The MSC *cnx1*: DLC establishment

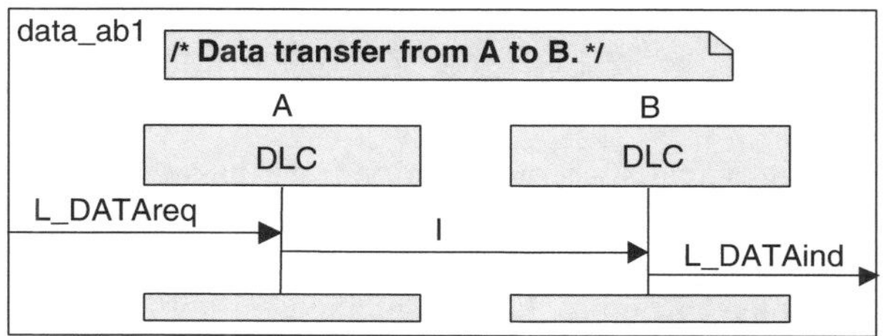

Figure 3.6 The MSC *data_ab1*: data transfer from *A* to *B*

transmit fits into a single I frame. This scenario can only occur after the DLC establishment represented in Figure 3.5.

The MSC in Figure 3.7 is symmetric with Figure 3.6.

The MSC in Figure 3.8 shows two DLC entities *A* and *B* performing a DLC release, as described in Section 3.2.5, initiated by DLC *A*. This scenario can only occur after the DLC establishment represented in Figure 3.5.

This protocol is symmetric, each side being identical: for example, the primitive L-RELEASE request can be received by DLC *A* or *B*, and the frame UA can be both sent or received by a DLC.

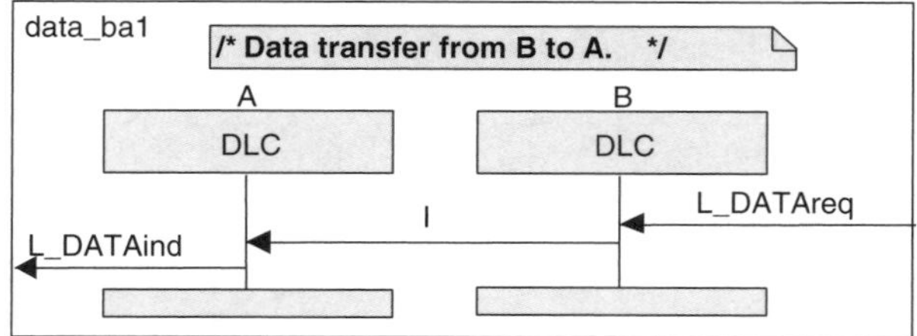

Figure 3.7 The MSC *data_ba1*: data transfer from *B* to *A*

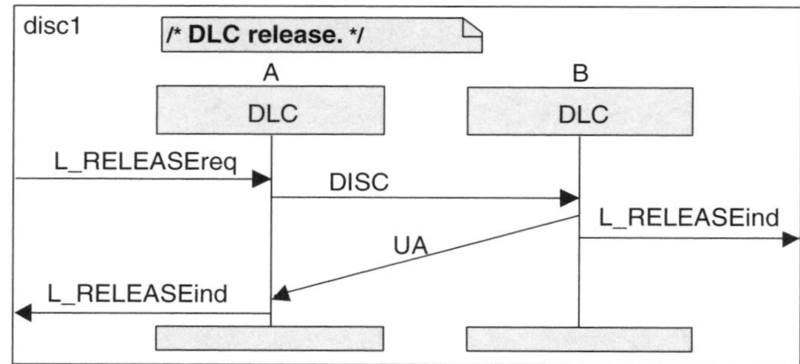

Figure 3.8 The MSC *disc1*: DLC release

3.4 THE SDL MODEL OF V.76

The version included in this section is the version before validation by simulation; it corresponds to Step 9 in [Doldi01] with a few minor changes. During the simulations performed in the next chapters, bugs will be detected and corrected.

3.4.1 The simulation configuration of V.76

To simulate our protocol in a realistic configuration[5], we have created the SDL system represented in Figure 3.9: two instances *DLCa* and *DLCb* of block type *V76_DLC* communicate through the block *dataLink*.

Block *dataLink*, represented in Figure 3.17, simulates a simplified data link layer.

Blocks *DLCa* and *DLCb* communicate with the service users, not modeled, through the channels *DLCaSU* and *DLCbSU*.

3.4.2 The package V76

The block type *V76_DLC* and its signal declarations are contained in the package *V76*, imported by the system *V76test* and shown in Figure 3.10.

The package *V76* also contains the declarations of data types used as signal parameters and the procedure *CRCok*.

[5] Realistic means that we simulate two V.76 peer entities instead of one alone, and we add the block *dataLink* to enable losing frames to test the retransmission mechanism.

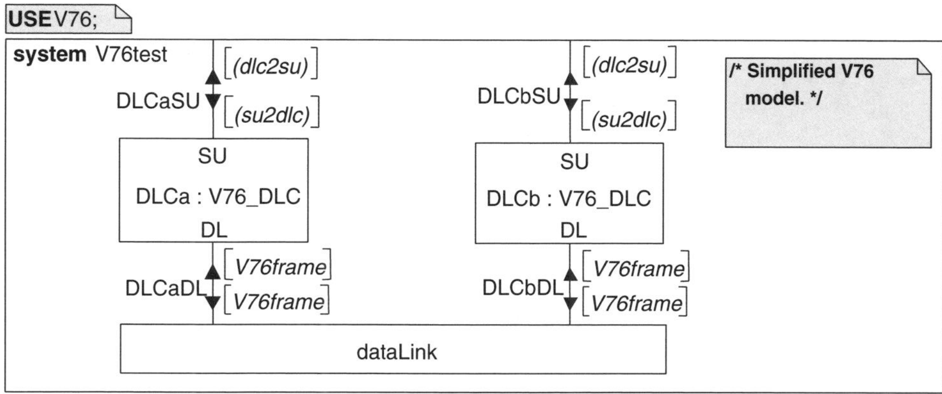

Figure 3.9 The simulation configuration of V.76

```
package V76

SIGNAL
  /* V.76 primitives L-ESTABLISH etc.*/
  L_EstabReq(DLCident),
  L_EstabInd(DLCident),
  L_EstabResp,
  L_EstabConf(DLCident),
  L_SetparmReq,
  L_SetparmInd,
  L_SetparmResp,
  L_SetparmConf,
  L_ReleaseReq(DLCident),
  L_ReleaseInd(DLCident),
  L_DataReq(DLCident, Integer),
  L_DataInd(DLCident, Integer);

SIGNAL
  /* V.76 commands and responses. */
  V76frame(V76paramTyp);

/* Service User to DLC: */
SIGNALLIST su2dlc=
  L_EstabReq,
  L_EstabResp,
  L_SetparmReq,
  L_SetparmResp,
  L_ReleaseReq,
  L_DataReq;

/* DLC to Service User: */
SIGNALLIST dlc2su=
  L_EstabInd,
  L_EstabConf,
  L_SetparmInd,
  L_SetparmConf,
  L_ReleaseInd,
  L_DataInd;
```

```
NEWTYPE V76paramTyp
 CHOICE{
   I Iframe,
   SABME SABMEframe,
   DM DMframe,
   DISC DISCframe,
   UA UAframe,
   XIDcmd XIDframe,
   XIDresp XIDframe }
ENDNEWTYPE ;

SYNONYM maxDLC Integer = 1;

/* DLC Identifier: */
SYNTYPE DLCident =
 Integer CONSTANTS 0 : maxDLC
ENDSYNTYPE;

NEWTYPE SABMEframe
 STRUCT
   DLCi DLCident;
ENDNEWTYPE;

NEWTYPE DMframe
 STRUCT
   DLCi DLCident;
ENDNEWTYPE ;

NEWTYPE DISCframe
 STRUCT
   DLCi DLCident;
ENDNEWTYPE ;

NEWTYPE UAframe
 STRUCT
   DLCi DLCident;
ENDNEWTYPE;
```

```
SYNTYPE XIDframe = Integer
ENDSYNTYPE ;

NEWTYPE Iframe
 STRUCT
   DLCi DLCident;
   data Integer;
   CRC Integer;
 OPERATORS
   fill_Iframe: DLCident, Integer, Integer
    -> Iframe;
 OPERATOR fill_Iframe;
   FPAR dlc DLCident, D Integer,
    Crc Integer;
   RETURNS res Iframe;
   START ;
   TASK res := (. dlc, D, Crc .);
   RETURN res;
 ENDOPERATOR ;
ENDNEWTYPE ;
```

V76_DLC — A block type

CRCok — This procedure checks the CRC

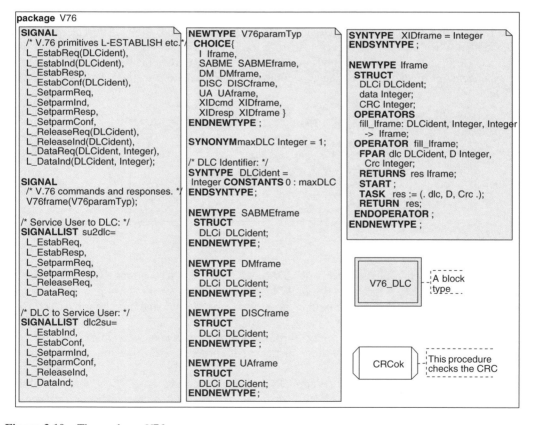

Figure 3.10 The package *V76*

As we will see later, we create one instance of process *DLC* on each side of the connection to handle each DLC; then, when a DLC is released, we stop the two corresponding process instances. For example, if three DLCs are created, we will have three instances of process *DLC* on each side of the connection; then three independent data transfers (through I frames) may occur several times before releasing one or more DLCs.

To simplify, we have not modeled the allocation of DLC numbers: we suppose that the user provides the DLC number as a parameter of the *L_EstabReq* signal when he wants to create a new DLC.

DLCident is the DLC number corresponding to a data stream multiplexed over the layer below V.76. As shown in Figure 3.10, *DLCident* is a SYNTYPE of Integer, ranging between 0 and *maxDLC. maxDLC* is a SYNONYM, equal to 1. *DLCident* may have the values 0 or 1. This parameter has been added to the signals *L_EstabReq*, *L_EstabInd*, *L_ReleaseReq*, *L_ReleaseInd*, *L_DataReq* and *L_DataInd*. For example, in *L_ReleaseReq, DLCident* is used to indicate which DLC must be released.

Then, to add *DLCident* to the *V76frame*, we have declared STRUCTs for *SABMEframe, DMframe* and so on. The unique field of those STRUCTs is *DLCi*. XID frames do not carry *DLCident*.

The procedure *CRCok*, shown in Figure 3.11 (in Tau, the header looks different, with extra ";"), is called when an I frame is received from the peer entity, to check that the carried data is correct. Here, it has been simplified, but normally the CRC (Cyclic Redundancy Check), a kind of checksum of the received data, is computed and compared to the CRC received with the data.

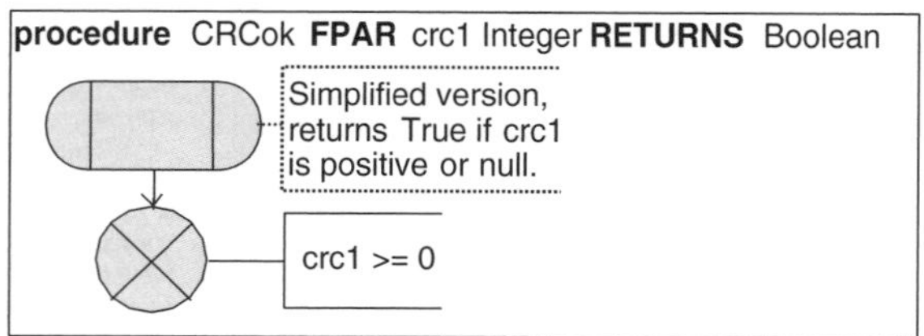

Figure 3.11 The procedure *CRCok*

Going down into the block type *V76_ DLC* we note that it contains two processes, as shown in Figure 3.12: a process *dispatch* having one instance at startup and maximum one instance, and a process *DLC* having no instance at startup and maximum *maxDLC+1* instances[6].

Process *DLC* receives signals only through process *dispatch: dispatch* forwards the received signals to the corresponding instance of process *DLC* (using the parameter DLC number).

Process *dispatch*, as shown in Figure 3.13, upon receiving an *L_EstabReq* (establish request) checks that the DLC to create, *DLCnum*, is not used and creates an instance of process *DLC*, passing *DLCnum* and *True* as parameters. Then[7] the Pid of the new instance of process *DLC* is stored into the array *DLCs*, at the index *DLCnum*. This array is declared

[6] The dashed arrow from *dispatch* to DLC indicates that *dispatch* may create instances of *DLC*.

[7] In a real situation, it should be safer to check that Offspring does not contain Null after the create request, meaning that the creation of the process instance failed.

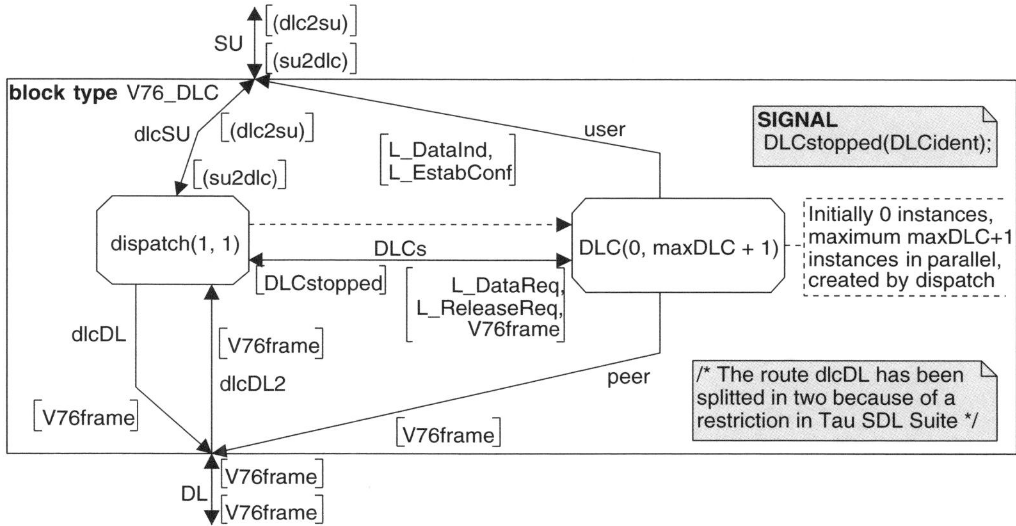

Figure 3.12 The block type *V76_ DLC*

process dispatch(1, 1)

ready

This is a priority input

L_DataReq (DLCnum, uData)

L_ReleaseReq (DLCnum)

L_EstabReq (DLCnum)

DLCs (DLCnum)

NULL /* DLCnum not setup */

ELSE

L_DataReq (DLCnum, uData) **TO** DLCs(DLCnum)

We pass the frame to the corresponding instance of process DLC

L_ReleaseReq (DLCnum) **TO** DLCs(DLCnum)

NULL

DLC (DLCnum, True)

DLCnumnot used, we create an instance of process DLC

DLCs(DLCnum) :=**OFFSPRING**

We store into the table the PID of the new instance

L_ReleaseInd (DLCnum)

waitUA

V76frame (V76para)

V76para ! **present**

UA

V76frame(V76para) **TO** DLCs(V76para ! UA ! DLCi)

ready

DLCstopped (DLCnum)

L_SetparmReq

L_ReleaseInd(DLCnum)

DLCs(DLCnum):= NULL

V76frame (XIDcmd : 0) **VIA** dlcDL

-

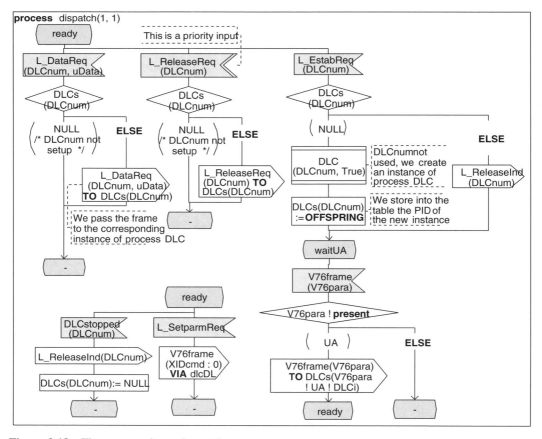

Figure 3.13 The process *dispatch* part 2

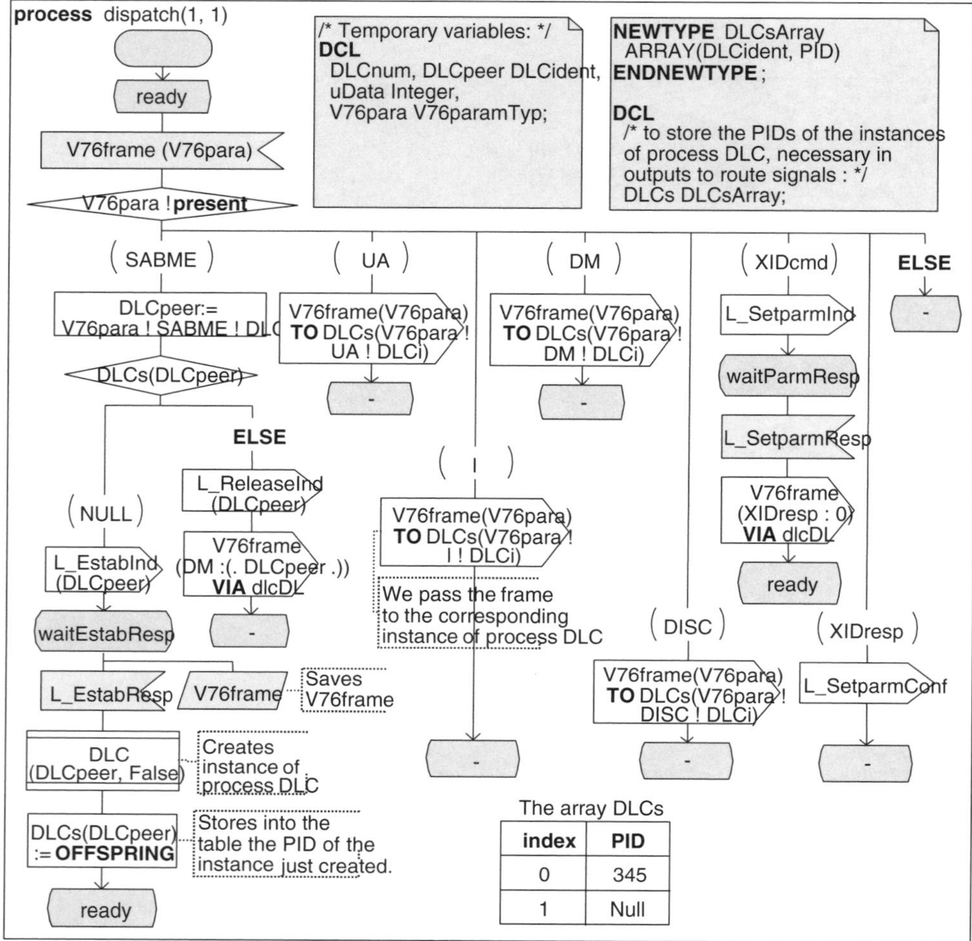

Figure 3.14 The process *dispatch* part 1

and illustrated in Figure 3.14. The table in the figure shows an example in which one DLC exists, of number 0.

This array is necessary to get the Pid of a *DLC* process instance from its DLC number, to forward a signal to it. After this creation, *dispatch* goes to state *waitUA*, waiting for the UA frame from its peer.

From the state *waitUA*, if a *V76frame* signal containing a *UA* arrives, we forward it to the corresponding instance of process *DLC*, using the array of Pids *DLCs* to get its Pid from the DLC number present in the *UA*, as shown in Figure 3.13.

Figure 3.14 shows the second part of process *dispatch*: when receiving a *V76frame* (from the peer), we use a decision to get the kind of element actually present in the received parameter *V76para*. If the field *present* is equal to *SABME*, we get the DLC number received in *V76para* and test if this number corresponds to a free DLC in the Array *DLCs*: Null in the Array means that no instance of process *DLC* exists. We send an *L_EstabInd* to the service user and wait for its response. On receiving the *L_EstabResp*, we create the instance of process *DLC*, passing parameters *DLCpeer* (its DLC number) and False (meaning that this side of the connection is not the originator of the *L_EstabReq*) to it. Finally, we store into the array *DLCs* the Pid of the instance just created and go to state *ready*.

Figure 3.15 and 3.16 show the contents of process *DLC*, dynamically created by process *dispatch*. When an instance of process *DLC* starts, if the parameter *originator* is True, it outputs a *SABME* (to the peer), else a *UA* (also to the peer). After transmitting a *SABME*, it outputs an *L_EstabConf* on receiving a *UA*: the connection is established. Symmetrically, after transmitting a *UA*, it goes to state *connected*.

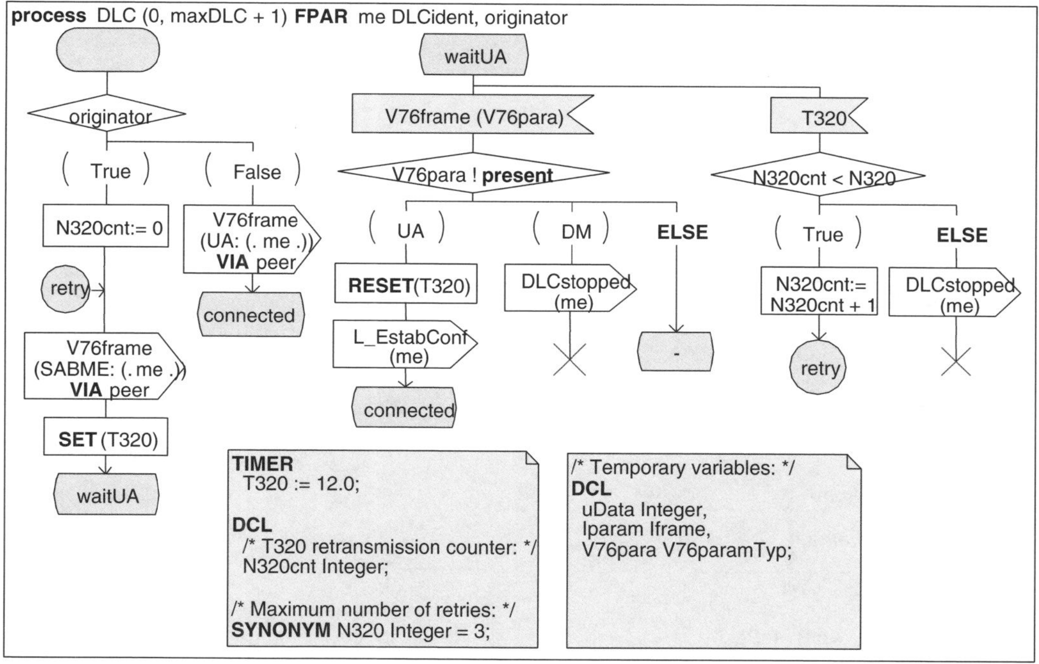

Figure 3.15 The process *DLC* part 1

From state *connected*, represented in Figure 3.16, we can see the *L_DataReq* input used for data transfer (through I frames), the *L_ReleaseReq* input to release the DLC (ending by a stop of the instance, symbol X), or the *V76frame* input containing either *DISC* or an I frame. Note that when a *DISC* is received, a signal *DLCstopped* is sent to process *dispatch* to update the array *DLCs*.

3.4.3 The block *dataLink*

Figure 3.17 depicts the contents of our data link layer block, whose role is just to retransmit (or to lose) the *V76frames* received from one DLC to the peer DLC. One process should have been enough to perform this function, but we would have faced addressing problems: in SDL,

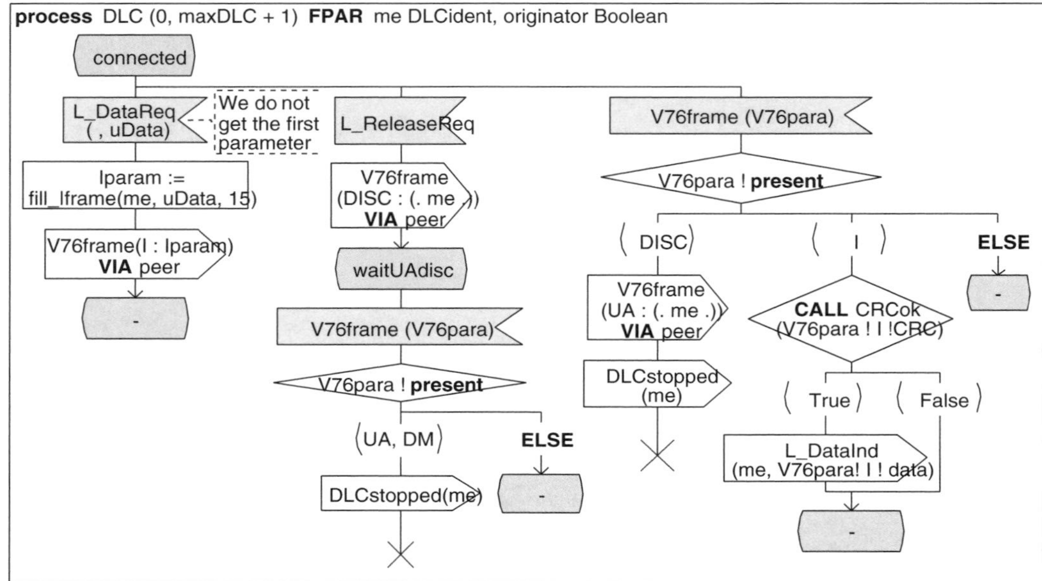

Figure 3.16 The process *DLC* part 2

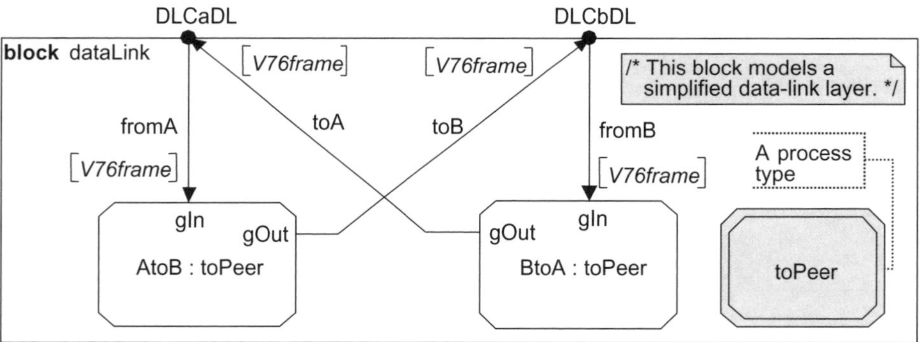

Figure 3.17 The block *dataLink*

no feature allows to send a received signal to a peer in a situation in which the two sides are identical. It would have been necessary to store Pids and so on.

A simple solution is to use one process for each direction:

- *AtoB* receiving signals from *DLCa* and forwarding them to *DLCb*,
- and *BtoA* receiving signals from *DLCb* and forwarding them to *DLCa*.

We have instrumented the process type *toPeer*, shown in Figure 3.18, to introduce faults, to test the robustness of our protocol, for example, to see its reaction when a signal is lost: in

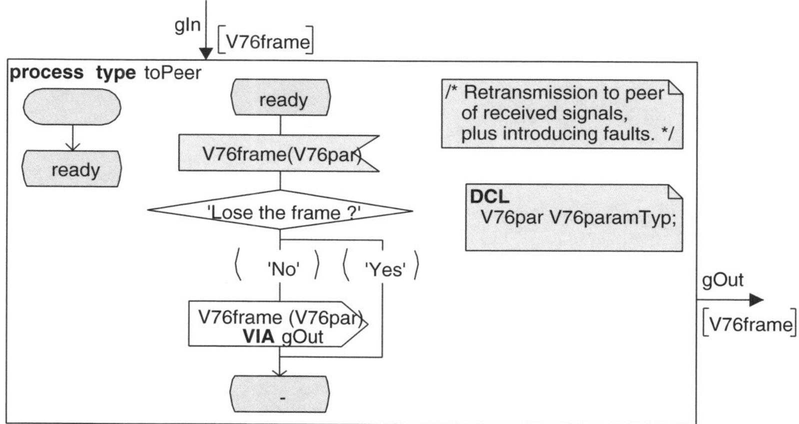

Figure 3.18 The process type *toPeer*

the informal decision *'Lose the frame?'*, the text between ' ' is not compiled. SDL simulators, upon reaching such a symbol, will prompt the user for the answer to the decision: if she or he selects 'No', signal *V76frame* is transmitted, otherwise it is not transmitted; it is lost.

4

Interactive Simulation

In this chapter, you will learn how to validate the V.76 SDL model using interactive simulation: compiling the SDL model, simulating the main scenarios such as connection establishment, generating MSC simulation traces, detecting, analyzing and correcting bugs, detecting and analyzing nonsimulated symbols and writing scripts to automatize the validation. Finally, the list of errors that can be detected by interactive simulation is presented, for the two Simulators used.

4.1 PRINCIPLES

As soon as an SDL model is terminated[1], it must be simulated interactively to check if its behavior is correct. The simulation allows the user to test the SDL model in an abstract and simplified world, and without waiting for the end of the coding or for the availability of the target hardware. In this simplified world, bugs are easier to find and correct than when the actual code is loaded into the target system, where the specification-level and design-level problems are complicated by coding-level issues.

During interactive simulation, the SDL model is executed one transition or one symbol at a time, like in a debugger. To be consistent with the SDL ITU-T Recommendation [SDL92], a transition must be terminated before starting another transition. In other words, the execution of a transition cannot be interrupted, except when calling a remote procedure or when waiting for the user to provide the answer to an informal decision.

Between two transitions or symbols, the SDL model is stopped and the contents of its variables can be displayed or modified.

After the execution of a transition, the simulators evaluate which transitions are ready to be executed. If several transitions are ready, they are presented to the user who selects one to execute (ObjectGeode) or the oldest transition is proposed (Tau SDL Suite Simulator). By default, all the process instances have the same priority.

During simulation, the SDL time progresses only if timers are started (in the real world, the SDL time progresses according to the real time, that is, even if timers are not used).

After each interesting simulation scenario, the corresponding MSC trace or script can be stored into a file to be replayed automatically, for example for nonregression testing after modifications.

Observers can be used to monitor the simulation, to automatically detect when the simulated behavior is or is not consistent with the expected behavior(s).

[1] In fact, an SDL model can be simulated even if some parts are not terminated, such as informal tasks, informal decisions or empty processes. Informal means that the text is placed between single quotes.

Validation of Communications Systems with SDL: The Art of SDL Simulation and Reachability Analysis.
Laurent Doldi © 2003 John Wiley & Sons, Ltd ISBN: 0-470-85286-0

Note that this kind of simulation does not verify the performances, such as response time, of the SDL model. For that, special libraries or extensions to SDL are provided by the simulators.

4.2 CASE STUDY WITH TAU SDL SUITE

The tools and platform used for the exercises are described in Section 1.5. The V.76 SDL model used for this case study is presented in Chapter 3.

As opposed to ObjectGeode, the simulation features of Tau SDL Suite are split in two tools:

- The Simulator: interactive simulation and automatic simulation, with ready-first[2] scheduling.
- The Validator: interactive, random and exhaustive simulation, with or without[3] ready-first scheduling.

4.2.1 Prepare the Simulator

4.2.1.1 Compile the SDL model

A. Download from the Internet (see the ftp address in Section 1.5) or create the SDL V.76 model, and launch Tau SDL Suite Organizer: double-click the icon *v76.sdt*[4], or for Unix: type *sdt v76.sdt*. The window represented in Figure 4.1 appears.

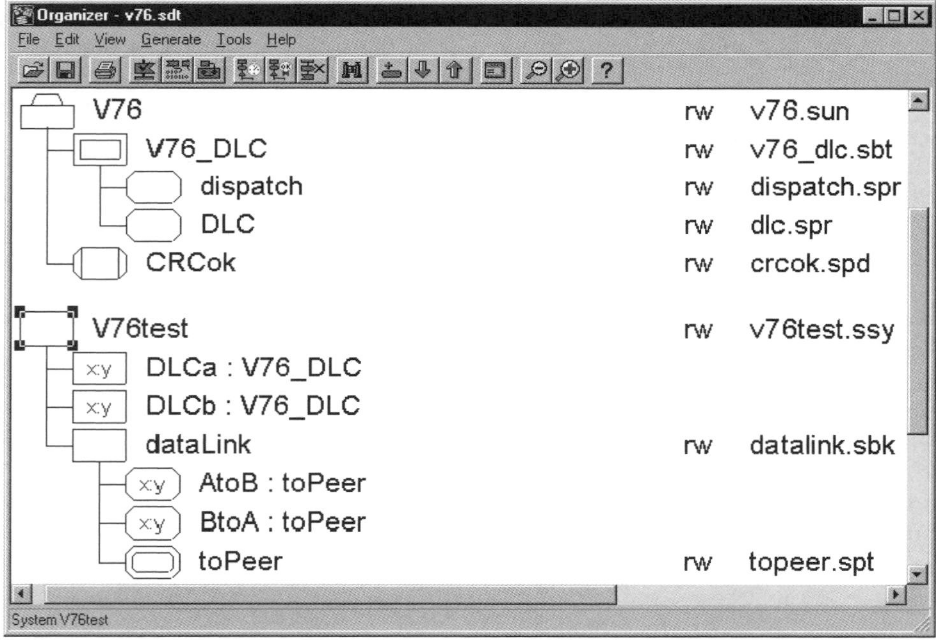

Figure 4.1 The Tau SDL Suite Organizer

[2] When two transitions are ready, the one that actually became ready first will be executed.

[3] When two transitions are ready, the two can be fired, even the last to become ready. The choice between the two transitions is made according to the simulation mode: either by the user (interactive) or according to a random number (random) or the two alternatives are explored (exhaustive).

[4] We recommend setting your Windows options to display the file extensions (in Windows NT4, open *Workstation*, then select *Display* > *Options*: in the *Display* tab, check *Display all files*, and uncheck *Mask the extensions for known file types*, then press *OK*).

B. In the Organizer, select the SDL system *V76test* and do *Generate > Make*: the window represented in Figure 4.2 appears.

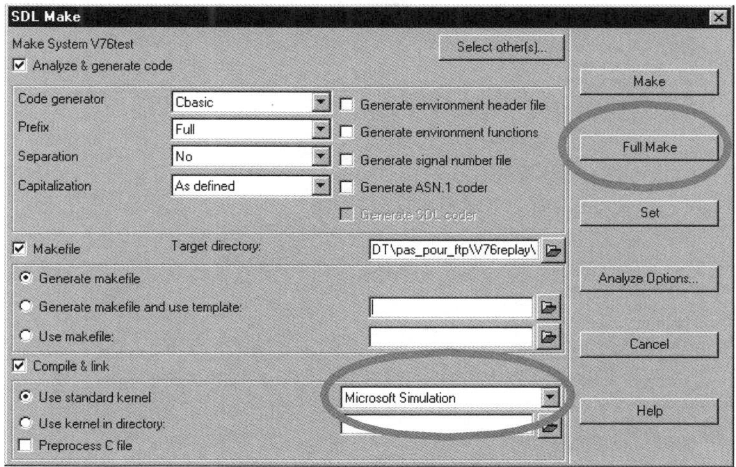

Figure 4.2 The SDL Make window

C. In the SDL Make window, check that the options are similar to the options selected in Figure 4.2, and depending on your C compiler, select *Microsoft Simulation* or *Borland Simulation*.

D. In the SDL Make window, press *Full Make*: the Organizer Log appears, as shown in Figure 4.3. If your SDL model is correct, Tau generates the C code, then compiles the C files and links them to the simulator library, to produce the executable file *v76test_smc.exe*[5], as depicted in Figure 4.4.

4.2.1.2 Start the Simulator

A. In the Organizer, press the *Simulate* button to start the Simulator[6]. The Simulator main window appears, as shown in Figure 4.5. Drag it to the upper left corner of your screen.

4.2.1.3 Start the SDL trace

For the moment, four start transitions can be fired, because the V.76 model contains four process instances: *AtoB:1, BtoA:1, << Block DLCa >> dispatch:1* and *<< Block DLCb >> dispatch:1*.

A. Press the button *SDL* in the *Trace* group as indicated in Figure 4.5.

B. Press the button *Transition* in the *Execute* group: the SDL Editor displays the next symbol ready to be executed, as illustrated in Figure 4.6.

[5] This is the name corresponding to Microsoft C compiler on Windows.

[6] Once you have set your options in the SDL Make window, you can directly compile the SDL model and launch the Simulator just by pressing the *Simulate* button.

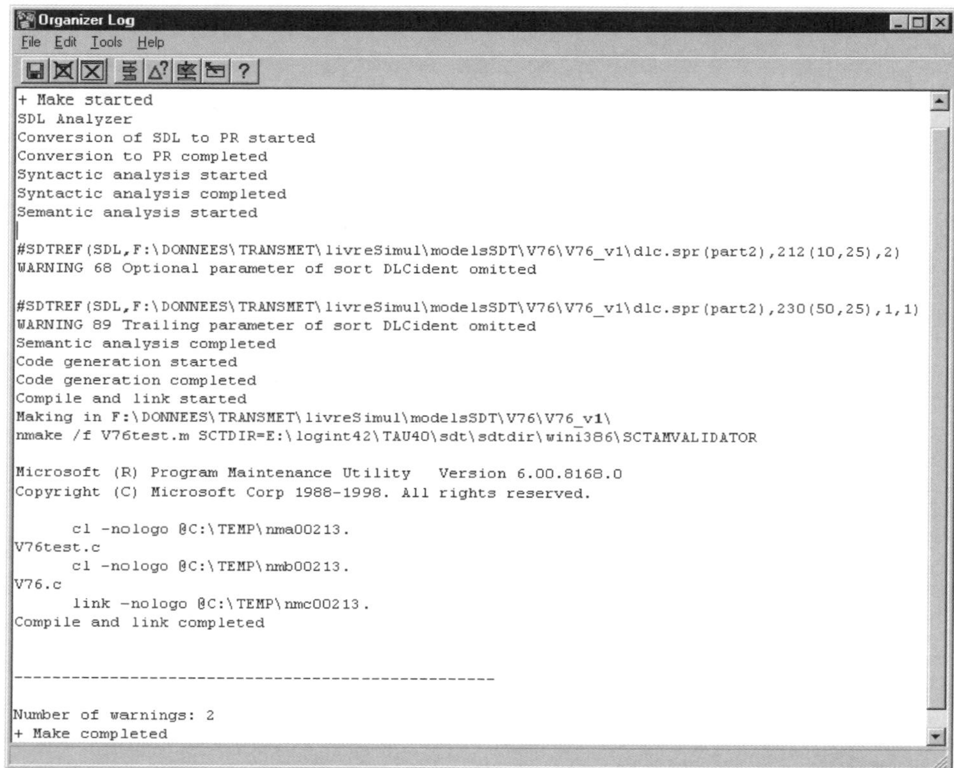

Figure 4.3 The Organizer Log after the Make command

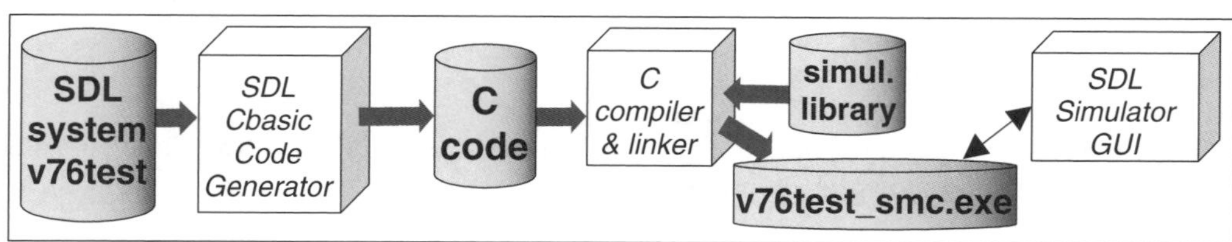

Figure 4.4 The Tau SDL Suite simulation building

The right part of the Simulator window contains:

```
*** TRANSITION START
*       PId     : AtoB:1
*       State   : start state
*       Now     : 0.0000
*** NEXTSTATE  ready
```

It means that the Simulator has executed the start transition in the instance 1 of process *AtoB* (based on process type *toPeer*), to reach the state *ready*.

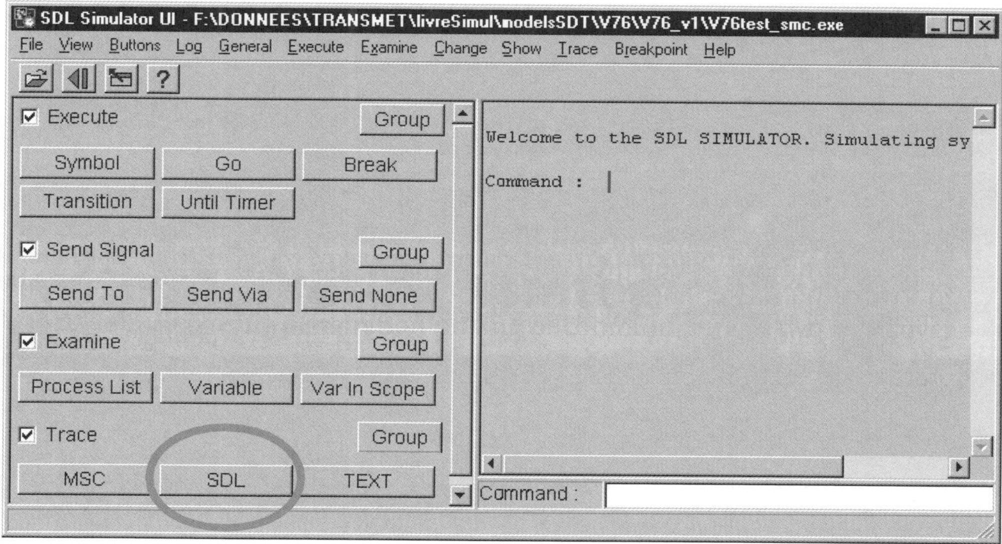

Figure 4.5 The Simulator is ready

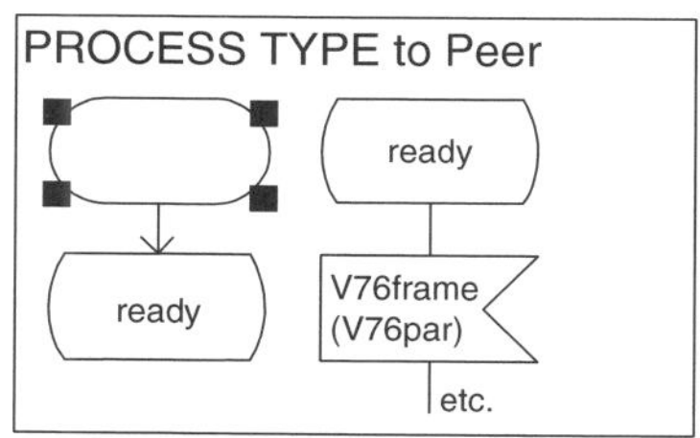

Figure 4.6 SDL trace: start symbol ready to be executed

C. In the Simulator, select *View > Command Window*: the *Command* Window appears, as shown in Figure 4.7; in its upper part you can see the ready queue of the Simulator, where three process instances are ready to be executed: *BtoA, dispatch* in block *DLCa* and *dispatch* in block *DLCb*.

D. In the Simulator, press the button *Transition* three times: the ready queue is now empty. If you press *Transition* again, the Simulator displays *No process instance scheduled for a transition*: the simulation is blocked because the SDL model now expects to receive some external signals.

4.2.1.4 Prepare the Simulator for MSC recording

A. In the Simulator, type the command *Define-MSC-Trace-Channels on* (or type *d-m-t-c on*): instead of drawing only one environment instance, the Simulator will generate one environment instance for each external SDL channel: *DLCaSU* and *DLCbSU*, as shown in Figure 4.8. This clarifies the MSC trace, separating signals linked to DLCs A and B.

B. In the Simulator, select *Trace > MSC Trace: Start*: a Select window appears.

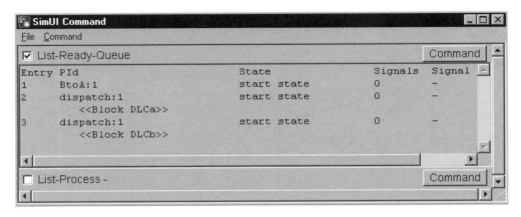

Figure 4.7 The Simulator *Command* Window

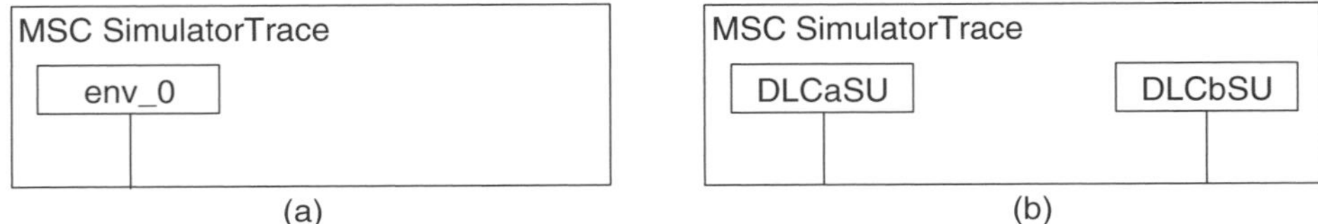

Figure 4.8 Environment: default (a) and *Define-MSC-Trace-Channels* (b)

C. In the Select window, select[7] *0 /* Basic MSC */* and press *OK*: the MSC Editor displays the MSC trace.

D. Arrange the Simulator and SDL and MSC Editors windows as in Figure 4.9.

4.2.2 Validate against the main scenarios

Now the Simulator is ready to check that the SDL model behaves as described in the MSCs drawn during the analysis step in Section 3.3. You can compare the analysis MSCs and the MSCs recorded during simulation.

4.2.2.1 Send a signal to the model

Figures 4.34 and 4.35 show that the V.76 SDL model requires the input of external signals, such as *L_EstabReq*, contained in the signal list *su2dlc* on the channels *DLCaSU* and *DLCbSU*. You will use the *Send Signal* command[8] to send such signals to the SDL model:

A. In the Simulator, press the *Send To* button: select *L_EstabReq* and press *OK*.
B. Enter the value *0* for the parameter of signal *L_EstabReq*, of type *DLCident*. Press *OK*.
C. Select the target process *<< Block DLCa >> dispatch* and press *OK*.

---

[7] If you select 1, the MSC also displays the state of each process instance, and if you select 2 it adds the SDL actions (assignments etc.).

[8] The ObjectGeode Simulator *feed* command has no equivalent here. The Tau *Send Signal* command corresponds to *Execute > Output* in ObjectGeode. Thus the external signals cannot be sent before the receiving processes reach a state where they are ready to input them, otherwise they are lost.

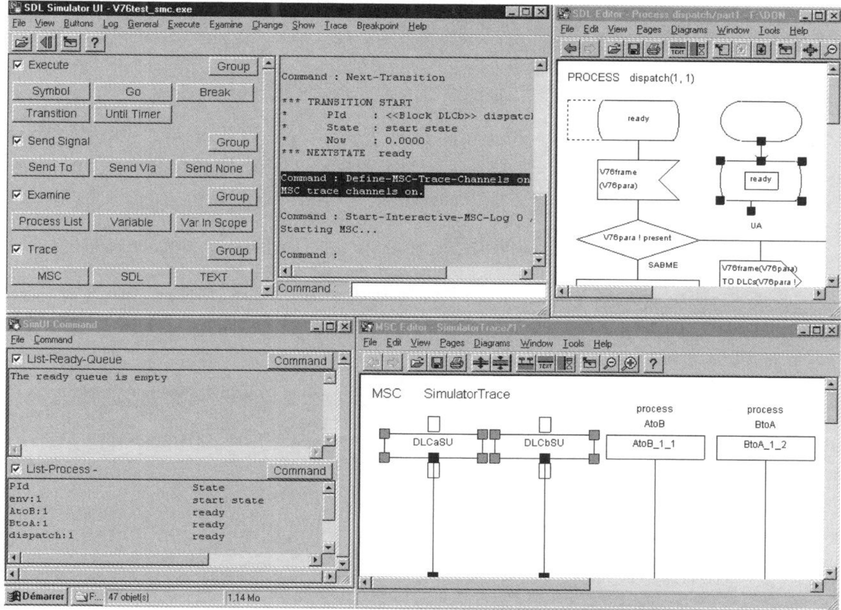

Figure 4.9 The Simulator is ready for MSC & SDL trace

In the *Command* Window, you can see that the ready queue now contains process *dispatch* from block *DLCa*, ready to input signal *L_EstabRe*q, as depicted in Figure 4.10.

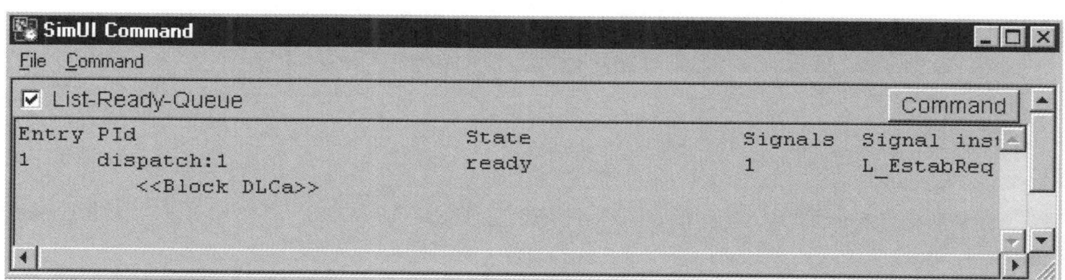

Figure 4.10 The Simulator ready queue

4.2.2.2 Play and record the scenario DLC establishment

You will simulate the SDL model, trying to play the scenario described in the MSC *DLC establishment*, shown in Figure 3.5. The Simulator will generate an MSC trace representing the actual SDL behavior, which you will compare with the expected MSC shown in Figure 3.5.

A. Press three times the *Transition*[9] button in the *Execute* group until you see the *Select - Enter path* window[10]: the Simulator executes the ready transitions (scheduled in the Simulator ready queue).

[9] If you had pressed on *Go*, the Simulator would have run automatically until the model is blocked.

[10] This window appears because the Simulator reaches a decision containing text between quotes, called an informal decision: it asks you which answer you want to execute.

B. In the *Select - Enter path* window, select *1 /* "No" */* and press *OK*. It means that the signal *v76frame* is not lost.

C. Press ONCE ONLY the *Transition* button.

In the MSC trace, you see that process *dispatch* on side B has just sent *L_EstabInd*, and in the SDL trace, you see that timer *T320* is ready to time-out. To avoid this, do NOT continue the execution: instead you will send an *L_EstabResp* to *dispatch* on side B.

D. In the Simulator, press the *Send To* button: select *L_EstabResp* and press *OK*.

E. Select the target process *<< Block DLCb >> dispatch* and press *OK*: the MSC Editor displays the trace[11] represented in Figure 4.11, arranged for clarity.

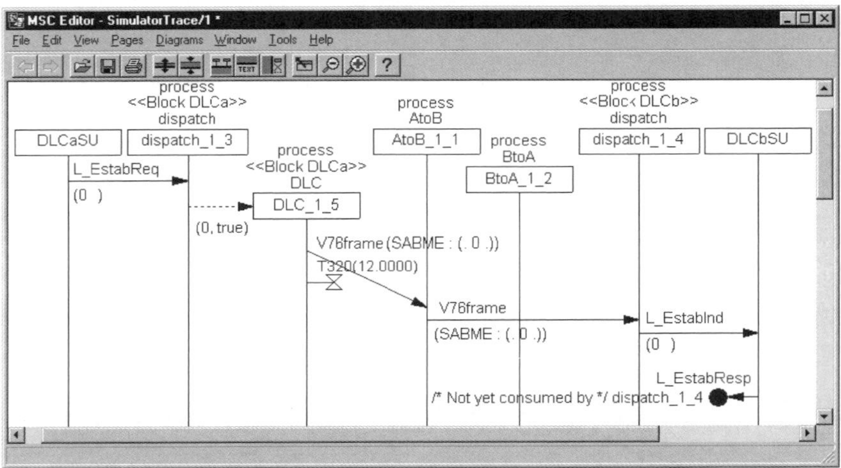

Figure 4.11 The MSC trace after sending *L_EstabResp*

The last executed SDL symbol, input of L_EstabResp, is selected, as illustrated in Figure 4.12(a).

F. Simulate symbol by symbol: press the *Symbol* button in the Simulator; the next SDL symbol, creation of a *DLC* instance, has been executed, as shown in Figure 4.12(b).

G. Terminate the scenario: in the Simulator, press again several times the *Transition* button answering *No* to the question, until you see the output of signal *L_EstabConf* in the MSC. The two DLCs are both in state *connected.*

The MSC trace generated by the Simulator, displayed in Figure 4.13, shows that the SDL model complies with the MSC in Figure 3.5. The MSC generated contains the values of the signal parameters, such as the DLC number, here equal to 0.

H. Save the MSC: in the MSC Editor, select *File > Save As*, navigate to the desired directory, enter *cnx1.msc* and press *OK*.

[11] The environment of the SDL model is represented here by the pseudo-instances *DLCaSU* and *DLCbSU*. In ObjectGeode, the environment is represented by the frame.

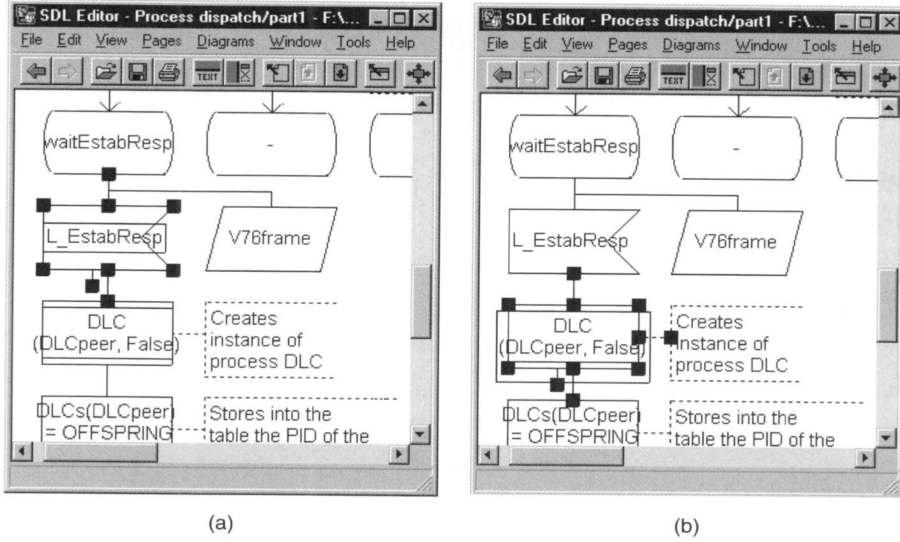

Figure 4.12 Simulating symbol by symbol

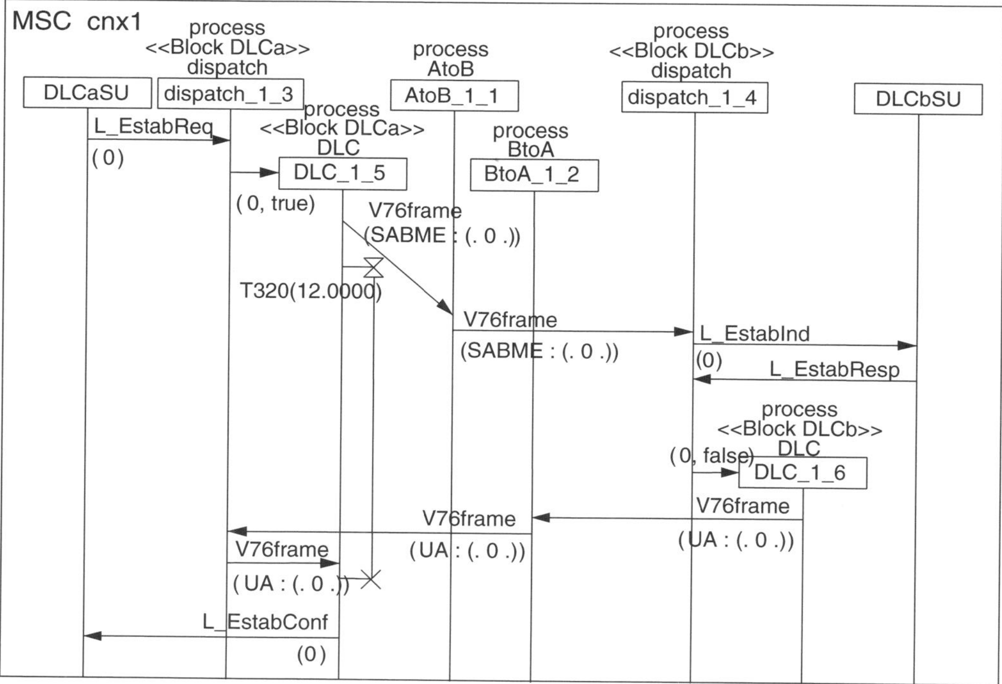

Figure 4.13 The MSC simulation trace

I. Save the simulation commands into a file: in the Simulator, select *Log > Save Input History*[12], select *All files*, enter *cnx1.com* and press *OK*.

J. Open the file *cnx1.com* with a text editor[13], and if necessary add a return in the two lines:

```
Next-Transition 1 /* "No" */
```

to get:

```
Next-Transition
1 /* "No" */
```

Remove the two lines:

```
Define-MSC-Trace-Channels on
Start-Interactive-MSC-Log 0
```

K. Save the file and close the text editor.

Later, you will do *File > Restart* and *Execute > Command Script* to replay the file *cnx1.com* automatically. Here is the content of *cnx1.com* (the first line starts the SDL trace):

```
Set-GR-Trace 1
Next-Transition
Next-Transition
Next-Transition
Next-Transition
Output-To L_EstabReq (0) <<Block DLCa>> dispatch
Next-Transition
Next-Transition
Next-Transition
1 /* "No" */
Next-Transition
Output-To L_EstabResp <<Block DLCb>> dispatch
Step-Symbol
Next-Transition
Next-Transition
Next-Transition
1 /* "No" */
Next-Transition
Next-Transition
```

Do not exit from the Simulator.

[12] *Save Command History* also works, but it also contains commands that have been undone or commands entered before Simulator restart.

[13] To easily open any file in Windows, we recommend you to create a shortcut to WordPad on your desktop. Then simply drag the file on the shortcut to open it.

4.2.2.3 Play the other scenarios

Once the DLC connection is established (if not, play the scenario described in Section 4.2.2.2), continue the simulation, playing the following three scenarios (important: play them in sequence, i.e. never reinitialize the Simulator):

A. Simulate XID, described in Figure 3.4: send signal *L_SetParmReq* to *<< Block DLCa >> dispatch*, press *Go* and select *1 /* "No" */*; send signal *L_SetParmResp* to *<< Block DLCb >> dispatch*, press *Go* and select *1 /* "No" */*; check that the MSC trace is compliant with Figure 3.4.

B. Simulate data transfer from A to B, described in Figure 3.6: send signal *L_DataReq* with parameters *0* and *86* to *<< Block DLCa >> dispatch*, press *Go*, select *1 /* "No" */*.

C. Simulate DLC release, described in Figure 3.8: send signal *L_ReleaseReq* with parameter *0* to *<< Block DLCa >> dispatch*, press *Go* and select *1 /* "No" */* two times.

D. Save the simulation commands into the file *test1.com,* as indicated previously (*Log > Save Input History*), because you will need it later.

E. Open the file *test1.com* with a text editor, and if necessary add a return in all the lines like:

```
Go 1 /* "No" */
Next-Transition 1 /* "No" */
```

to split them:

```
Go
1 /* "No" */
Next-Transition
1 /* "No" */
```

Remove the two lines:

```
Define-MSC-Trace-Channels on
Start-Interactive-MSC-Log 0
```

Here is the content of *test1.com*:

```
Set-GR-Trace 1
Next-Transition
Next-Transition
Next-Transition
Next-Transition
Output-To L_EstabReq (0) <<Block DLCa>> dispatch
Next-Transition
Next-Transition
Next-Transition
1 /* "No" */
Next-Transition
Output-To L_EstabResp <<Block DLCb>> dispatch
Next-Transition
```

```
Next-Transition
Next-Transition
1 /* "No" */
Next-Transition
Next-Transition
Output-To L_SetparmReq <<Block DLCa>> dispatch
Go
1 /* "No" */
Output-To L_SetparmResp <<Block DLCb>> dispatch
Go
1 /* "No" */
Output-To L_DataReq (0, 86) <<Block DLCa>> dispatch
Go
1 /* "No" */
Output-To L_ReleaseReq (0) <<Block DLCa>> dispatch
Go
1 /* "No" */
1 /* "No" */
```

Save the file and close the text editor. Do not exit from the Simulator.

4.2.2.4 Generate a black box MSC

The MSC generated in Figure 4.13 contains all the model process instances. You will now generate a more abstract "black box" MSC, containing only the blocks *(DLCa, dataLink* and *DLCb*), to validate the external behavior of the system:

A. In the Simulator, select *File > Restart ...* and press *OK*.

B. Type the command *Define-MSC-Trace-Channels on* (or type *d-m-t-c on*) to get two environment instances.

C. Select *Trace > MSC Level: Set*: a *Select Unit name* window appears, select *System V76test* and press *OK*: a *Select Trace value* window appears, select *3 /* Block trace */* and press *OK*.

D. Select *Trace > MSC Trace: Start*: select 0 /* Basic MSC */ and press OK.

E. Select *Execute > Command Script*, choose *test1.com* and press *OK*: the Simulator executes the commands contained in *test1.com*, and the MSC trace contains only blocks, instead of processes, as illustrated in Figure 4.14.

F. Rename the MSC: in the MSC Editor, type *test1* instead of *SimulatorTrace*.

G. Save the MSC: in the MSC Editor, select *File > Save As*, navigate to the desired directory, enter *test1.msc* and press *OK*.

H. Exit from the Simulator.

4.2.3 Detect a bug in the SDL model

4.2.3.1 Check the number of retransmissions

You will check that the retransmission of SABME occurs a maximum of three times after its first transmission, as specified in Section 3.2.3. For that, you will generate an MSC trace

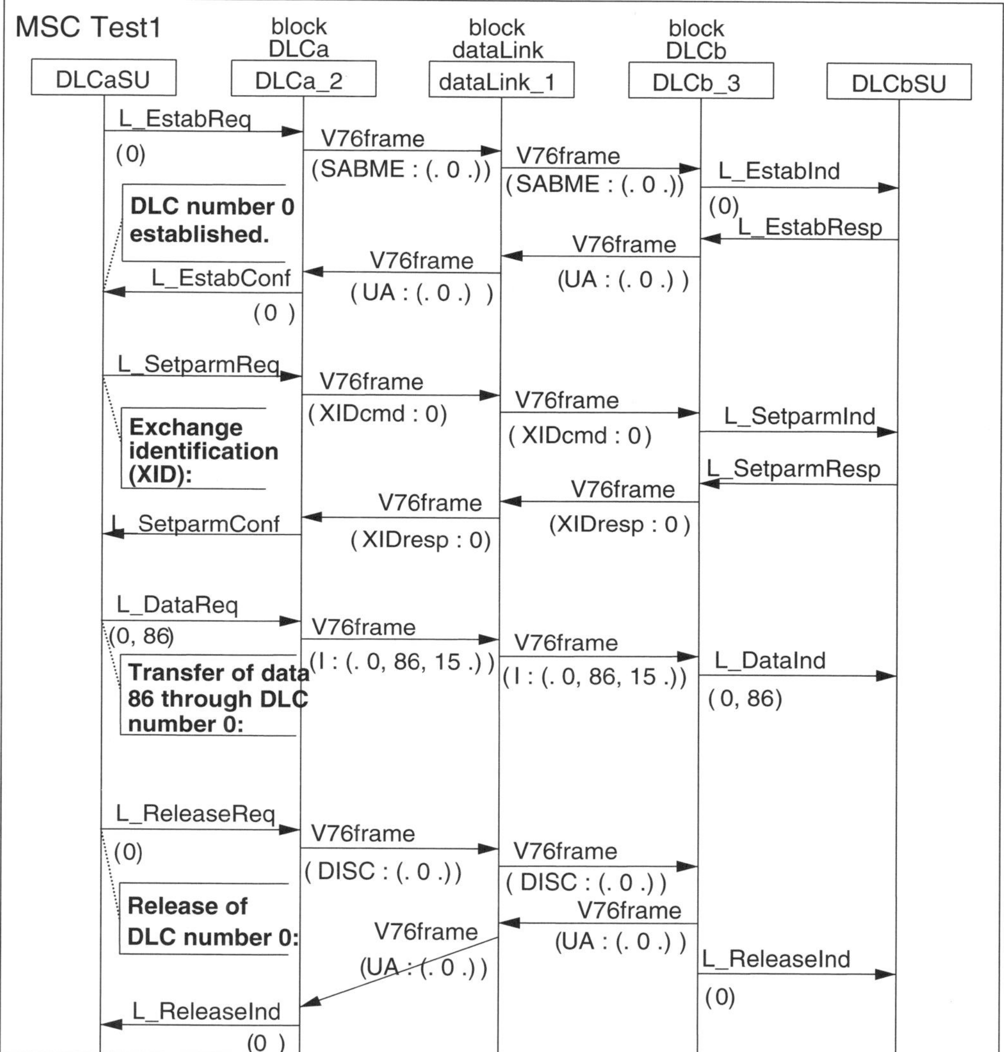

Figure 4.14 The black box MSC trace created by the Simulator

containing only the process *DLC* in block *DLCa* and display the value of the retransmission counter *N320cnt*.

A. Launch (or restart) the Simulator.

B. Press the button *SDL* in the *Trace* group.

C. Select *Trace > MSC Level: Set*: a *Select Unit name* window appears, select *System V76test* and press *OK*: a *Select Trace value* window appears, select *0 /* No MSC trace */* and press *OK*.

D. Select *Trace > MSC Level: Set*: a *Select Unit name* window appears, select *Process << Block DLCa >> DLC* and press *OK*: a *Select Trace value* window appears, select *2 /* Unconditional MSC trace */*[14] and press *OK*.

E. Select *Trace > MSC Level: Show* and check that you have:

```
Default          1 = Conditional MSC trace
System  V76test    0 = No MSC trace
Process <<Block DLCa>> DLC 2 = Unconditional MSC trace
```

F. Select *Trace > MSC Trace: Start*: a Select window appears, select *2 /* MSC with states and actions */* and press *OK*: the MSC Editor displays the MSC trace.

G. In the Simulator, press the *Send To* button: select *L_EstabRe*q and press *OK*, enter the value *0* for the parameter and press *OK*, select the target process *<< Block DLCa >> dispatch* and press *OK*.

H. Press *Go* in the *Execute* group.

I. Each time the *Select - Enter path* window appears, select *2 /* "Yes" */* and press *OK*.

As expected, the generated MSC, represented in Figure 4.15, contains only process *DLC* of block *DLCa*, and displays *N320cnt* plus the state of the process. You can see that the number of retransmission of SABME is correct.

But now if you press on *Transition*, the Simulator answers: *No process instance scheduled for a transition*. It means that the signal *DLCstopped* that you see at the bottom of the MSC trace is not expected by the process *dispatch*: it goes into the input queue of process *dispatch* in block *DLCa*, but unfortunately under the current state of *dispatch* there is no input of signal *DLCstopped*.

J. Save the MSC: in the MSC Editor, select *File > Save As*, navigate to the desired directory, enter *retry1.msc* and press *OK*.

K. Save the simulation commands into a file: in the Simulator, select *Log > Save Input History*, select *All files*, enter *retry1.com* and press *OK*.

L. Open the file *retry1.com* with a text editor, and if necessary add a return before each *2* in the *Go* line, to get the following file:

```
Set-GR-Trace 1
Set-MSC-Trace System V76test 0
Set-MSC-Trace Process <<Block DLCa>> DLC 2
List-MSC-Trace-Values
Start-Interactive-MSC-Log 2
Output-To L_EstabReq (0) <<Block DLCa>> dispatch
Go
2 /* "Yes" */
2 /* "Yes" */
2 /* "Yes" */
```

[14] Conditional means that an entity – for example, a process – will be traced only if it exchanges signals with entities present in the MSC trace. This avoids having entities present in the MSC trace but not exchanging any signal.

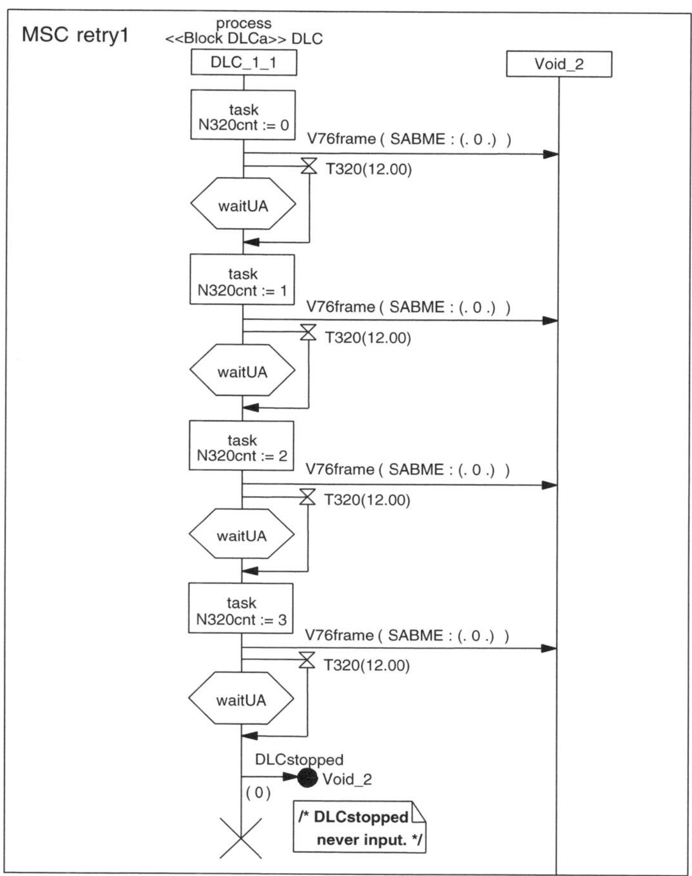

Figure 4.15 The three SABME retransmissions in the MSC generated by the Simulator

```
2 /* "Yes" */
Next-Transition
```

Do not exit from the Simulator.

4.2.3.2 Analyze the bug

To understand the bug, you will search in which state process *dispatch* (in block *DLCa*) was when process *DLC* transmitted the signal *DLCstopped* to it.

A. In the Simulator, select *View > Command Window*: the *Command* Window appears, as shown in Figure 4.7; check the box *List-Process*: you can see that *dispatch* in block *DLCa* is in state *waitUA*.

B. In the Simulator, select *Show > Prev Symbol*: as shown in Figure 4.16, in process *DLC* the Editor selects the stop symbol, below the output of *DLCstopped*, which caused the problem.

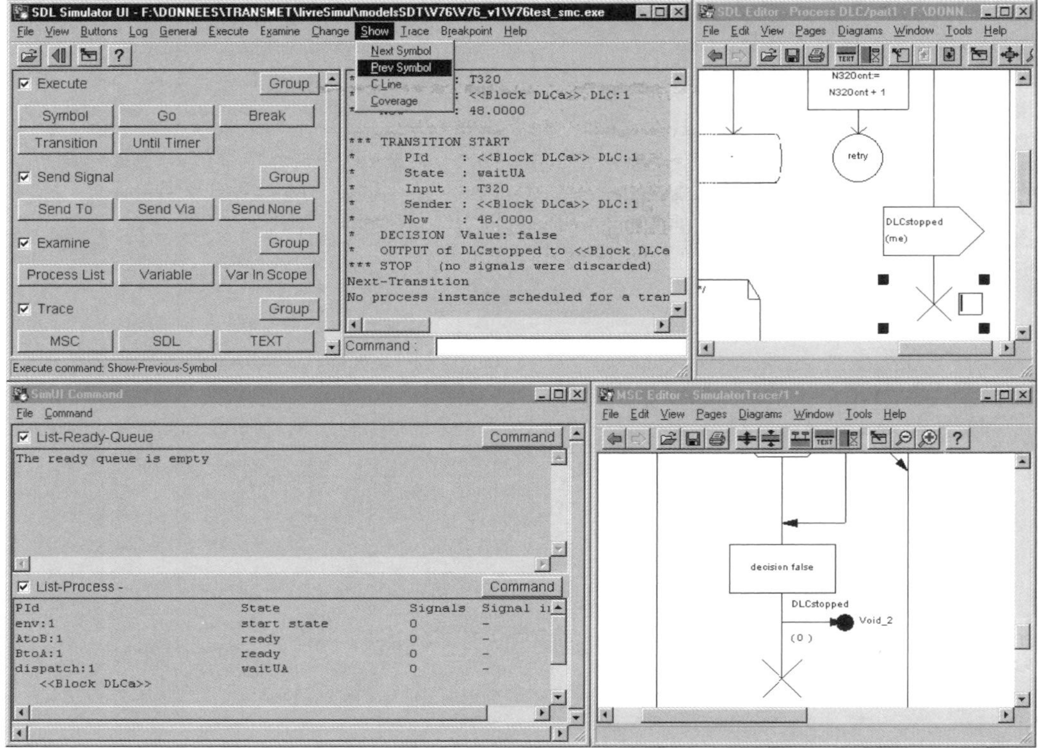

Figure 4.16 Finding the stop symbol that caused the problem

C. In the SDL Editor, select *Page > Edit Reference Page*: the block type *V76_DLC* is now displayed, as in Figure 3.12; you see that signal *DLCstopped* goes to process *dispatch*, through the signal route *DLCs*.

D. In the SDL Editor, double-click on the process *dispatch* to open it; press the *Next Page* button to display *part2*: you see that under state *waitUA*, the input of signal *DLCstopped* is missing.

E. Exit from the Simulator.

4.2.3.3 Correct the bug

You will add the missing input of signal *DLCstopped* under state *waitUA* in process *dispatch*.

A. In Windows (or Unix), make a copy of the file *dispatch.spr*, containing the process *dispatch*, into *dispatch_v1.spr*. Continue editing *dispatch.spr*.

B. In process *dispatch*, select the input of *DLCstopped* and the next three symbols under the state *ready*, copy them, paste them near state *waitUA* and connect the state to the input as shown in Figure 4.17.

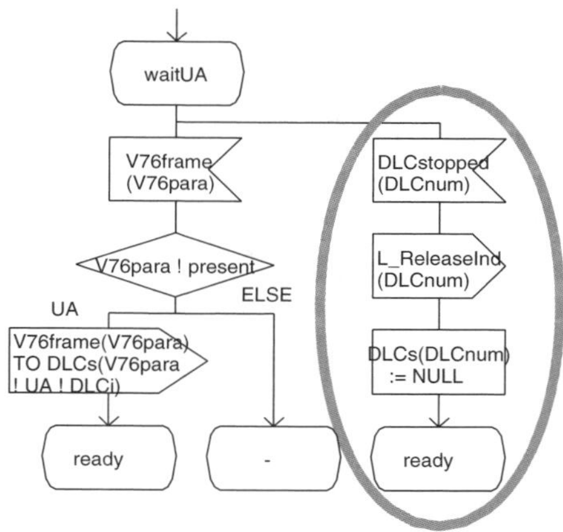

Figure 4.17 Missing *DLCstopped* input added under state *waitUA*

C. In the pasted transition, don't forget to replace - in the nextstate symbol by *ready*, otherwise you would be stuck in state *waitUA*.

D. Save the SDL model.

4.2.3.4 Simulate to check the bug correction

To check that the bug has been corrected, you will load and automatically replay the scenario stored in Section 4.2.3.1.

A. In the Organizer, select the SDL system *V76test* and press the *Simulate* button: Tau compiles the SDL model and starts the Simulator.

B. Select *Execute > Command Script*, choose *retry1.com* and press *OK*: the Simulator executes the commands contained in *retry1.com*.

This time, the bottom of the MSC generated by the Simulator looks like Figure 4.18(b): signal *DLCstopped* has been consumed by process *dispatch*. The bug is corrected.

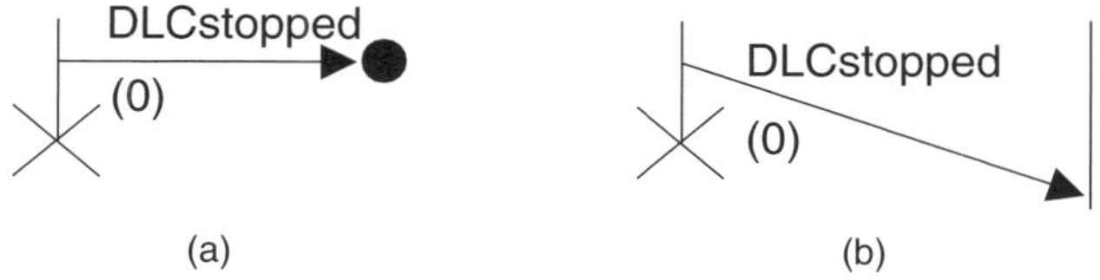

Figure 4.18 Signal *DLCstopped* in the process queue (a) and consumed (b)

4.2.4 Detect nonsimulated parts

After a simulation session, the SDL Coverage Viewer displays the SDL symbol coverage[15]. This shows you which parts of the SDL model have not been simulated.

[15] A counter is associated with each SDL symbol: every time a symbol is simulated, its counter is incremented. A value of 0 means never simulated.

Then you can simulate the SDL model again until you reach 100% symbol coverage. After playing all possible scenarios (which is easier using exhaustive simulation, if the model does not have too many states), the symbols not simulated are considered as "dead" parts: they can be removed, after careful inspection.

A. If the Simulator is already running, restart it and go to C.

B. In the Organizer, select the SDL system *V76test* and press the *Simulate* button: Tau compiles the SDL model and restarts the Simulator.

C. Select *Execute* > *Command Script*, choose *test1.com* and press *OK*: the Simulator executes the commands contained in *test1.com*.

D. Select *Show* > *Coverage*: the SDL Coverage Viewer appears, as illustrated in Figure 4.19, displaying the 21 SDL symbols that have never been executed since the Simulator startup.

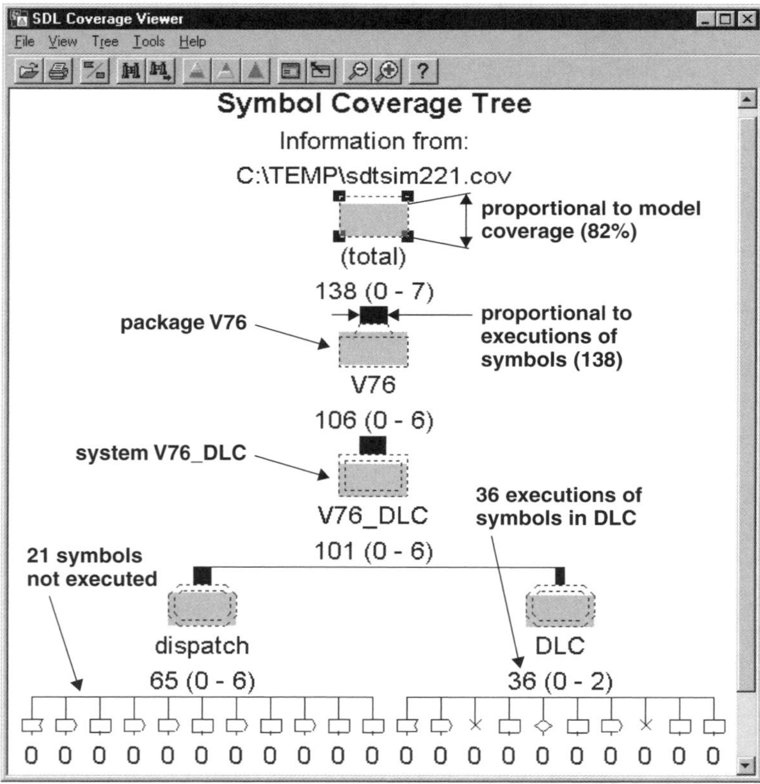

Figure 4.19 The SDL Coverage Viewer

E. In the Coverage Viewer, double-click the first symbol below *DLC* (an input): the SDL Editor opens the process *DLC* and selects the uncovered symbol, as depicted in Figure 4.20.

F. In the Coverage Viewer, select the top symbol above *total* and press the *Details* button: the SDL coverage details window appears, as shown in Figure 4.21. There you see that:

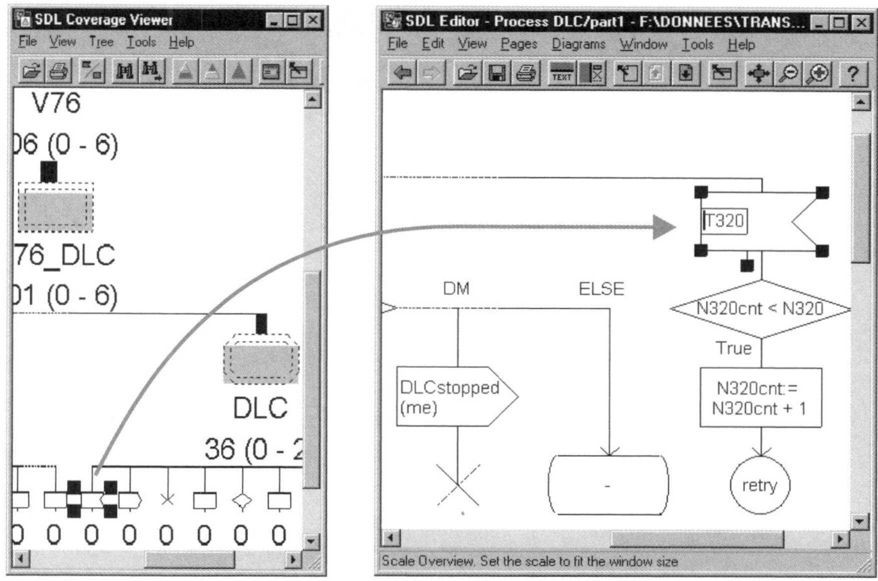

Figure 4.20 A symbol not simulated automatically located

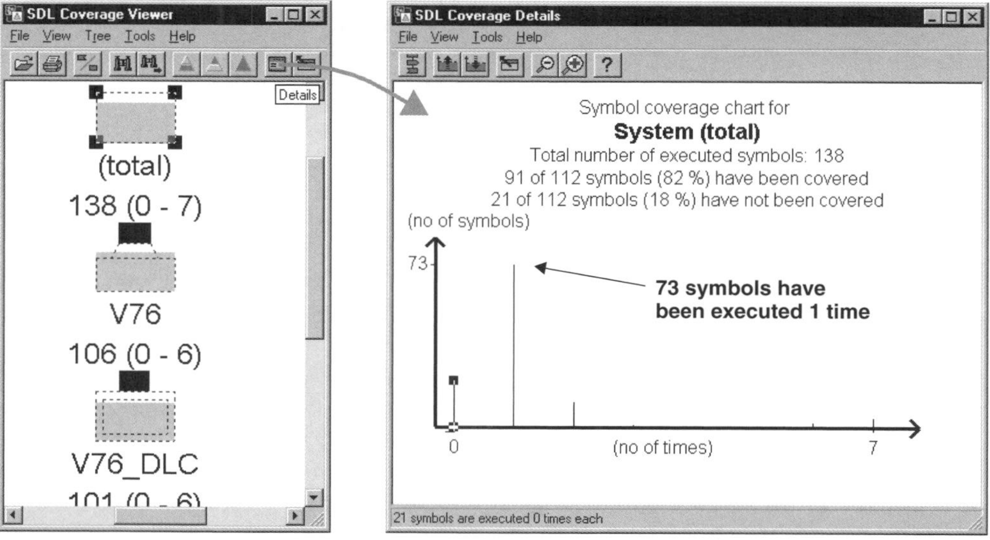

Figure 4.21 The SDL coverage details

- 73 symbols have been executed 1 time,
- the total symbols coverage for the SDL model is 82%,
- 21 symbols, displayed at the bottom of Figure 4.19, have not been executed

G. Close the coverage details window.

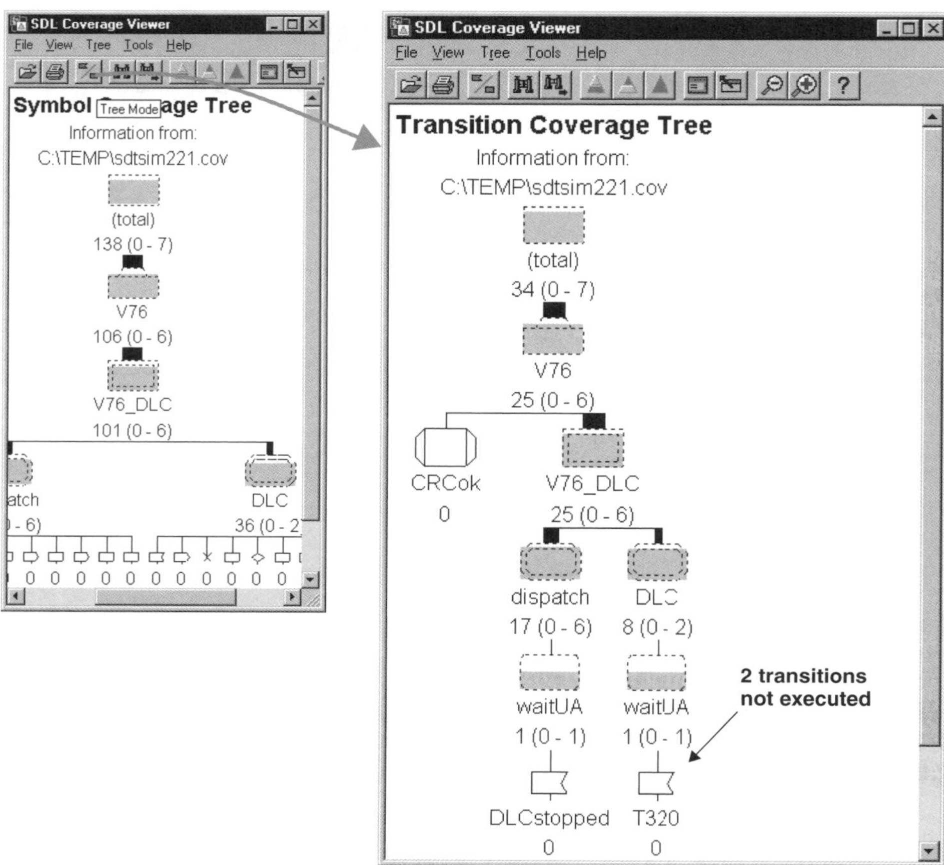

Figure 4.22 The SDL Coverage Viewer in transition tree mode

H. In the Coverage Viewer, press the *Tree Mode* button: the SDL Coverage Viewer switches from symbol to transition coverage tree, as shown in Figure 4.22.

The information displayed here is more synthetic as it only shows the transitions not executed. But even if it indicates that all the transitions have been executed, a branch after an answer to a decision could remain uncovered: this can only be detected by the symbol coverage tree.

As we have only replayed the scenario *test1.scn*, it is normal that two transitions that are not executed are displayed in Figure 4.22, one in process *dispatch* corresponding to the input of *DLCstopped*, the other in process *DLC* corresponding to the input of timer *T320*; to simulate them, we also need to replay *retry1.com*, and then press on *Transition* to execute one more transition.

4.2.5 Validate against more scenarios

After simulation of the main scenarios, described in Section 4.3.2, it is wise to play more scenarios to check the reaction of the SDL model. Those scenarios can be

- more complex: for example, two simultaneous connections,
- beyond limits: for example, creation of more connections than allowed.

4.2.5.1 Simulate two simultaneous connections

You will simulate to check that the SDL model can handle two connections in parallel.

A. In the Simulator, select *File > Restart...* and press *OK*.

B. Type the command *Define-MSC-Trace-Channels on* (or type *d-m-t-c on*) to get two environment instances.

C. Select *Trace > MSC Trace: Start*: a Select window appears. In the Select window, select *0 /* Basic MSC */* and press *OK*.

D. Select *Execute > Command Script*, choose *cnx1.com* and press *OK*; the Simulator executes the commands contained in *cnx1.com*: one instance of process *DLC* exists on each side of the system.

   Now establish one more connection:

E. Press the *Send To* button: select *L_EstabReq* and press *OK*. Enter the value *1* for the parameter of signal *L_EstabReq* and press *OK*. Select the target process *<< Block DLCa >> dispatch* and press *OK*.

F. Press several times the *Transition* button until you see the *Select – Enter path* window. A new instance of process *DLC* is created.

G. In the *Select - Enter path* window, select *1 /* "No" */* and press *OK*.

H. Press ONCE ONLY the *Transition* button.

I. Press the *Send To* button: select *L_EstabResp* and press *OK*, select the target process *<< Block DLCb >> dispatch* and press *OK*.

J. Press several times the *Transition* button until you see the *Select – Enter path* window.

K. In the *Select - Enter path* window, select *1 /* "No" */* and press *OK*.

L. Press two times the *Transition* button: signal *L_EstabConf* is transmitted.

   The new connection has been established between sides A and B.

M. To check the state of the two instances of process *DLC* on each side, select *View > Command Window*; in the List-Process part, you see that all four instances of process *DLC* exist and are in state *connected*:

```
DLC:2                  connected
   <<Block DLCa>>
DLC:1                  connected
   <<Block DLCa>>
DLC:2                  connected
   <<Block DLCb>>
DLC:1                  connected
   <<Block DLCb>>
```

N. To test that the new connection[16] works, let us transfer data through it: send signal *L_DataReq* with parameters *1* and *39* to *<< Block DLCa >> dispatch*, press *Go*, select *1 /* "No" */* and press *Go*.

The generated MSC, represented at blocks level by Figure 4.23, shows that block *DLCb* transmitted signal *L_DataInd(1, 39)* to *env_b* (representing Service User B): the data *39* has been successfully transferred from A to B through DLC *1*.

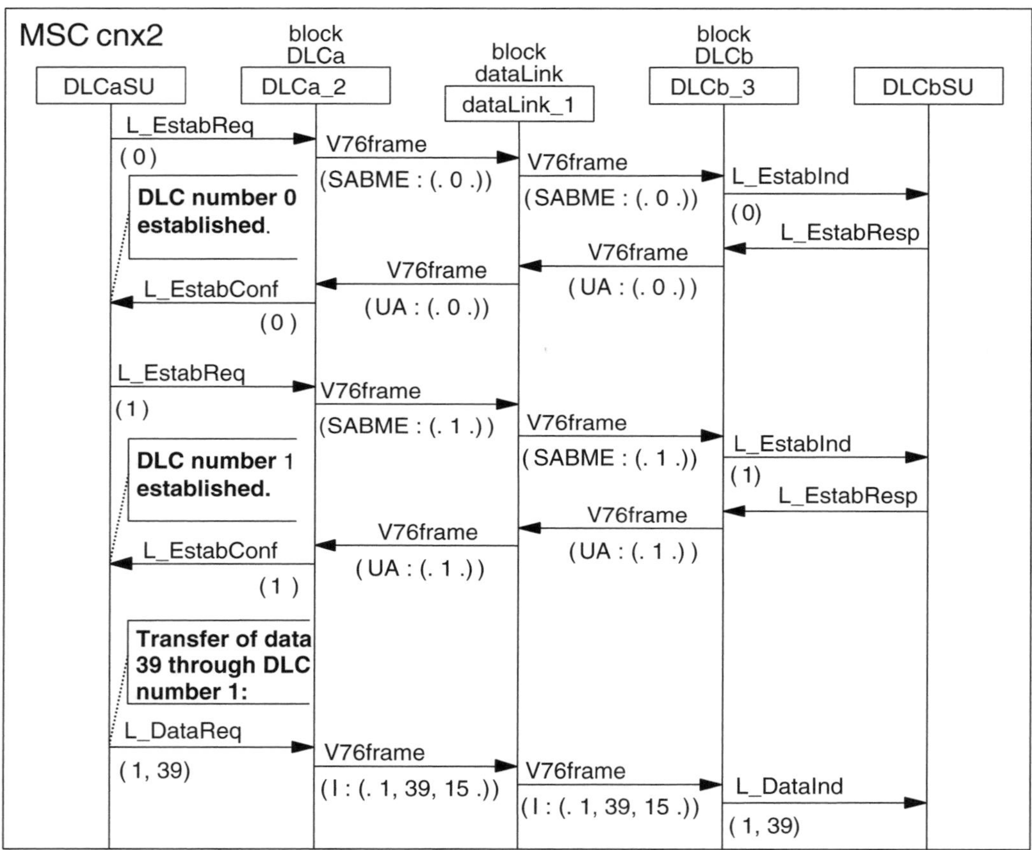

Figure 4.23 Two DLCs 0 and 1 connected in parallel

O. Save the simulation commands into a file: in the Simulator, select *Log* > *Save Input History*, select *All files*, enter *cnx2.com* and press *OK*.

P. Open the file *cnx2.com* with a text editor, and if necessary add a return in the two lines:

```
Next-Transition 1 /* "No" */
```

[16] The DLC number (of type *DLCident*) of the new connection is 1, and the corresponding instance number of process *DLC* (given by the Simulator) is 2.

to get:

```
Next-Transition
1 /* "No" */
```

Add a return in the line containing *Go* if necessary.

Transform the absolute file names into relative file names (remove pathnames such as *C:* etc. before `cnx1.com`).

Remove the following line, redundant with `Include-File cnx1.com`:

```
execute-command-script cnx1.com
```

The file *cnx2.com* should now contain:

```
Define-MSC-Trace-Channels on
Start-Interactive-MSC-Log 0 /* Basic MSC */
Include-File cnx1.com
Output-To L_EstabReq (1) <<Block DLCa>> dispatch
Next-Transition
Next-Transition
Next-Transition
1 /* "No" */
Next-Transition
Output-To L_EstabResp <<Block DLCb>> dispatch
Next-Transition
Next-Transition
Next-Transition
1 /* "No" */
Next-Transition
Next-Transition
Output-To L_DataReq (1, 39) <<Block DLCa>> dispatch
Go
1 /* "No" */
```

Q. Save the MSC: in the MSC Editor, select *File > Save As*, enter *cnx2.msc* and press *OK*.

4.2.5.2 Simulate an attempt to create too many connections

You will simulate to see what happens if you try to create more connections than allowed. The maximum number of parallel connections in our model is $maxDLC + 1 = 2$. Figure 3.12 shows that this number corresponds to the maximum number of instances of process DLC, identical to the size of the array *DLCs*, declared in Figure 3.14.

A. If you exited from the Simulator since Section 4.2.5.1, launch the Simulator and replay the command script *cnx2.com*: two instances of process *DLC* exist on each side of the system, the maximum is reached.

B. Press the *Send To* button: select *L_EstabReq* and press *OK*, enter the value *0* for the parameter of signal *L_EstabReq* and press *OK*.

C. Select the target process << *Block DLCa* >> *dispatch* and press *OK*.

D. Press the *Transition* button: you see in the MSC trace that the system answers with an *L_RelelaseInd(0)*: it means that no more connections can be established.

   But if you look at the *Command* Window, you discover that process *dispatch* in block *DLCa* is stuck in state *waitUA*: this is a modeling bug.

E. Exit from the Simulator.

   You will correct process *dispatch* to remain in state *ready* instead of going to state *waitUA* after transmitting *L_RelelaseInd*:

F. In Windows (or Unix), make a copy of the file *dispatch.spr* into *dispatch_v2.spr*

G. In process *dispatch*, page *part2*, select the output of *L_RelelaseInd* under the *ELSE* branch of the decision and double-click on the *state* palette symbol; enter – in the newly created symbol, as shown in Figure 4.24.

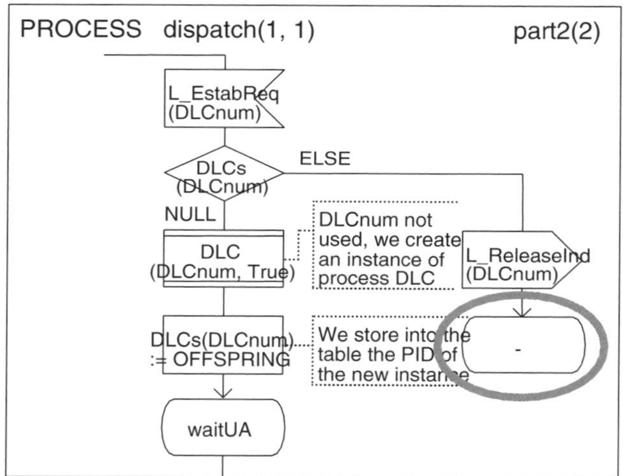

Figure 4.24 After the bug correction with the SDL Editor

H. Save the process.

I. Simulate again to check that the bug has disappeared: launch the Simulator, replay the command script *cnx2.com* as before, send an *L_EstabReq* to the model and execute its input: you see that process *dispatch* now stays in state *ready*.

4.2.6 Write a script for automatic validation

In an actual project, to test a more complex SDL model, you would produce, for example, 40 simulation scripts. After a change in the SDL model, you must check that it still works correctly (nonregression) : the Simulator command language[17] enables you to write a script to

[17] A few commands such as *add-macro* are interpreted by the SimUI (Simulator User Interface), not by the Simulator itself.

automatically replay the 40 scripts (stored in . *com* files) and write the test results into a log file, and write the MSC trace. After the execution of the script, looking into the log file and into the MSC trace tells you if the 40 scripts have been replayed correctly or not.

Here is a Simulator script to replay our two scripts *test1.com* and *cnx2.com*:

```
Log-On test_res1.wri
Define-MSC-Trace-Channels on
Start-Interactive-MSC-Log 0 /* Basic MSC */

Include-File test1.com
/* The two arrays DLCs must contain (: null, null :): */
Examine-Variable ( <<Block DLCa>> dispatch DLCs
Examine-Variable ( <<Block DLCb>> dispatch DLCs

Include-File cnx2.com
List-Process - /* to check that all DLC inst. are connected */

Log-Off
```

Here are a few more explanations on the script:

- `Log-On test_res1.wri`: all simulation traces will be written into the file *test_res1.wri*.
- `Define-MSC-Trace-Channels on`: to get two environment instances in the MSC trace.
- `Start-Interactive-MSC-Log 0`: starts the MSC trace (0 means MSC without states or actions).
- `Include-File test1.com`: executes the commands contained in the file *test1.com*.
- `Include-File cnx2.com`: executes the commands contained in the file *cnx2.com*.
- `Log-Off`: stops the logging, otherwise the file cannot be opened while the Simulator is running.

Once you have typed the script into the file *script1.com*, execute it:

A. Launch the Simulator.

B. Select *Execute > Command Script*, choose *script1.com* and press *OK*: the Simulator executes the commands contained in *script 1.com*. If the model (and the script!) is correct, you get a file *test_res1.wri* containing the simulation traces, plus an MSC displayed by the MSC Editor[18].

You will learn in Chapter 5 how to check that your SDL model behavior complies with an MSC, for example. It will be even more powerful to test an SDL model automatically: for example, we will be able to check the value of a parameter in a signal transmitted.

4.2.7 Other Simulator features

This section describes features of the Simulator that are not absolutely essential to validate an SDL model, but which can be very helpful and save a lot of time on an actual system validation.

[18] To speed up the replay for a large number of long scenarios, you can turn off the SDL trace.

4.2.7.1 Macros

You can define macros (a kind of aliases), for example, to give your own name to a Simulator command. To define the macro *trans*:

```
Command: add-macro trans Next-Transition
```

To list the current defined macros:

```
Command: list-macros
Number of macros defined: 1
Macro name:  trans
Macro value: Next-Transition
```

Then, you have only to enter *$trans* instead of *Next-Transition* (or *n-t*):

```
Command: $trans
Next-Transition

*** TRANSITION START
*       PId    : AtoB:1
*       State  : start state
*       Now    : 0.0000
*** NEXTSTATE  ready
```

4.2.7.2 Automatic script execution at Simulator startup

When the Simulator starts, the file *siminit.com*, if present in the current directory, is automatically executed, as shown in Figure 4.25. For example, *siminit.com* could contain the following commands to automatically start the MSC and SDL traces:

```
Start-Interactive-MSC-Log 0
Set-GR-Trace 1
```

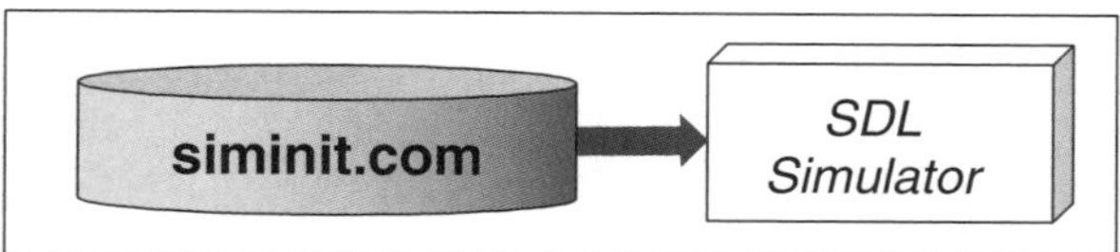

Figure 4.25 The Simulator startup file

4.2.7.3 Running the Simulator without its graphical interface

To perform automatic tasks such as automatic replay of scenarios, the Simulator can be launched without its graphical interface; this is called batch mode. To run in batch mode, after the build, type the following command (we suppose the SDL model is in the file *v76.pr*) in a DOS shell (containing *V76test_smc.exe*) or in a Unix shell:

```
F:\DONNEES\V76>V76test_smc.exe
No connection with the Postmaster. Running stand-alone.
```

```
Welcome to the SDL SIMULATOR. Simulating system V76test.
Command :
```

To see the Ready Queue, enter *List-Ready-Queue*, or *l-r-q*: the Simulator displays the list of processes ready to execute a transition:

```
Command : l-r-q
Entry PId                State         Signals  Signal instance
1     AtoB:1             start state   0        -
2     BtoA:1             start state   0        -
3     dispatch:1         start state   0        -
         <<Block DLCa>>
4     dispatch:1         start state   0        -
         <<Block DLCb>>
```

To execute a transition, enter *Next-Transition* or *n-t*:

```
Command : n-t
*** TRANSITION START
*       PId     : AtoB:1
*       State   : start state
*       Now     : 0.0000
*** NEXTSTATE  ready
```

Then you can enter any Simulator textual command:

```
Command : Output-To L_EstabReq(0) <<Block DLCa>> dispatch

Signal L_EstabReq was sent to <<Block DLCa>> dispatch:1 from env:1
Process scope : <<Block DLCa>> dispatch:1
```

The commands to execute can be inserted in the file *siminit.com*, automatically played when the Simulator starts. For example, the automatic simulation script *script1.com* described in Section 4.2.6 can be executed automatically if the file *siminit.com* contains:

```
Include-File script1.com
quit
```

4.2.7.4 Commands history

The Simulator automatically stores the last 100 commands executed. To reenter a Simulator command, simply scroll by pressing the up arrow key until you see the desired command.

4.2.7.5 Examining the SDL model: the Command window

Select *View > Command Window*: the *Command* Window appears, as shown in Figure 4.26.

The upper part of the *Command* Window displays the ready queue of the Simulator: the next process instance to be executed is the first process in the ready queue (entry number 1),

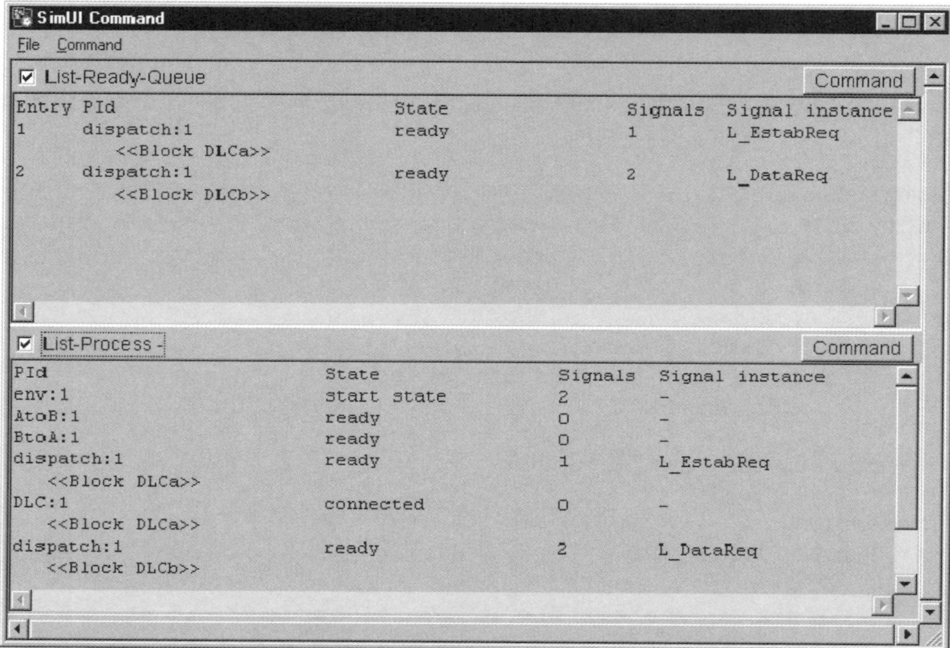

Figure 4.26 The Command Window

here instance 1 of process *dispatch* in block *DLCa*. This instance is in state *ready* and has one signal in its queue, *L_EstabReq*.

The lower part of the *Command* Window displays, for each process instance,

- its current state,
- the number of signals in its input queue,
- the name of the first signal in its input queue.

4.2.7.6 Examining the SDL model: Examine-Variable

The *Examine-Variable* command displays the contents of any SDL variable. The Simulator provides a button *Variable* in the *Examine* group to avoid entering the textual command manually.

To examine variable *DLCs* in process *dispatch* in block *DLCa*, press the button *Variable*, select << *Block DLCa* >> *dispatch* and select *DLCs*; the Simulator displays the equivalent textual command followed by its result:

```
Command : Examine-Variable ( <<Block DLCa>>  dispatch DLCs
DLCs (DLCsArray) = (: <<Block DLCa>> DLC:1, null :)
```

The result indicates that *DLCs* is of type *DLCsArray* and that the first element of the array contains the Pid of the first instance of process *DLC* in block *DLCa*, the second element containing Null.

4.2.7.7 Examining the SDL model: List-Input-Port

To display the input queue contents of process *dispatch* in block *DLCb*, select *Examine > Set Scope* to set the scope to the desired process:

```
Command : Set-Scope <<Block DLCb>> dispatch
Process scope : <<Block DLCb>> dispatch:1
```

then select *Examine > Input Port*:

```
Command : List-Input-Port
Input port of <<Block DLCb>> dispatch:1
Entry   Signal name          Sender
*1      L_DataReq            env:1
 2      L_EstabResp          env:1
```

The queue contains two signals, *L_DataReq* and *L_EstabResp*. To display the parameters of signal *L_DataReq*, select *Examine > Signal* and enter the entry number (here 1):

```
Command : Examine-Signal-Instance 1
Signal name : L_DataReq
Parameter(s) : 0, 97
```

4.2.7.8 Examining the SDL model: the watch window

The *Edit > Watch Window* command opens a watch window displaying the contents of any SDL variable, refreshed every time its value changes.

In the *Watch* window, select *Watch > Add . . .* and enter the name of the SDL variable to watch. For example, to watch variable *N320cnt* in process *DLC* in block *DLCa*, type the syntax:

```
( <<Block DLCa>> DLC N320cnt
```

The easiest way to get the correct syntax is to examine the variable by pressing the *Variable* button in the *Examine* group; then you can copy the argument of the *Examine-Variable* command and paste it into the *Add Watch* field.

To avoid entering the variable names again, the next time you launch the Simulator, you can select *File > Save As* in the watch and enter the file name *def.vars*. The Simulator will automatically load the variable names into the watch window as shown in Figure 4.27.

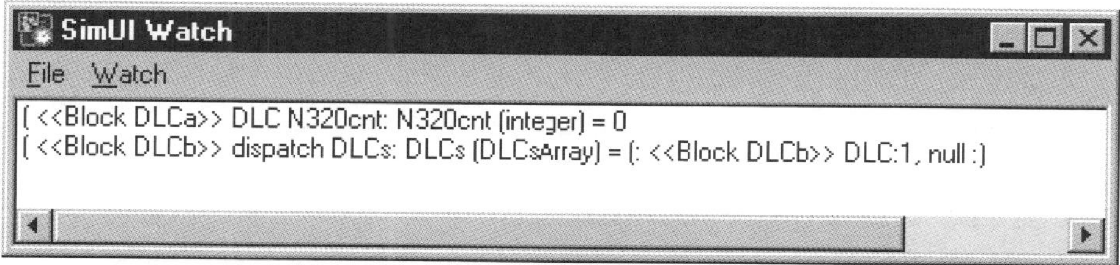

Figure 4.27 The Simulator watch window

4.2.7.9 Modifying the SDL model: Rearrange-Ready-Queue

To test a special case impossible to simulate with the current Simulator process scheduling, you may want to rearrange the Simulator ready queue. To do that, select *Change > Ready Queue* and enter successively the queue entry numbers to swap. Here is an example where we have swapped entries 3 and 2:

```
Command : List-Ready-Queue
Entry PId                       State           Signals  Signal
                                                         instance
1     AtoB:1                    start state     0         -
2     BtoA:1                    start state     0         -
3     dispatch:1                start state     0         -
         <<Block DLCb>>

Command : Rearrange-Ready-Queue 3 2
Rearranged!
Process scope : AtoB:1

Command : List-Ready-Queue
Entry PId                       State           Signals  Signal
                                                         instance
1     AtoB:1                    start state     0         -
2     dispatch:1                start state     0         -
         <<Block DLCb>>
3     BtoA:1                    start state     0         -
```

Then process *dispatch* will be executed immediately after process *AtoB*.

4.2.7.10 Modifying the SDL model: variables

To change the value of a variable, you must first set the scope to the desired process instance, if it is not in the current scope.

For example, to enter 2 into the variable *N320cnt* in block *DLCa*, use *Examine > Set Scope* and select *<< Block DLCa >> DLC*. Then select *Change > Variable*, choose *N320cnt* and enter 2.

```
Command : Set-Scope <<Block DLCa>> DLC
Process scope : <<Block DLCa>> DLC:1

Command : Assign-Value N320cnt 2
Value assigned
N320cnt (integer) = 2
```

To change easily the values of complex structures such as arrays, struct or ASN.1 values, the Simulator automatically prompts you for each required field value.

4.3 CASE STUDY WITH OBJECTGEODE

The tools and platform used for the exercises are described in Section 1.5. The V.76 SDL model used for this case study is presented in Chapter 3.

4.3.1 Prepare the Simulator

4.3.1.1 Compile the SDL model

A. Download from the Internet (see the ftp address in Section 1.5) or create the SDL V.76 model, and launch the ObjectGeode SDL Editor: double-click the icon *v76.pr*[19], or for Unix type *geodedit v76&*.

B. In the SDL Editor, select *Tools > SDL & MSC Simulator*. The window represented in Figure 4.28 appears.

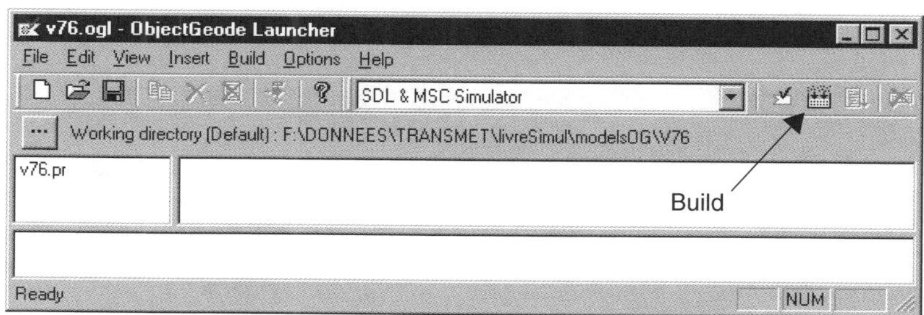

Figure 4.28 The ObjectGeode Launcher

The left area[20] contains *v76.pr*, the name of the file containing the SDL model. Remember to save your SDL model each time you modify it in the Editor, otherwise the file *v76.pr* will not contain your changes.

C. Select *File > Save* to create the file *v76.ogl*[21] in your current directory (this feature may not be available in the Unix version).

D. Press the *Build* button: this checks your SDL model, and if it is correct translates it into C code[22], stored in the subdirectory *.geodesm*. Then the C files are compiled[23] and linked to the simulator library, to produce the executable file *v76.sim*, as depicted in Figure 4.29.

4.3.1.2 Start the Simulator

A. Press the *Execute* button indicated in Figure 4.30 to start the Simulator. The Simulator[24] main window appears, as shown in Figure 4.31. Drag it to the upper left corner of your screen.

---

[19] We recommend setting your Windows options to display the file extensions (in Windows NT4, open *Workstation*, then select *Display > Options*: in the *Display* tab, check *Display all files* and uncheck *Mask the extensions for known file types*, then press *OK*).

[20] If this area contains other models you do not want to simulate, remove them: select them and do *Edit > Remove*. By default, the files you see in this area are all the files loaded in the SDL Editor when you invoked the command *Tools > SDL & MSC Simulator*, MSCs included.

[21] This file saves the current simulation builder options and SDL file name, and enables launching the simulator (or the Target C Code Generator) without the SDL Editor.

[22] The code generated here is specific to simulation. Another code optimized for real-time targets is generated when using the tool ObjectGeode SDL C Code Generator.

[23] If you do not see the file names *b0.c* and so on, as in Figure 4.30, check that a C compiler is available.

[24] In fact when you start the Simulator, two processes start: *ogsmgui.exe*, the Graphical User Interface, and *v76.sim*, the command line interface Simulator.

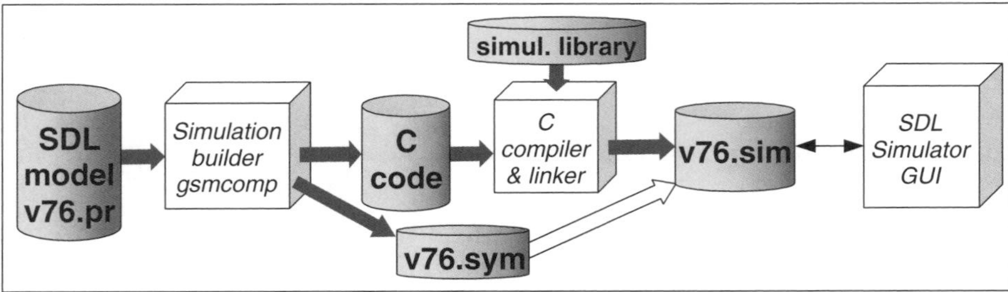

Figure 4.29 The ObjectGeode simulation building

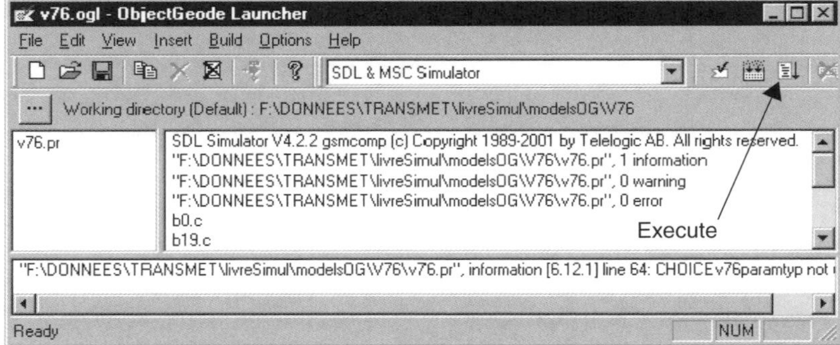

Figure 4.30 The Simulator is ready to be launched

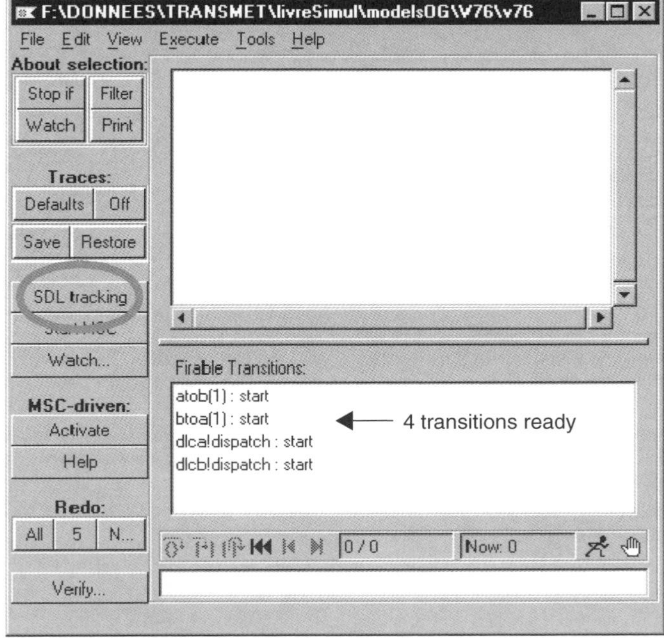

Figure 4.31 The Simulator main window

Four transitions can be fired, because the V.76 model contains four process instances: *AtoB(1)*[25], instance number one of process *AtoB*, *BtoA(1)*, *DLCa!dispatch* and *DLCb!dispatch*.

4.3.1.3 Start the SDL tracking

A. Press the button *SDL tracking* shown in Figure 4.31: this opens a window in the SDL Editor, named *Default tracking*. Close (do not minimize) the other SDL windows in the Editor, if any.

B. Select the first ready transition in the list shown in Figure 4.31: the SDL Editor opens process *AtoB(1)* and displays in bold its start symbol, as shown in Figure 4.32(a).

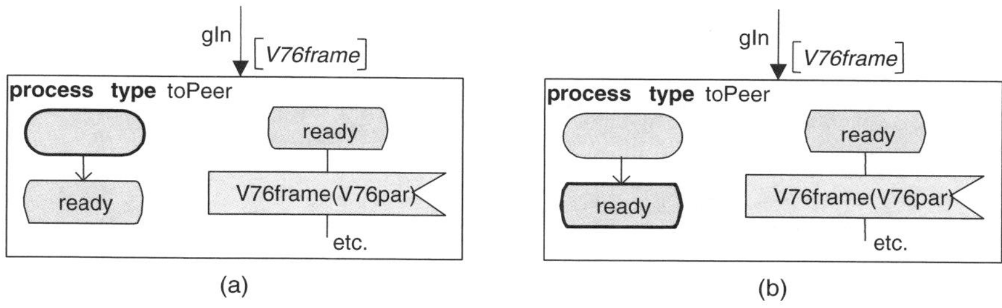

Figure 4.32 SDL tracking: start transition ready (a) and executed (b)

C. Double-click[26] on the same transition (*atob(1):start*): this executes the transition from the start symbol to the state *ready* which is now in bold, as depicted in Figure 4.32(a).

D. Double-click on the three transitions remaining in the list to execute them. As you can see in Figure 4.33, no transition is ready to be fired: this is a deadlock. In fact, the simulation is blocked because the SDL model now expects to receive some external signals.

E. Save the current simulation scenario into the file *start.scn* by typing the following command in the Simulator:

```
save start.scn
```

4.3.1.4 Send signals to the model

Figures 4.34 and 4.35 show that the V.76 SDL model requires the input of external signals, such as *L_EstabReq*, contained in the signallist *su2dlc* on the channels *DLCaSU* and *DLCbSU*. We will tell the Simulator to send such signals to the SDL model, using the *feed* command[27].

---

[25] The Simulation Builder converts the SDL names to lower case; thus the actual names you see in the tool do not contain capitals.

[26] If your mouse has three buttons, it is easier to click once on its middle button.

[27] The Simulator *output* command (in the *Execute* menu) can also be used, but it is less convenient as it sends the signal only once: to send the signal again, the *output* command must be repeated, which is not possible during an automatic simulation (such as random or exhaustive).

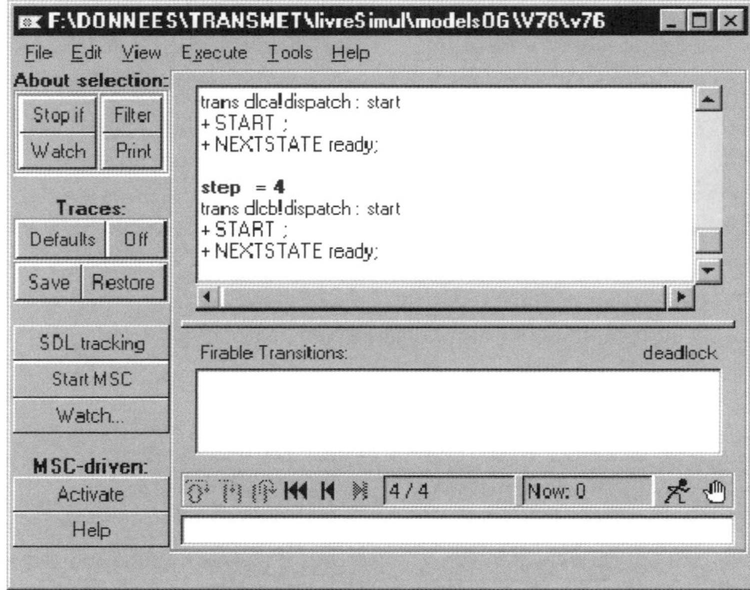

Figure 4.33 The simulation is blocked

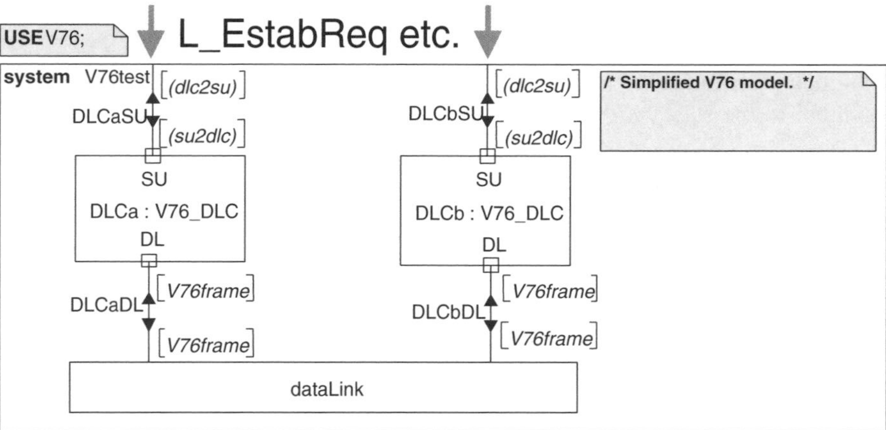

Figure 4.34 The V.76 model waits for external signals

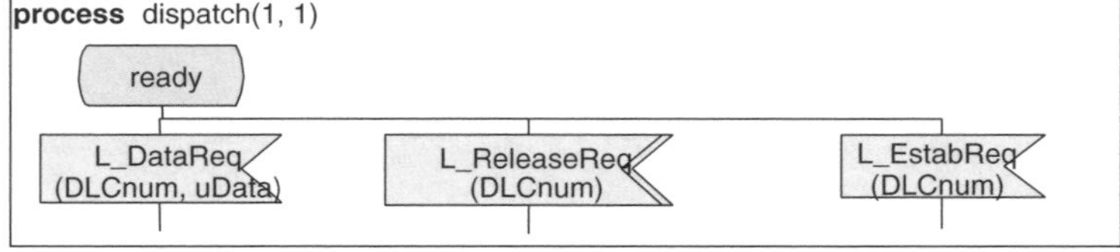

Figure 4.35 Some of the external signals required in the blocks *DLCa* and *DLCb*

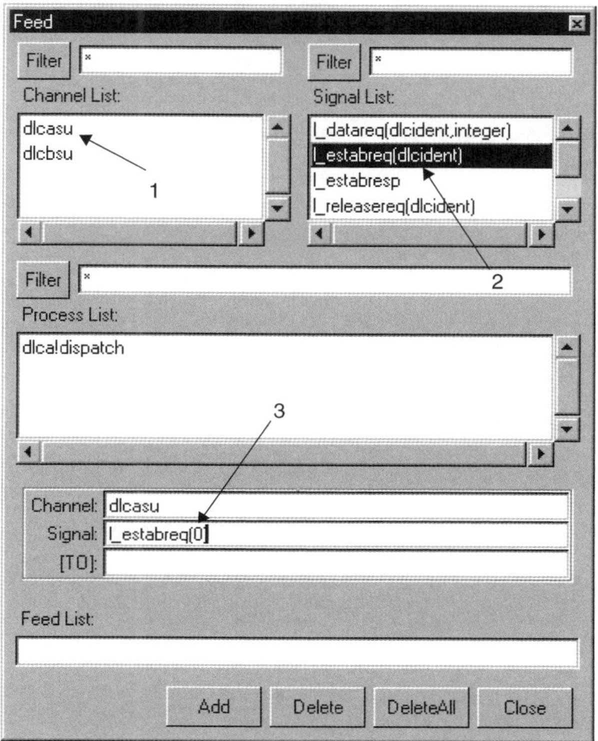

Figure 4.36 Adding feed commands

A. In the Simulator, select *Edit > Feed*: the window shown in Figure 4.36 appears. Select a channel (1), a signal (2), enter the value of the parameter(s) (3) and press on *Add*: you have created one *feed*.

   This means that every time the process *dispatch* in the block *DLCa* is in a state where the input of *L_EstabReq* is possible (i.e. input queue empty and state = *ready*, see Figure 4.35), the Simulator will propose to execute the input of *L_EstabReq* by *dispatch*[28].

B. Continue in the same way to create the following *feed* list, then press on *Close*.

```
dlcasu l_estabreq(0)
dlcasu l_estabreq(1)
dlcasu l_setparmreq()
dlcasu l_releasereq(1)
dlcasu l_datareq(0,86)
dlcasu l_datareq(1,39)
dlcbsu l_estabresp()
```

[28] As opposed to actual signal outputs, signals sent by Simulator *feed* commands do NOT go into the input queue of the target process (here *dispatch*). The behavior is equivalent, but not using the queue reduces the number of global states during exhaustive simulation.

```
dlcbsu l_setparmresp()
dlcbsu l_releasereq(0)
```

C. The feed commands created here will be necessary for all the exercises in the book. To avoid entering them manually repeatedly, save them: in the Simulator window, type as shown in Figure 4.37, terminated by return:

```
list feed >> v76_feed.wri
```

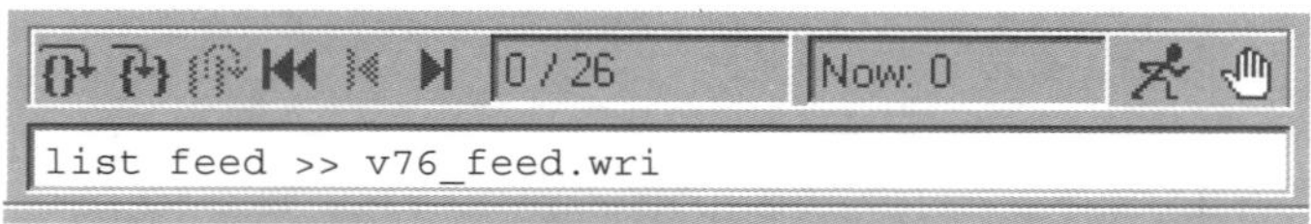

Figure 4.37 Saving the *feed* list

This creates in the current directory a file *v76_feed.wri*[29] containing the feed commands

```
feed dlcasu l_estabreq(0)
feed dlcasu l_estabreq(1)
etc.
```

4.3.1.5 Create a startup file

To avoid entering feed commands manually and executing the four process start transitions each time we launch the Simulator, you will create a startup file.

Remember to always invoke the ObjectGeode tools from your working directory. If you launch the ObjectGeode Editor and then navigate to another directory to open the SDL model, the Simulator might not run in the directory where the SDL model is; thus the startup file will not be found and so on. This is especially true for older versions.

A. With a text editor[30], create the file[31] *v76.startup* (and NOT *v76.startup.wri*) in the current directory, containing

```
source v76_feed.wri
source start.scn
```

The files *v76_feed.wri* and *start.scn* were described in the previous sections.

The next time you run the Simulator, it will automatically execute the file *v76.startup*. This will load your feed commands and execute the four process start transitions. See details on startup files in Section 4.3.7.2.

[29] The purpose of the . *wri* extension is to provide automatic opening of the file by WordPad in Windows.

[30] To easily open any file in Windows, we recommend that you create a shortcut to WordPad on your desktop. Then simply drag the file on the shortcut to open it.

[31] Warning: in Unix if you name the file *V76.startup*, the Simulator will not load it, because it does not expect the capital V.

4.3.2 Validate against the main scenarios

Now the Simulator is ready to check that the SDL model behaves as described in the MSCs drawn during the analysis step in Section 3.3. You can compare[32] the analysis MSCs and the MSCs recorded during simulation.

4.3.2.1 Prepare the simulator for MSC recording

You would have carried out the steps described in the previous sections (SDL tracking, feed, etc.).

A. In the Simulator, press the *Start MSC* button: the Editor opens a new window containing an empty MSC[33].

B. In the Editor, close the Framework window.

C. In the Editor, check that you have only two open windows: the empty MSC and the SDL tracking (named *Default tracking*). Close (do not minimize) any other window.

D. In the Editor, Select *Window > Tile Horizontally*, and arrange the Simulator and Editor windows as in Figure 4.38.

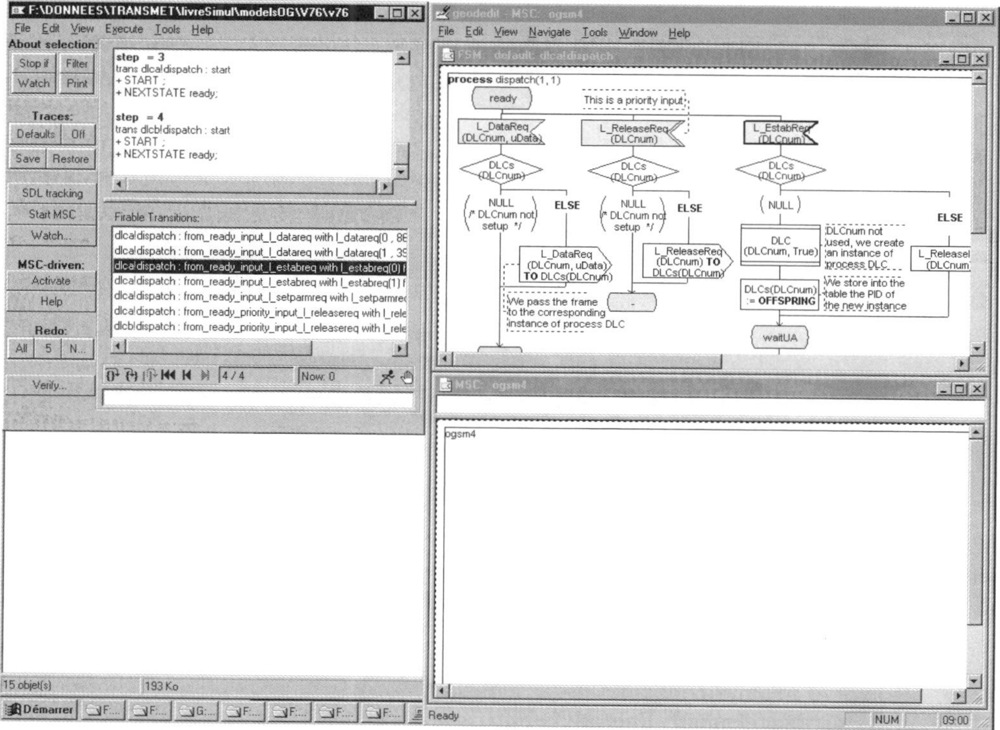

Figure 4.38 The Simulator ready for SDL animation and MSC recording

---

[32] For the moment, you will check this "manually", we will see later how to automatically validate the SDL model against a set of MSCs.

[33] To see the state of each process in the MSC, type *trace state* in the Simulator command line.

E. If you do not get firable transitions as in the Simulator window in Figure 4.38, see Section 4.3.1.4.

F. Select the third transition in the *Firable transitions* Simulator area, named *trans dlca!dispatch: from_ready_input_l_estabreq with l_estabreq(0) from env_dlcasu*: the Editor displays in bold the SDL input symbol corresponding to this transition, as depicted in Figure 4.38.

4.3.2.2 Play and record the scenario DLC establishment

You will simulate the SDL model, trying to play the scenario described in the MSC *DLC establishment*, shown in Figure 3.5. The Simulator will generate an MSC trace representing the actual SDL behavior, which you will compare with the expected MSC shown in Figure 3.5.

A. In the Simulator, double-click the transition[34]:

```
trans dlca!dispatch : from_ready_input_l_estabreq with
l_estabreq(0) from env_dlcasu
```

The Simulator transmits the signal *L_EstabReq* to the SDL model, which inputs it and executes the transition to the state *waituUA,* as illustrated in Figure 4.39.

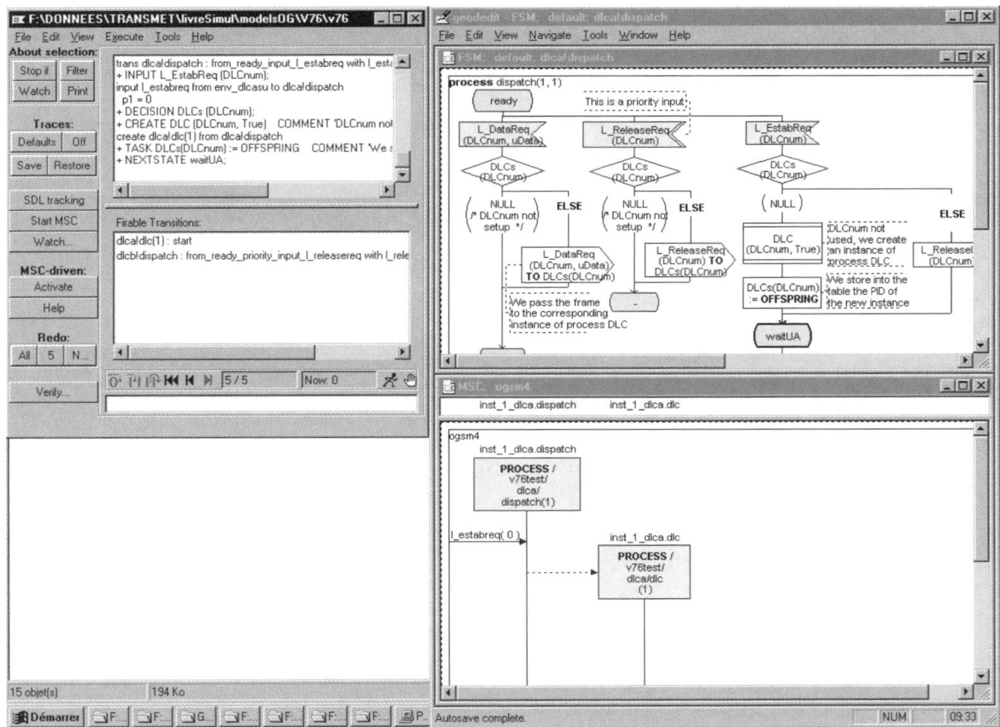

Figure 4.39 After sending *L_EstabReq(0)* to the SDL model

[34] If you execute a wrong transition, return to the previous simulation step by pressing the *undo* ⏮ button (see Section 4.3.7.1).

In the MSC trace, you see that the process *dispatch* on side A has received *L_EstabReq* with *0* as parameter and that *dispatch* has created one instance of the process *DLC*.

B. Simulate symbol by symbol: in the Simulator, select the transition *trans dlca!dlc(1): start*: the start SDL symbol of process *DLC* is in bold. Press the *Step Into* button shown in Figure 4.40: the first SDL symbol in the transition, *decision originator*, is executed and is displayed in bold, as illustrated in Figure 4.41(a).

Figure 4.40 The *Step Into* button

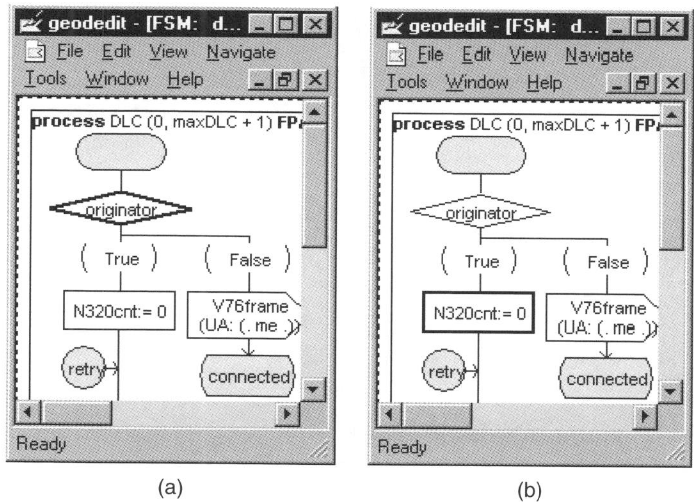

Figure 4.41 Symbol by symbol simulation

C. Press the *Step Into* button again: the next SDL symbol, *N*320*cnt* := 0, is executed, as shown in Figure 4.41(b).

D. Press the *Step Out* button to terminate the transition: the process *DLC* now reaches the state *waitUA*.

E. Terminate the scenario: in the Simulator, double-click the following transitions:

```
trans atob(1) : from_ready_input_v76frame
trans atob(1) : decision_lose_the_frame('No')
trans dlcb!dispatch : from_ready_input_v76frame
trans dlcb!dispatch : from_waitestabresp_input_l_estabresp ...
trans dlcb!dlc(1) : start
trans btoa(1) : from_ready_input_v76frame
```

```
trans btoa(1)  : decision_lose_the_frame('No')
trans dlca!dispatch : from_waitua_input_v76frame
trans dlca!dlc(1)  : from_waitua_input_v76frame
```

After executing those transitions, the MSC trace generated by the Simulator, arranged and displayed in Figure 4.42, shows that the SDL model complies with the MSC in Figure 3.5. The MSC generated contains the values of the signal parameters, such as the DLC number, here equal to 0.

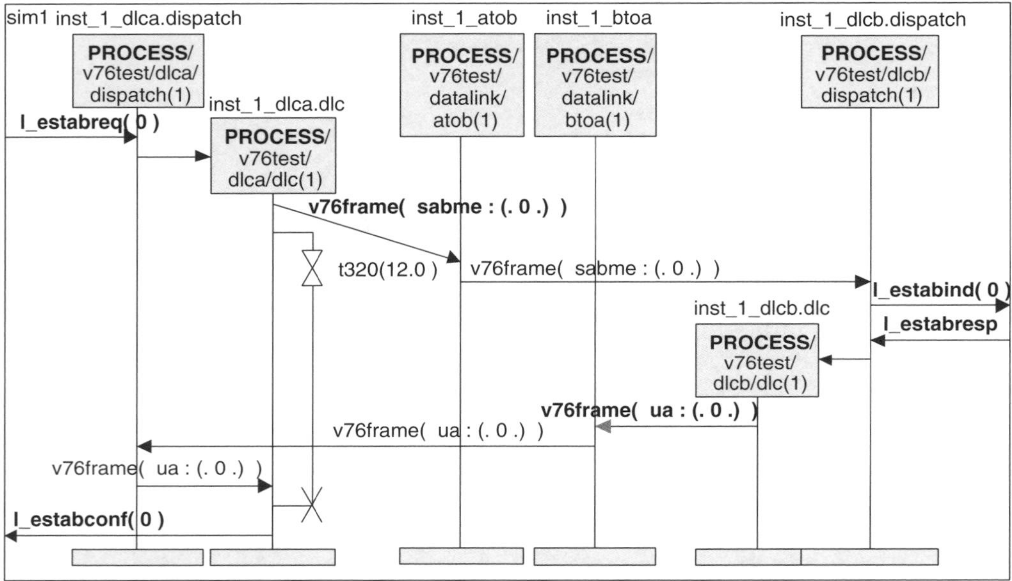

Figure 4.42 The MSC generated by the Simulator

F. Type the following command into the Simulator to save the current simulation scenario[35] into the file *cnx1.scn*:

```
save cnx1
```

Later, you can load *cnx1.scn* into the Simulator to automatically replay the recorded scenario.

G. While the Simulator is running, you cannot modify or save the MSCs. To avoid exiting from the Simulator, you can enter the following command into the Simulator to save the current MSC into the file *cnx1.msc*:

```
msc cnx1
```

Do not exit from the Simulator.

[35] Do not confuse the Simulator scenario, which is the recording of the transition execution, and the graphical simulation trace stored in the form of an MSC scenario.

4.3.2.3 Play the other scenarios

Once the DLC connection is established (if not, play again the scenario described in Section 4.3.2.2), continue the simulation, playing three more scenarios (important: play them in sequence, that is, never reinitialize the Simulator, and always select *decision_lose_the_frame('**No**')*):

A. Simulate XID, described in Figure 3.4.

B. Simulate data transfer from A to B, described in Figure 3.6. Select the correct transition (i.e. use DLC 0, not DLC 1, which was not established).

C. Simulate DLC release, described in Figure 3.8. Select the correct transition (i.e. release DLC 0, not DLC 1, which was not established).

D. The current step number should be 41. Save the current Simulator scenario into the file *test1.scn*, using the command: *save test1.scn*.

   Do not exit from the Simulator.

4.3.2.4 Generate a black box MSC

The MSC generated in Figure 4.42 contained all the model process instances. You will now generate a more abstract "black box" MSC, containing only the blocks *(DLCa, dataLink* and *DLCb*), to validate the external behavior of the system:

A. In the Simulator, select *File > Generate MSC...*

B. In the *Generate MSC* window, press the *Add...* button under *For entities:*

C. Select blocks *datalink, dlca* and *dlcb* (the selection is multiple) and press *OK*.

D. In the *Generate MSC* window, press *Apply* and *Cancel* (do NOT press *Save*).

E. In the Simulator, press on the buttons *init* and *Redo: All*.

F. Enter the following Simulator command to generate an MSC in the file *test1.msc*:

```
msc test1
```

   The Editor loads automatically *test1.msc*.

G. In the Editor, double-click on the box *test1* to display the generated MSC, similar to the one represented in Figure 4.43, edited to add comments.

   Exit from the Simulator.

4.3.3 Detect a bug in the SDL model

4.3.3.1 Check the number of retransmissions

You will check that the retransmission of SABME occurs maximum three times after its first transmission, as specified in Section 3.2.3.

A. Prepare the Simulator as indicated in Section 4.3.2.1.

B. Execute the 4 start transitions if not already done.

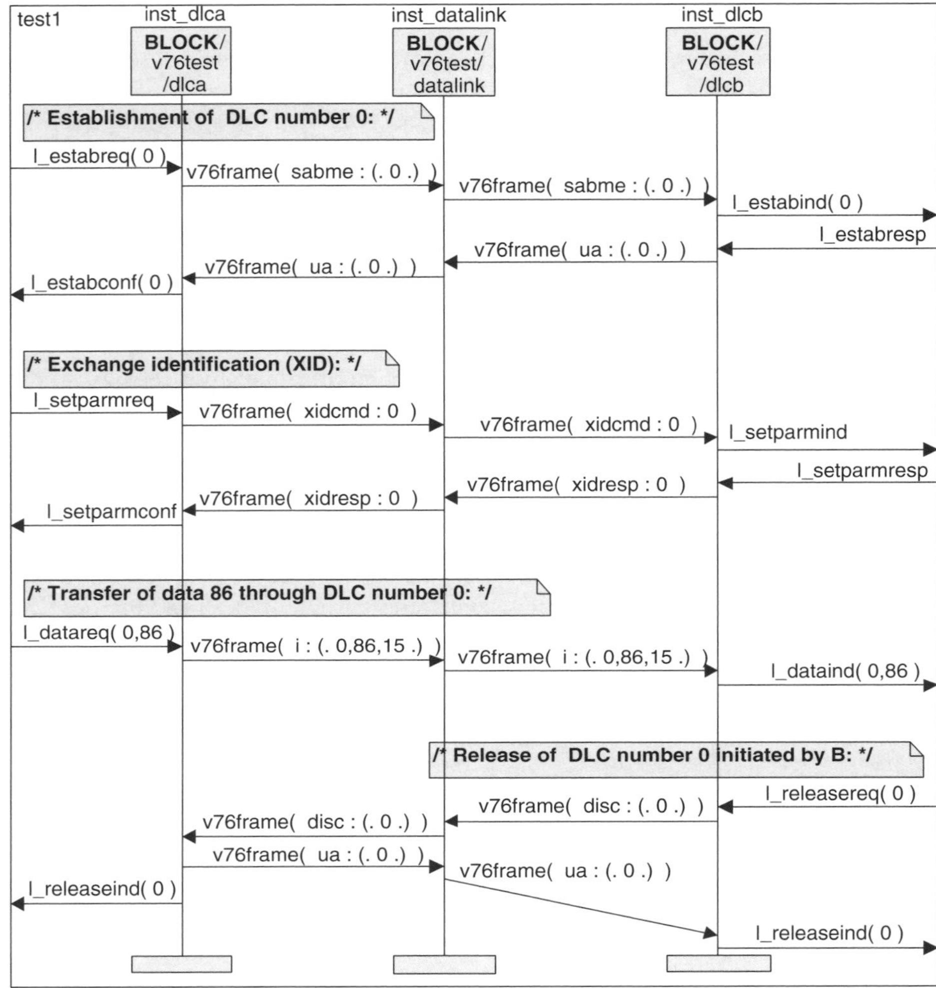

Figure 4.43 The black box MSC generated by the Simulator

C. Execute the transition:

```
trans dlca!dispatch : from_ready_input_l_estabreq with
                        l_estabreq(0) from env_dlcasu
```

You are in the situation shown in Figure 4.39[36].

D. In the Simulator command line, type as in Figure 4.44:

```
print n320cnt
```

[36] If you do not see the processes in the generated MSC, type *msc for all* in the Simulator command line and press the *init* and *Redo: All* Simulator buttons.

Figure 4.44 The Simulator command line

The Simulator displays the value of the retransmission counter *N320cnt* of the instance number 1 of process *DLC* in block *DLCa*:

```
process dlcb!dlc not created or stopped
dlca!dlc(1) ! n320cnt = 0
```

E. To display the value of *N320cnt* at every step, type in the Simulator command line:

```
trace dlca!dlc(1) ! n320cnt
```

the Simulator will display the value of the retransmission counter *N320cnt* in the generated MSC, when its value changes.

F. In the Simulator, double-click the following transitions:

```
trans dlca!dlc(1) : start
trans atob(1) : from_ready_input_v76frame
trans atob(1) : decision_lose_the_frame('Yes')
```

G. As we have not transmitted the signal *V76frame*, the Simulator proposes to make the time progress until the time-out of timer *T320*; double-click the transition:

```
trans time(12.0)
```

H. The current SDL time, *Now*, displayed in the Simulator, has progressed from 0 to 12: the Simulator proposes to time-out timer *T320*; double-click the transition:

```
trans dlca!dlc(1) : timeout_t320
```

I. The Simulator proposes to input the timer *T320*; double-click the transition:

```
trans dlca!dlc(1) : from_waitua_input_t320
```

As the decision *N320cnt* < *N320* has returned *True*, *N320cnt* has been incremented and the execution has jumped to the label *retry*, where the SABME has been retransmitted and *T320* has been set.

J. Execute the sequence of transitions given below three times:

```
trans atob(1) : from_ready_input_v76frame
trans atob(1) : decision_lose_the_frame('Yes')
trans time(12.0)
trans dlca!dlc(1) : timeout_t320
trans dlca!dlc(1) : from_waitua_input_t320
```

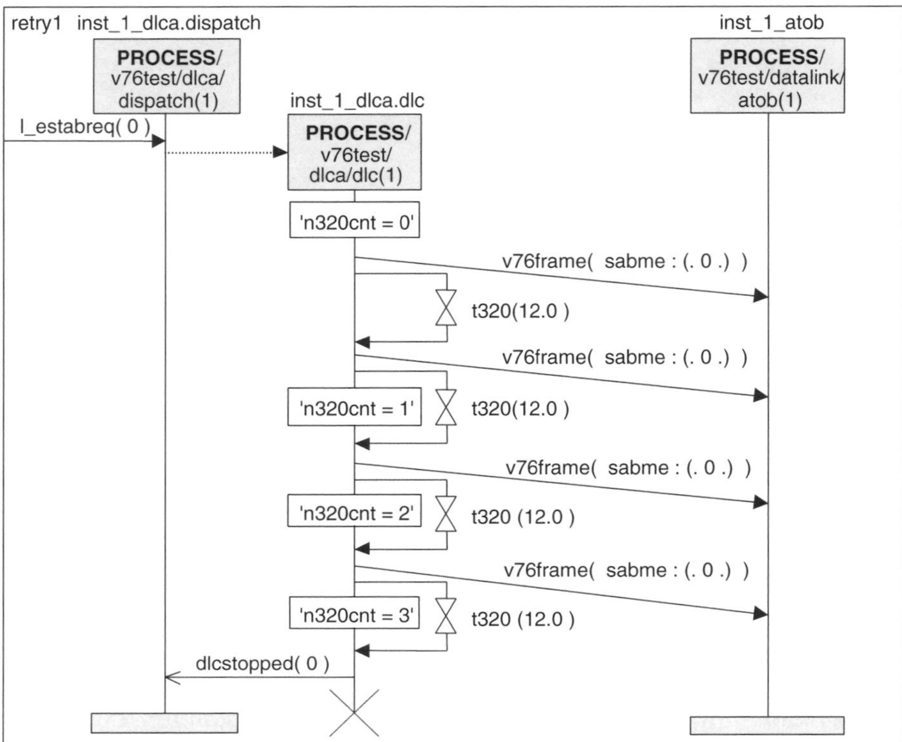

Figure 4.45 MSC generated by the Simulator showing three *SABME* retransmissions

In the generated MSC, represented in Figure 4.45, you see that the number of retransmission of SABME is correct, but the Simulator indicates[37]: *exception state, Unexpected signal dlcstopped in dlca!dispatch.*

It means that the signal *dlcstopped* transmitted by *DLC* goes into the input queue of process *dispatch* in block *dlca*, but unfortunately, in the current state of *dispatch*, the input of signal *dlcstopped* is not specified.

K. Save the Simulator scenario leading to the bug: in the Simulator, select *File > Scenario > Save As*, enter *retry1* and press *save*.

L. To save the current MSC into the file *retry1.msc*, type in the Simulator:

```
msc retry1
```

Do not exit from the Simulator.

4.3.3.2 Analyze the bug

To understand the bug, you will search in which state process *dispatch* (in block *dlca*) was when process *DLC* transmitted to it the signal *dlcstopped*.

---

[37] If you do not see this error, in the Simulator select *Edit > Configuration* and check the box *Trap unexpected signals*.

A. Return to the previous simulation step by pressing the *undo* ⏮ Simulator button.

B. In the Simulator, select the transition[38] *trans dlca!dlc(1): from_waitua_input_t320*: as shown in Figure 4.46, the Editor displays in bold the corresponding input *(T320)*. In the ELSE branch, you see the output of *DLCstopped*, which caused the problem.

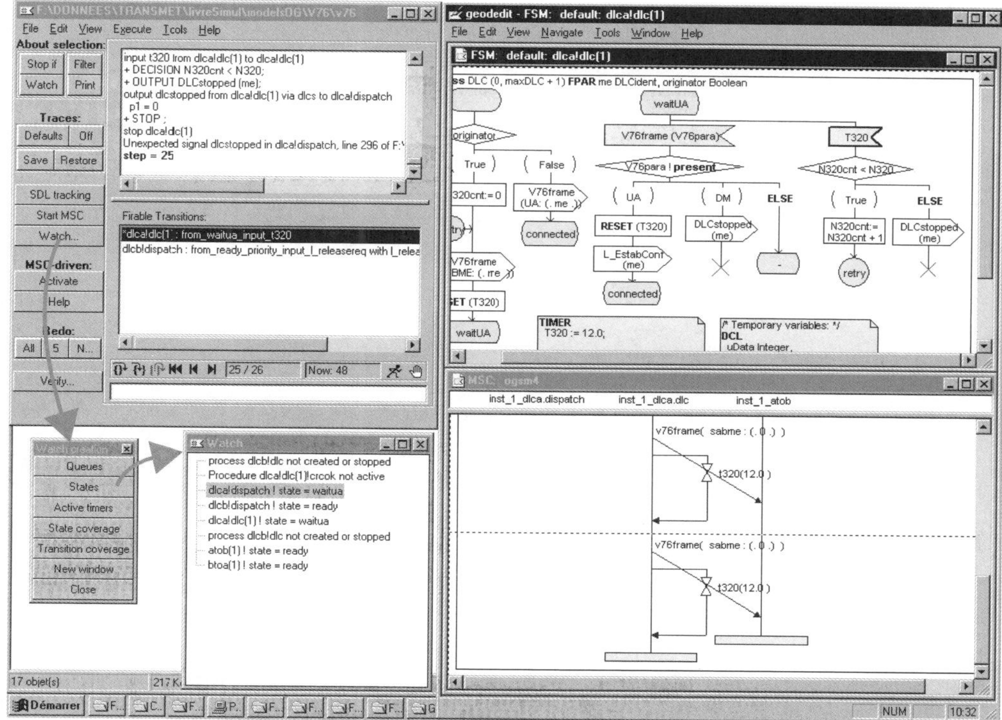

Figure 4.46 Searching for the unexpected signal bug

C. In the Editor, select *Navigate > Up*: the block type *V76_DLC* is now displayed, as in Figure 3.12; you see that signal *DLCstopped* goes to process *dispatch*, through the signal route *DLCs*.

D. In the Editor, select process *dispatch* and do *Navigate > Down* and *Navigate > Next Partition*: you see that under state *waitUA*, the input of signal *DLCstopped* is missing.

4.3.3.3 Correct the bug

You will add the missing input of signal *DLCstopped* under state *waitUA* in process *dispatch*.

A. When the Simulator is running, the Editor prevents you from modifying the SDL model: exit from the Simulator (answering *No* to the question) to enable the modification features of the Editor. Do not exit from the Editor.

---

[38] The firable transition *trans dlca!dlc(1): from_waitua_input_t320* is preceded by a *: it reminds you that you executed it before the undo.

B. In Windows (or Unix), make a copy of the file *v76.pr* into *v76_v1.pr* (but continue working on *v76.pr*, which becomes version 2).

C. In process *dispatch*, partition *part2*, select the input of *DLCstopped* under state *ready*, copy it, select the state *waitUA* and paste: the whole transition is inserted, as shown in Figure 4.47.

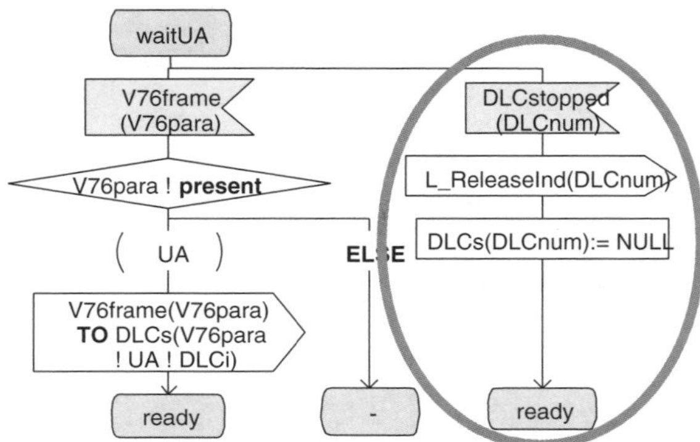

Figure 4.47 Missing input of signal *DLCstopped* added under state *waitUA*

D. In the pasted transition, don't forget to change the *nextstate* - into *nextstate ready*, otherwise you will be stuck in state *waitUA*.

E. Save the SDL model.

4.3.3.4 Simulate to check the bug correction

To check that the bug has been corrected, you will load and automatically replay the scenario stored in Section 4.3.3.1. See Section 4.3.1 for details on restarting the Simulator.

A. In the SDL Editor, unload all files except *v76.pr*.

B. If the ObjectGeode Launcher is not running, in the Editor select *Tools > SDL & MSC Simulator*.

C. In the ObjectGeode Launcher, Press the *Build* button, then if you do not get any SDL errors, press the *Execute* button.

D. The Simulator starts: press on *SDL Tracking* and on *Start MSC*.

E. In the Editor, close all windows except *Default tracking* and *ogsm4*, close the *Framework* view and select *Window > Tile Horizontally*, to obtain the screen shown in Figure 4.38.

F. In the Simulator, select *File > Scenario > Load*, and open *retry1.scn*: after *end of scenario loading*, you see *0/26* in the lower part of the Simulator, as shown in Figure 4.48: it means that you are at Step 0, and the loaded scenario has 26 steps.

G. Press the button *All* located under *Redo:* (or press 26 times the *redo* ⏯ Simulator button): when you see *end of scenario execution* and *26/26*, it means that the scenario loaded from the file *retry1.scn* has been replayed entirely[39].

---

[39] If the scenario does not replay until the end, check that your feeds are loaded: see Section 4.3.1.4.

Figure 4.48 The current and maximum step numbers after loading the scenario

H. The bottom of the MSC generated by the Simulator looks like Figure 4.49(a): the signal *dlcstopped* has been transmitted, but it is in the queue of process *dispatch*.

Figure 4.49 MSC trace: signal *dlcstopped* in the process queue (a) and consumed (b)

I. To watch the input FIFO queues of the model, press the *Watch...* button; in the *Watch creation* window, press *Queues*: as depicted in Figure 4.50, a watch window appears, displaying the contents of the queues. As expected, you can see the signal *dlcstopped* in the queue of process *dispatch*.

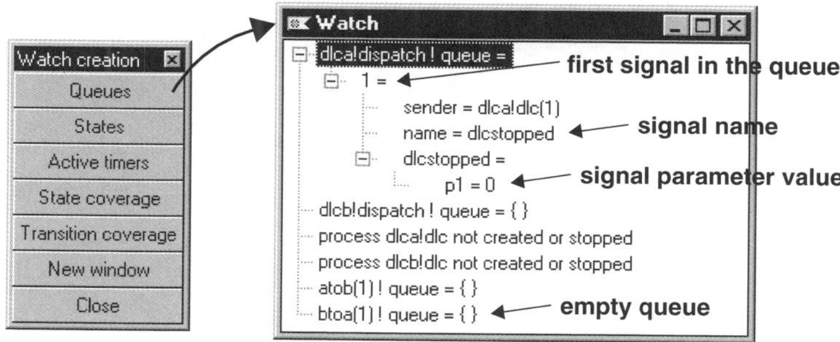

Figure 4.50 Watching the input queues

J. In the Simulator, double-click the transition

```
trans dlca!dispatch : from_waitua_input_dlcstopped
```

the signal *dlcstopped* disappears from the watch and the MSC shows a filled arrowhead as in[40] Figure 4.49(b).

We have returned to the initial model state, from where we can simulate other scenarios.

[40] The arrow is inclined, because the actual input of *dlcstopped* (the arrowhead) occurred after the process stop (the X symbol). A horizontal arrow would mean output *dlcstopped*, followed by input *dlcstopped* and then by process stop, which is not the actual behavior.

4.3.4 Detect nonsimulated parts

After a simulation session, the Simulator indicates[41] which parts of the SDL model have not been simulated:

- states
- transitions
- basic blocks[42]

An example of state, transition and basic block is provided in Figure 4.51.

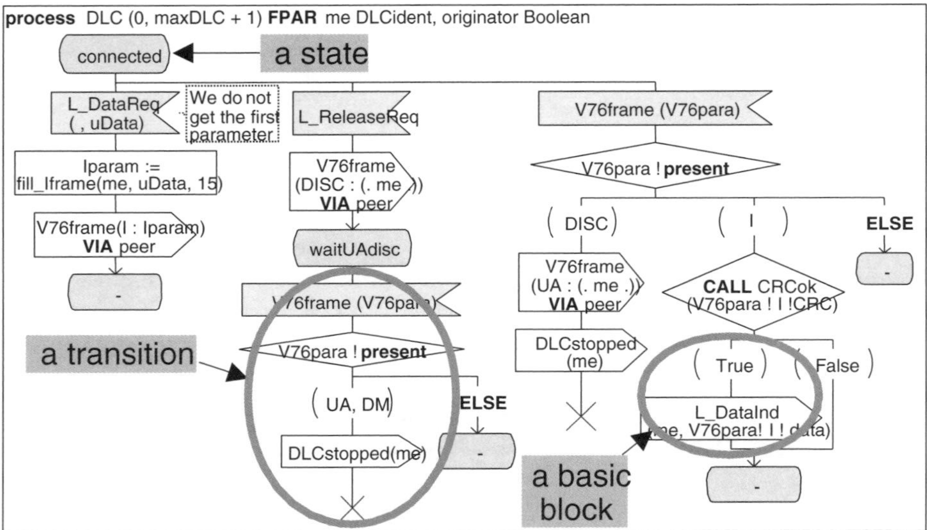

Figure 4.51 Examples of state, transition and basic block

Then you can simulate again the SDL model until you reach 100% coverage for the states, transitions and basic blocks. After playing all possible scenarios (which is easier using exhaustive simulation, if the model does not have too many states), the states, transitions and basic blocks not simulated are considered as "dead" parts: they can be removed, after careful inspection.

A. If the Simulator is already running, jump to E.

B. In the SDL Editor, unload all files except *v76.pr*.

C. If the ObjectGeode Launcher is not running, in the Editor select *Tools > SDL & MSC Simulator*.

D. In the ObjectGeode Launcher, Press the *Execute* button: the Simulator starts.

---

[41] A counter is associated to each state, each transition and each basic block: every time a state, transition or basic block is simulated, its counter is incremented. A value of 0 means it has never been simulated.

[42] In fact, basic blocks include transitions (except the implicit transition corresponding to discarding an unexpected signal).

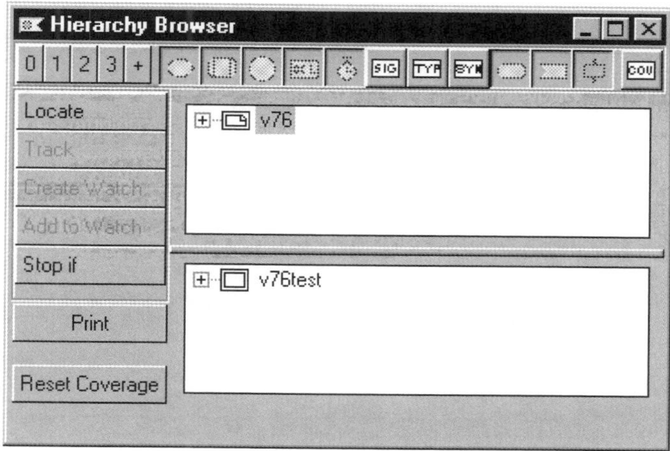

Figure 4.52 The Simulator Hierarchy Browser

E. In the Simulator, select *View > Hierarchy*: the Hierarchy Browser appears, as shown in Figure 4.52.

F. In the Hierarchy Browser, press the button *Reset Coverage*: this[43] sets to 0 the coverage counters, in case you did not restart the Simulator just before this exercise.

G. In the Hierarchy Browser, select *v76test* and press the button+ : this displays the coverage rates of the SDL model, as illustrated in Figure 4.53.

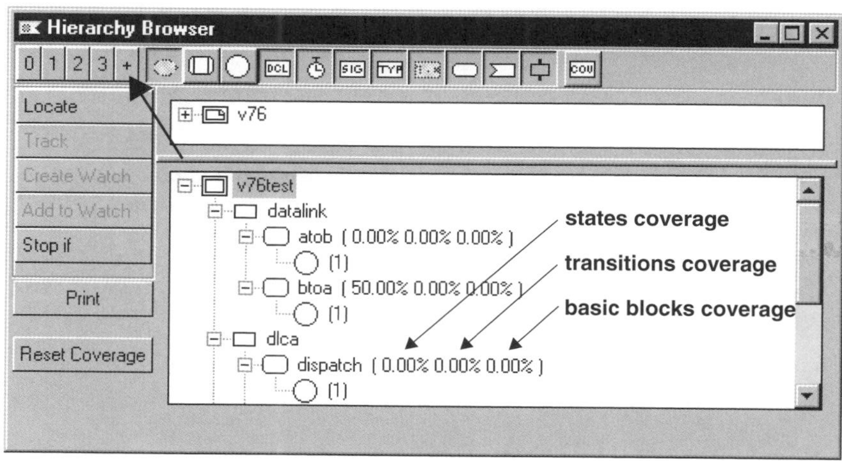

Figure 4.53 The coverage rates

[43] It seems that there are two minor problems in the version 4.2.2 of the Simulator: just after starting the Simulator, without executing any transition, the start symbols of processes *atob*, *btoa* and *dispatch* are marked as executed 2 times instead of 0 (type *cover state all*), and if you press the button *Reset Coverage* in the Hierarchy Browser, the coverage counters are all correctly set to 0, but the Hierarchy Browser does not display 0 concerning *btoa* and *dlcb!dispatch*. In fact the coverage for *btoa* and *dlcb!dispatch* should not be displayed, as they are instances of the same entities than *atob* and *dlca!dispatch*. As they concern only start transitions, which will be executed in any case, these problems are not dangerous.

The first number is the percentage of states simulated, the second number is the percentage of transitions simulated and the last number is the percentage of basic blocks simulated.

H. In the Simulator, select *File > Scenario > Load*, and open *test1.scn* (see Section 4.3.2.3).

I. When you see *end of scenario loading*, press the button *All* located under *Redo*.

The Hierarchy Browser now displays the new values of the coverage rates, as shown in Figure 4.54.

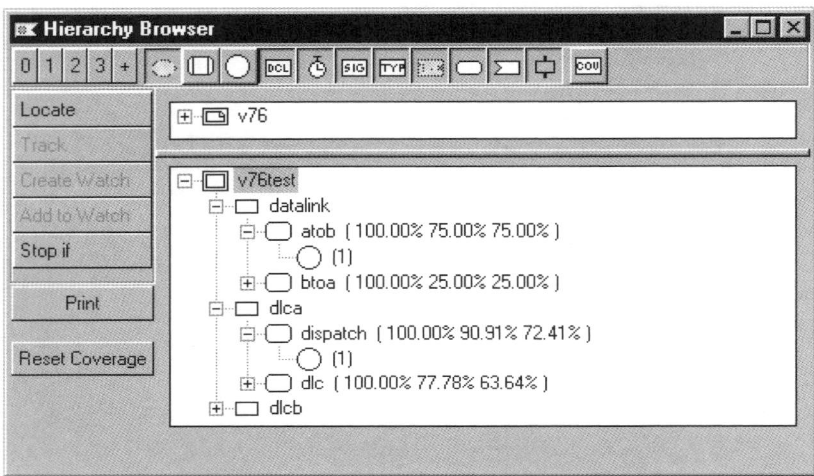

Figure 4.54 The coverage rates after replaying *test1*

We see, for example, that

- all the states in process *atob, dispatch* and *dlc* have been simulated,
- 90.9% of the transitions in process *dispatch* have been simulated,
- 72.4% of the basic blocks in process *dispatch* have been simulated.

J. Select *dispatch* in *dlca* in the Hierarchy Browser and press the two buttons shown in Figure 4.55: one displays all the transitions, the other displays only nonexecuted transitions; you see that the only transition not simulated is *from_waitua_input_dlcstopped.*

K. In the Hierarchy Browser, select *from_waitua_input_dlcstopped* and press *Locate*: the Editor opens process *dispatch* and selects the uncovered transition, as shown in Figure 4.56.

As we have only replayed the scenario *test1.scn*, it is normal for this transition to be detected as not simulated. To simulate it, we need to replay (without resetting the coverage counters) *retry1.scn* also and execute the transition *trans dlca!dispatch: from_waitua_input_dlcstopped.*

4.3.5 Validate against more scenarios

After simulation of the main scenarios described in Section 4.3.2, it is wise to play more scenarios to check the reaction of the SDL model. Those scenarios can be:

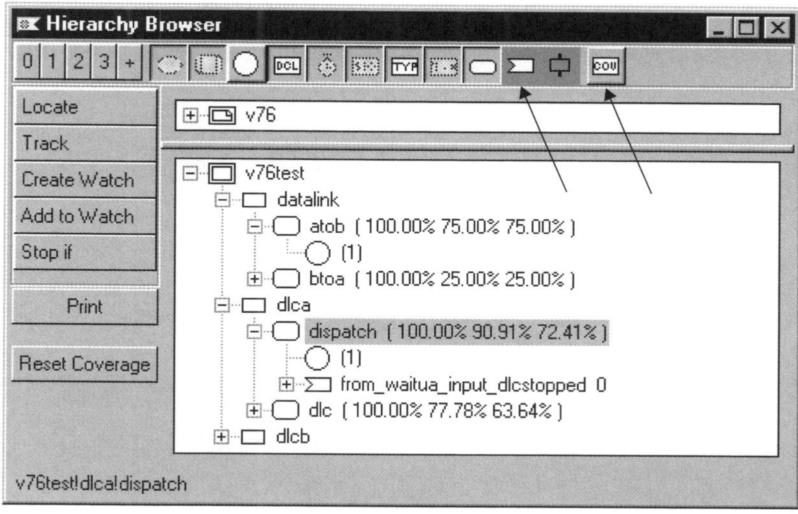

Figure 4.55 The transition not simulated in *dispatch*

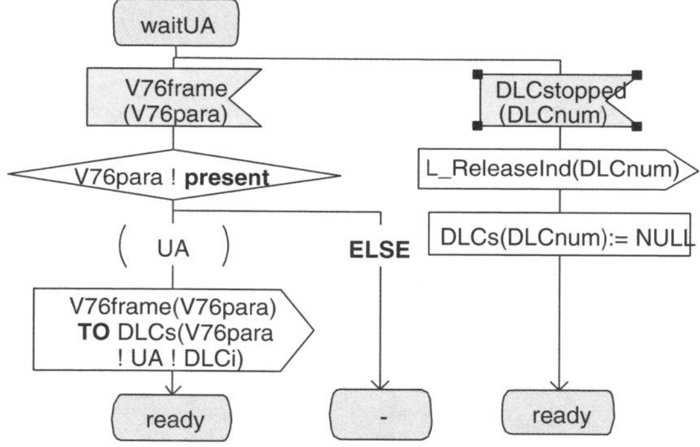

Figure 4.56 The transition not simulated is automatically located

- more complex: for example, two simultaneous connections,
- beyond limits: for example, creation of more connections than allowed.

4.3.5.1 Simulate two simultaneous connections

You will simulate to check that the SDL model can handle two connections in parallel.

A. If the Simulator is already running, jump to E.

B. In the SDL Editor, unload all files except *v76.pr*.

C. If the ObjectGeode Launcher is not running, in the Editor select *Tools > SDL & MSC Simulator.*

D. In the ObjectGeode Launcher, press the *Build* button, then if you do not get any SDL errors, press the *Execute* button.

E. The Simulator starts: press on *SDL Tracking* and on *Start MSC*.

F. In the Editor, close the *Framework* view and select *Window > Tile Horizontally*, to obtain the screen shown in Figure 4.38.

G. Load and replay the scenario *cnx1.scn*, as indicated for *retry1.scn* in Section 4.3.3.4: now one instance of process *DLC* exists on each side of the system.

   Now establish one more connection:

H. Execute the transition *dlca!dispatch*: *from_ready_input_l_estabreq with l_estabreq(1) from env_dlcasu* (double-click it in the Simulator list, as shown in Figure 4.57). You see in the MSC trace that a new instance of process *DLC* is created, numbered 2.

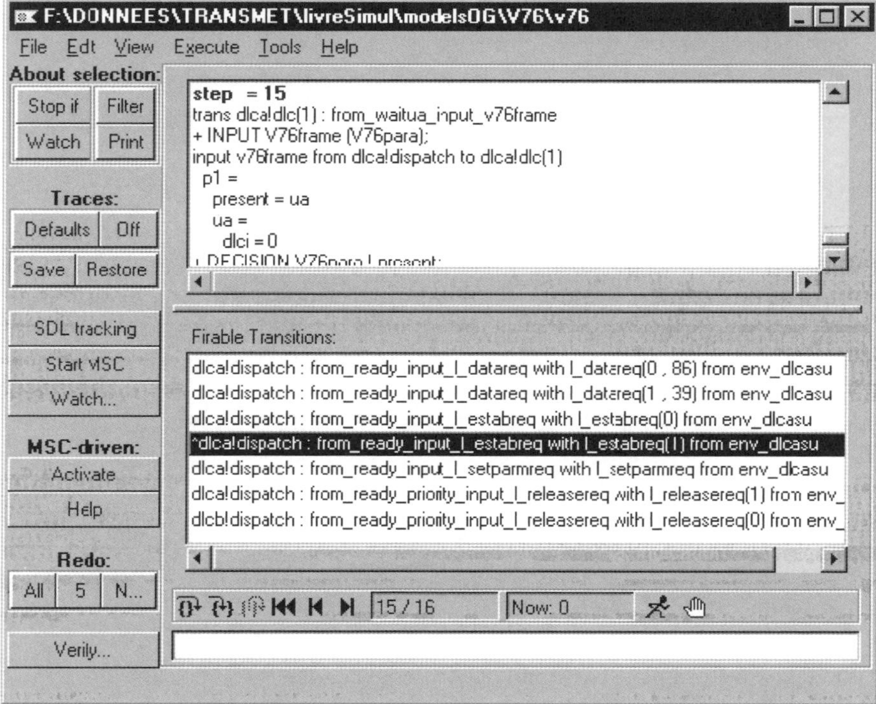

Figure 4.57 The transition to fire

I. In the same way, execute the following transitions:

```
dlca!dlc(2) : start
atob(1) : from_ready_input_v76frame
atob(1) : decision_lose_the_frame('No')
dlcb!dispatch : from_ready_input_v76frame
dlcb!dispatch : from_waitestabresp_input_l_estabresp with
                l_estabresp from env_dlcbsu

dlcb!dlc(2) : start
```

```
btoa(1)  : from_ready_input_v76frame
btoa(1)  : decision_lose_the_frame('No')
dlca!dispatch : from_waitua_input_v76frame
dlca!dlc(2)  : from_waitua_input_v76frame
```

The new connection has been established between sides A and B.

J. To check that all four instances of process *DLC* exist and are in state *connected*, enter in the Simulator:

```
print dlc!state
```

The Simulator displays:

```
dlcb!dlc(1) ! state = connected
dlcb!dlc(2) ! state = connected
dlca!dlc(1) ! state = connected
dlca!dlc(2) ! state = connected
```

K. To test that the new connection[44] works, let's transfer data through it; execute the following transitions:

```
dlca!dispatch : from_ready_input_l_datareq with l_datareq(1,
                  39) from env_dlcasu
dlca!dlc(2) : from_connected_input_l_datareq
atob(1)  : from_ready_input_v76frame
atob(1)  : decision_lose_the_frame('No')
dlcb!dispatch : from_ready_input_v76frame
dlcb!dlc(2) : from_connected_input_v76frame
```

The generated MSC, represented at blocks level in Figure 4.58, shows that block *DLCb* transmitted signal *L_DataInd(1, 39)* to the environment (representing Service User B): the data *39* has been successfully transferred from A to B through DLC *1*.

L. Save the Simulator scenario: in the Simulator, select *File* > *Scenario* > *Save As*, enter *cnx2* and press *save*.

M. To save the current MSC into the file *cnx2.msc*, enter the following command into the Simulator:

```
msc cnx2
```

4.3.5.2 Simulate an attempt to create too many connections

You will simulate to see what happens if you try to create more connections than allowed. The maximum number of parallel connections in our model is $maxDLC + 1 = 2$. Figure 3.12 shows that this number corresponds to the maximum number of instances or process *DLC*, which is equal to the size of the array *DLCs*, declared in Figure 3.14.

[44] The DLC number (of type *DLCident*) of the new connection is 1, and the corresponding instance number of process *DLC* (given by the Simulator) is 2.

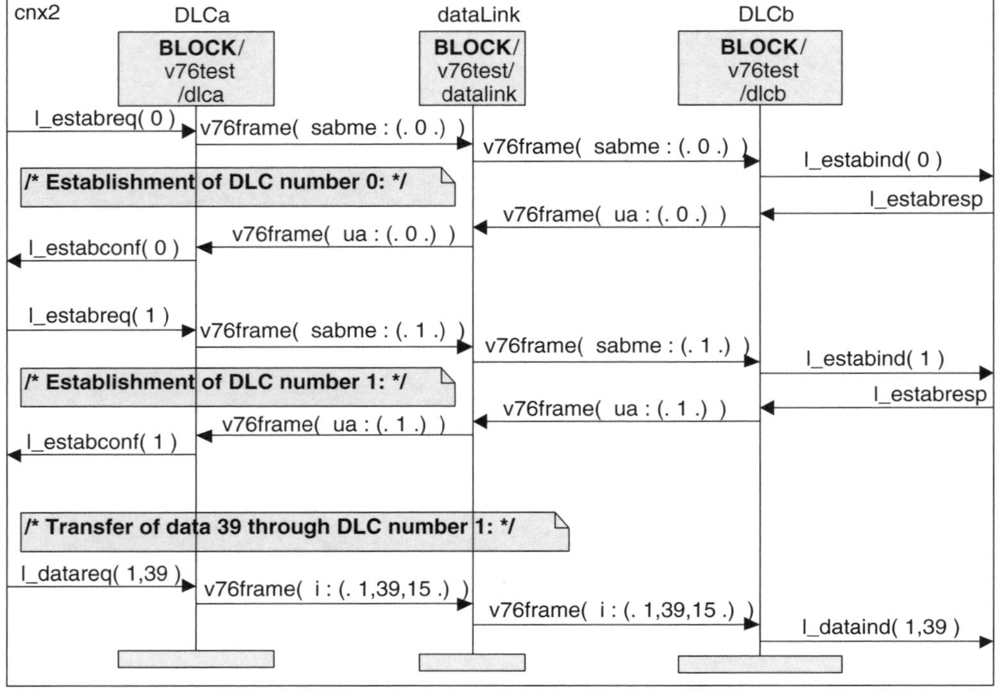

Figure 4.58 Two connections 0 and 1 in parallel

A. If you exited the tools since Section 4.3.5.1, launch the Simulator and replay the scenario *cnx2.scn*: two instances of process *DLC* exist on each side of the system, the maximum is reached.

B. Execute the transition *dlca!dispatch*: *from_ready_input_l_estabreq with l_estabreq(0) from env_dlcasu*. You see in the MSC trace that the system answers with an *L_RelelaseInd(0)*: it means that no more connection can be established.

   But if you look at the SDL trace (or enter *print state*), you discover that process *dispatch* is stuck in state *waitUA*: this is a modeling bug. Also, transitions are missing in the list of firable transitions.

   You will correct process *dispatch* to go to state *ready* instead of state *waitUA* after transmitting *L_RelelaseInd*.

C. Exit from the Simulator (answering *No* to the question) to enable the modification features of the Editor. Do not exit from the Editor.

D. In Windows (or Unix), make a copy of the file *v76.pr* into *v76_v2.pr*.

E. In process *dispatch*, partition *part2*, select the output of *L_RelelaseInd* under the *ELSE* branch of the decision and click on the *nextstate* palette symbol; enter – in the newly created symbol, as shown in Figure 4.59.

F. Save the SDL model.

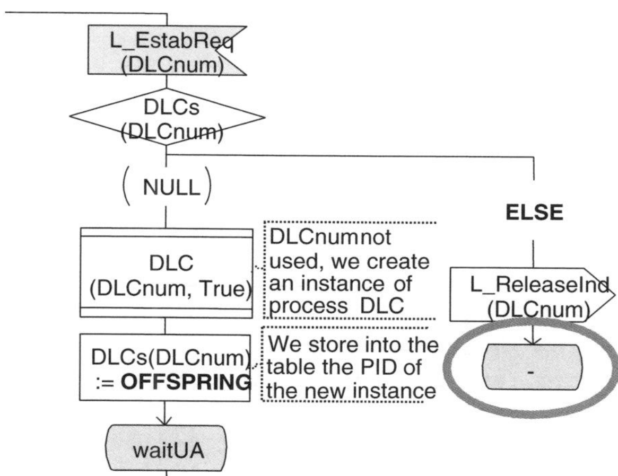

Figure 4.59 After the bug correction

G. Simulate again to check that the bug has disappeared: replay the scenario *cnx2.scn*, and execute the transition *dlca!dispatch*: *from_ready_input_l_estabreq with l_estabreq(0) from env_dlcasu*: process *dispatch* now stays in state *ready*.

4.3.6 Write a script for automatic validation

In an actual project, to test a more complex SDL model, you would produce, for example, 40 simulation scenarios. After a change in the SDL model, you must check that it still works correctly (nonregression) : the Simulator command language enables you to write a script to automatically replay the 40 scenarios (stored in *.scn* files) and write the test results into a file. By just looking if the file contains *TEST FAILED* or *TEST PASSED*, you will know if the 40 scenarios replayed correctly or not.

Here is a Simulator script to replay three of our scenarios:

```
-- This geodesim script replays 3 scenarios to test V.76
-- and writes the results into test_res1.wri

define test_failed False

echo ' '
echo 'Execution of script1.wri:' || test_res1.wri
echo '' ||| test_res1.wri

init -- return to step 0
source test1.scn -- replays test1.scn
-- now check that test1.scn executed till the end (step 41)
-- and that both DLCs arrays contain Null:
if (step=41) and (dlca!dispatch ! dlcs = (. Null .)) and \
   (dlcb!dispatch ! dlcs = (. Null .)) ;
  echo 'Replay test1.scn OK' ||| test_res1.wri;
```

```
17  else ;
18    define test_failed True;
19    echo 'Error found in test1.scn' ||| test_res1.wri;
20  fi
21
22  init -- return to step 0
23  source cnx2.scn
24  -- now check that cnx2.scn executed till the end (step 32)
25  -- and that the 4 inst. of process DLC are in state connected:
26  if (step=32) and (dlca!dlc(1)!state = connected) and \
27     (dlca!dlc(2)!state = connected) and \
28     (dlcb!dlc(1)!state = connected) and \
29     (dlcb!dlc(2)!state = connected) ;
30    echo 'Replay cnx2.scn OK' ||| test_res1.wri;
31  else ;
32    define test_failed True;
33    echo 'Error found in cnx2.scn' ||| test_res1.wri;
34  fi
35
36  init -- return to step 0
37  source retry1.scn
38  -- now check that retry1.scn executed till the end (step 26) and
39  -- that queue of process dispatch contains signal DLCstopped:
40  if (step=26) and (dlca!dispatch ! queue(1)!name = DLCstopped) ;
41    echo 'Replay retry1.scn OK' ||| test_res1.wri;
42  else ;
43    define test_failed True;
44    echo 'Error found in retry1.scn' ||| test_res1.wri;
45  fi
46
47  echo ' ' ||| test_res1.wri
48  if $test_failed ;
49     echo '** TEST FAILED; see test_res1.wri ***' ||| test_res1.wri
51  else ; echo 'TEST PASSED' ||| test_res1.wri
52  fi
```

Here are a few explanations on the script:

- The character \ means continue on next line.
- Line 4: `define test_failed False`: defines the value *test_failed* and sets it to *False*.
- Line 7: `echo 'Execution of script1.wri:'` || `test_res1.wri`: writes 'Execution of script1.wri:' in the simulator trace and into the file *test_res1.wri*.
- Line 11: `source test1.scn`: same as loading scenario *test1.scn* and redoing all.
- Line 14: `if (step=41)`: step contains the current step number; we know that the scenario in *test1.scn* has 41 Steps; if after replay the step number is lower than 41, it means that the scenario did not run completely.

- Line 14: `DLCa!dispatch!DLCs = (. Null.)`: *DLCs* is an array declared in process *dispatch*; after the scenario replay, we know that all its elements must contain Null.
- Line 16: if the conditions above are true, we write *'Replay test1.scn OK'* into the file *test_res1.wri*, otherwise we write *'Error found in test1.scn'* and we set *test_failed* to *True*.
- Then we repeat a similar code for scenarios *cnx2.scn* (line 23) and *retry1.scn* (line 37), with different conditions.
- Line 48: finally, if *test_failed* contains *True*, we write *TEST FAILED*, otherwise *TEST PASSED*.

Once you have typed the script into the file *script1.wri*[45] (do not type the line numbers), execute it:

A. Launch the Simulator.

B. Click the Simulator button *Traces: Off* to see only the script traces.

C. Enter *source script1.wri* in the Simulator. If the model (and the script!) is correct, you see

```
> source script1.wri

Execution of script1.wri:

scenario in initial state
Replay test1.scn OK
scenario in initial state
Replay cnx2.scn OK
scenario in initial state
Replay retry1.scn OK

TEST PASSED
```

The file *test_res1.wri* contains the same results, informing you of which scenario(s) failed. By just reading the last line, you are sure that all the scenarios passed.

If you want to generate the MSC trace and store it into the file *test5.msc* after replaying a scenario, you can add the following command into the script:

```
msc test5
```

You will learn in Chapter 5 how to check that your SDL model behavior complies with an MSC. This will be even more powerful to test automatically an SDL model: for example, we will be able to check the value of a parameter in a signal transmitted, which is not performed here.

4.3.7 Other Simulator features: watch, trace, filter etc.

This section describes features of the Simulator not absolutely essential to validate an SDL model, but which can be very helpful and save a lot of time on an actual system validation.

[45] The file extension *.wri* ensures opening the file with WordPad just by double-click, if you are running Microsoft Windows™.

4.3.7.1 Aliases

To shorten textual commands, you can create aliases, a kind of macros. Alias definitions are generally stored in the file *geodesm.startup* (see Section 4.3.7.2) to be executed automatically. Here are some examples of aliases you could write:

```
alias h 'history'
alias p 'print'

-- Basic blocks (paths) never executed:
alias bb0 'cover bblocks all 0:0'
```

Then, you have only to enter *h* instead of *history*, *bb0* instead of *cover bblocks all 0:0*, and so on.

Also, aliases can have parameters; for example, to print the state of instance $2 of process *DLC* in block *DLC*$1:

```
> alias p_dlc 'print dlc$1!dlc($2)!state'
> p_dlc a 1
dlca!dlc(1) ! state = waitua
```

4.3.7.2 Automatic scripts execution at simulator startup

When the Simulator starts, three scripts are automatically executed, if they exist, as shown in Figure 4.60:

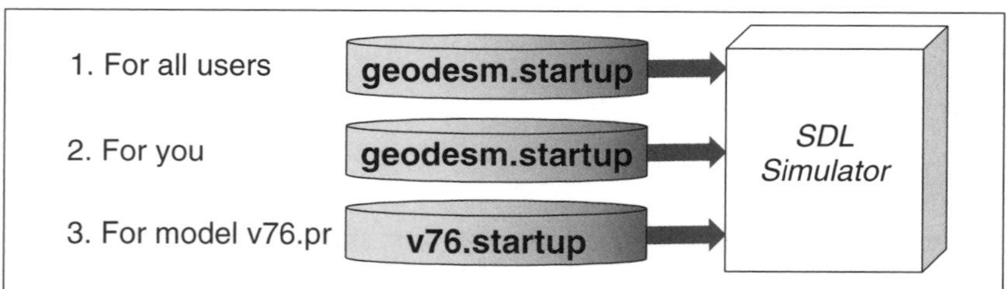

Figure 4.60 The three Simulator startup files

1. The file *geodesm.startup* common to all the Simulator users, rarely used, located in the installation directory of ObjectGeode.
2. The file located for Unix in your home directory, *$HOME/geodesm.startup*, or for Windows NT in your profile, *C:\winnt\profiles\your_name\ObjectGeode\geodesm.startup*.
3. The file *name.startup*, where *name* is the name of the file containing the SDL system. For example, if your model is stored in *v76.pr*, the startup file must be named *v76.startup*, located in the current directory[46]. It contains commands specific to a certain model, such as firing the start transitions and loading feed commands.

[46] In Unix, beware of navigating (using *cd*) to the directory containing the SDL model **before** launching ObjectGeode. Otherwise, the Simulator will run in the wrong directory: files will not be generated where you expect them and the model startup file will not be found.

A fourth script is executed if the Simulator is invoked with the *-startup* option. For example,

```
geodesim v76 -startup my_file.wri
```

Example of script *geodesm.startup* to put in your home directory (Unix) or in your profile directory (Windows NT):

```
-- Abbreviated commands :
alias h 'history'
alias p 'print'
alias u 'undo'
alias r 'redo'
alias t 'trace'
alias ve 'verify'
alias vo 'verify options'

-- Coverage rates:
alias cov 'print state_cover_rate(system); \
print trans_cover_rate(system); \
print bb_cover_rate(system)'

-- States never executed:
alias cs0 'cover state all 0:0'

-- Transitions never executed:
alias ct0 'cover trans all 0:0'

-- Basic blocks (paths) never executed:
alias bb0 'cover bblocks all 0:0'

-- More compact generated MSCs: (does not work on PC)
define msc_xspace '140'
define msc_yspace '20'

-- To store only 2 .scn files of each kind (for verify),
-- (default is 10):
stop limit 2
deadlock limit 2
exception limit 2
error limit all 2
success limit all 2
define scc_sink_limit '2'

-- To display states number of every process and queue
-- after exhaustive simulation:
define verify_stats 'true'
```

```
-- To be able to time-out timers more easily:
define loose_time 'true'

-- List the alias definitions:
alias
```

4.3.7.3 Running the Simulator without its graphical interface

To perform automatic tasks such as automatic replay of scenarios, the Simulator can be launched without its graphical interface; this is called batch mode. To run in batch mode, after the build (textual command: *gsmcomp v76*), type the following command (suppose the SDL model is in the file *v76.pr*) in a DOS or Unix shell (containing *v76.sim*):

```
F:\DONNEES\V76> geodesim -b v76
SDL Simulator V4.2.2 Copyright 1989-2001 by Telelogic AB.
1 : atob(1) : start
2 : btoa(1) : start
3 : dlca!dispatch : start
4 : dlcb!dispatch : start
```

The Simulator displays the list of ready transitions. To fire a transition, just enter its number:

```
> 2
step = 1
trans btoa(1) : start
+ START ;
+ NEXTSTATE ready;
1 : atob(1) : start
2 : dlca!dispatch : start
3 : dlcb!dispatch : start
>
```

Then you can enter any Simulator textual command:

```
> print now
now = 0.0
```

The commands to execute can be inserted in the file *v76.startup*, automatically played when the Simulator starts. For example, the automatic simulation script *script1.wri* described in Section 4.3.6 can be executed automatically if the file *v76.startup* contains

```
source script1.wri
quit
```

4.3.7.4 Commands history

To reenter a Simulator command like in a Unix shell, type *history*, and the 10 last executed commands will be displayed:

```
> history
```

```
1 : let x = 34
2 : print state
3 : alias f 'feed'
4 : print now
5 : print trans_cover_rate(v76test)
6 : print dlcs
7 : trace
8 : redo 6
9 : mode depth
10 : history
```

Then you can reexecute a command by typing *!n*, where *n* is the number in the list. For example, *!4* reexecutes *print now*, and !! reexecutes the last command.

4.3.7.5 Examining the SDL model: print

The *print* command displays the contents of any SDL or simulation element. Here are some examples, to type in the Simulator command line, shown in Figure 4.44.

To display the current time:

```
> print now
now = 48.0
```

To display the synonym *maxDLC*:

```
> print maxDLC
maxdlc = 1
```

To display all the array variables named *DLCs*:

```
> print DLCs
dlcb!dispatch ! dlcs =    -- in DLCb
  0 = null
  1 = null
dlca!dispatch ! dlcs =    -- in DLCa
  0 = dlca!dlc(2)
  1 = null
```

To display only the variable *DLCs* contained in process *dispatch* in block *DLCa*:

```
> print dlca!dispatch ! dlcs
dlca!dispatch ! dlcs =
  0 = dlca!dlc(2)
  1 = null
```

To display the input queue contents of each process instance:

```
> print queue
btoa(1) ! queue = { }      -- means queue empty
atob(1) ! queue = { }
```

```
process dlcb!dlc not created or stopped
process dlca!dlc not created or stopped
dlcb!dispatch ! queue = { }
dlca!dispatch ! queue =
  1 =                          -- first queue element
  sender = dlca!dlc(1)         -- who sent it
  name = dlcstopped            -- signal name
  dlcstopped =
    p1 = 0                       -- first signal parameter
```

To display the current state of each process instance:

```
> print state
btoa(1) ! state = ready
atob(1) ! state = ready
process dlcb!dlc not created or stopped
process dlca!dlc not created or stopped
dlcb!dispatch ! state = ready
dlca!dispatch ! state = waitua
```

To display all the model contents (current global state):

```
> print all
atob(1) =
  substate = none
  state = ready
  sender = dlca!dlc(1)
  parent = null
etc.
```

To test if process *dispatch* in block *DLCa* is in state *waitUA* and *Now* < 10:

```
> print (DLCa!dispatch!state=waitUA) and (Now < 10)
(dlca!dispatch ! state = waitua) and (now < 10.0) = false
```

To call the operator *fill_Iframe* to test it:

```
> print fill_Iframe(1, 39, 15)
fill_iframe(1, 39, 15) =
  dlci = 1
  data = 39
  crc = 15
```

To call the ObjectGeode-specific predefined SDL operator *hexadecimal* to convert a number to hexadecimal:

```
> p hexadecimal(15)
hexadecimal(15) = 0xf
```

To call the ObjectGeode-specific predefined SDL operator *integer* to convert a hexadecimal number:

```
> p integer(0xF)
integer(0xf) = 15
```

4.3.7.6 *Examining the SDL model: watch windows*

The *watch* command opens one or more watch windows displaying the contents of any SDL or simulation element, refreshed every time its value changes.

The button *Watch* allows to create predefined watch windows to display the contents of all the process input queues, process (and procedure or services) states, and so on.

To watch a specific element, you must select *Edit > Create watch*, then right-click to open a shortcut menu, where the choice *Add Item* prompts you to enter the name of the entity you want to put in the watch. The syntax for names you enter here is the same as that for the textual command *print*. For example, you can enter a variable name, such as *DLCs*, to watch all the occurrences of variable *DLCs*, as shown in Figure 4.61.

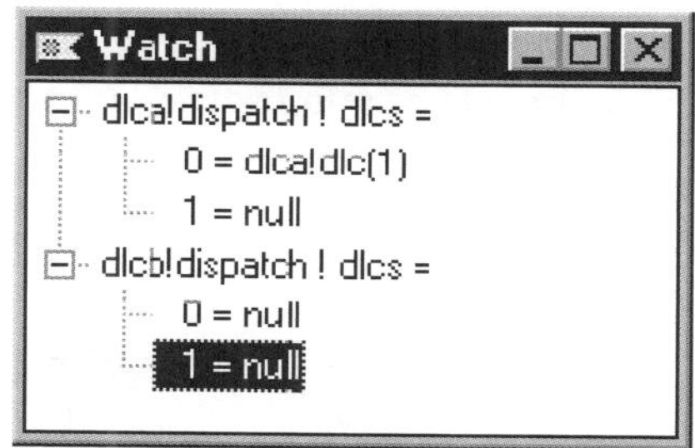

Figure 4.61 A watch window

As each *DLCs* is an array with two elements, the watch shows the content of each element.

To get the same watch windows the next time you launch the Simulator, you can select *File > Watches > Save As* and enter a file name such as *test1.watch*. Then you can add the command *source test1.watch* into your model startup file (example: *v76.startup*) to create the watch windows automatically.

4.3.7.7 *Examining the SDL model: trace*

The *trace* command displays in the textual trace the content of any SDL or simulation element, after every simulation step. It also writes this content in the MSC trace, every time its value changes. The syntax for names you enter here is the same as that for the textual command *print*. For example, to trace the value of the variable *N320cnt* in block *DLCa*, type in the Simulator command line

```
trace dlca!dlc(1) ! n320cnt
```

The effect in the MSC trace is depicted in Figure 4.45. Trace is handy when searching the origin of a hard to find bug: you can trace the state of each process instance, of variables, of process queues, and so on.

4.3.7.8 Modifying the SDL model

Two commands allow you to modify during simulation the content of an element in the SDL model: the *let* textual command, and the *change value* graphical command, available in the watch windows.

The *let* command accepts the same arguments as the *print* command (except that you must specify the path if more than one occurrence of the element exists in the SDL model). For example, to enter *2* in the variable *N320cnt* in block *DLCa*, type in the Simulator command line:

```
let DLCa!DLC(1)!N320cnt = 2
```

To change easily the value of complex structures such as arrays, struct or ASN.1 values the Simulator provides the *change value* command in the shortcut menu available in watch windows.

For example, to change the value of the second element in the array *DLCs* declared in process *dispatch* contained in block *DLCb*,

A. Create the watch as in Figure 4.62: select *Edit > Create watch*, choose *Add Item* in the shortcut menu and type *DLCs*.

B. Select the line corresponding to the element to change, as in Figure 4.62a.

C. Right-click to open the shortcut menu, and select *Change Value*: the Change Value window appears.

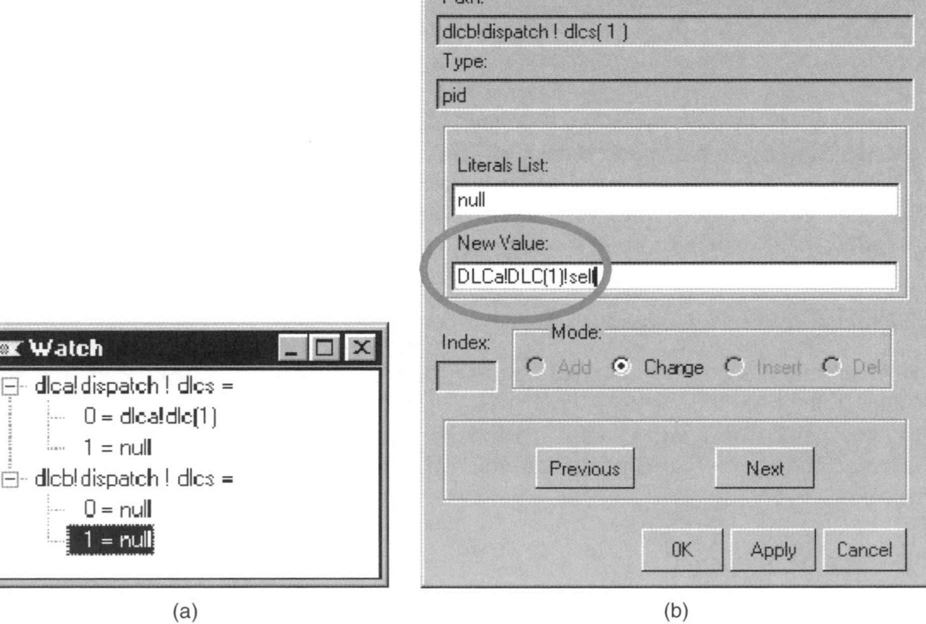

Figure 4.62 (a) Variables *DLCs* in a watch and (b) the Change Value window

D. In the field *New Value*, enter the desired value (such as *DLCa!DLC(1)!self*, the Pid of the instance *1* of process *DLC* contained in block *DLCa*) and press *Apply*.

The array *DLCs* in *DLCb* is changed; note that the SDL model is now in an inconsistent state and is no longer supposed to work.

4.3.7.9 Removing some behaviors: filters

The Simulator *filter* command allows you to freeze some parts of the SDL model. For example, to block the answer *'Yes'* in the informal decision represented in Figure 4.63, type the following command in the Simulator:

```
filter trans atob(1):decision_lose_the_frame('Yes')
```

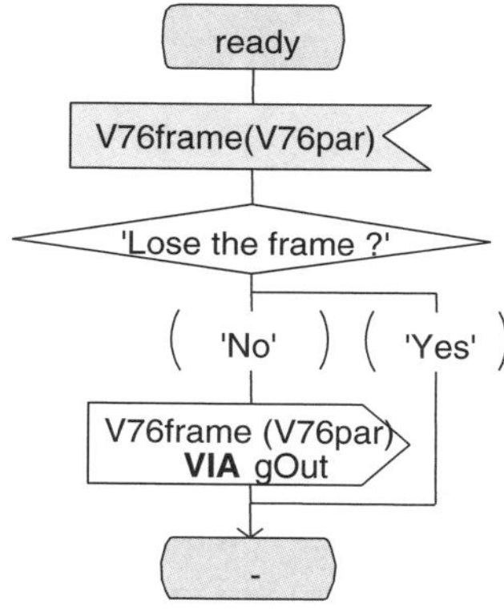

Figure 4.63 The informal decision in process type *topeer*

When you reach the decision, the *Firable Transition* area of the Simulator will only contain the transition corresponding to *'No'*:

```
trans atob(1) : decision_lose_the_frame('No')
```

instead of:

```
trans atob(1) : decision_lose_the_frame('No')
trans atob(1) : decision_lose_the_frame('yes')
```

You can add several filter conditions and manage them with *Edit > Filter Conditions*. To enter a filter condition with this command, do not type the keyword *filter* in front of the condition.

Filter conditions are also active during random and exhaustive simulation, where they are very handy to guide the simulation automatically. In such modes, the transitions fired are selected by the Simulator, not by you; thus, you must filter a certain transition if you do not want the Simulator to fire it.

We have seen how to filter a certain transition (*filter trans*...), but the argument of the *filter* command can also be an expression or an event. Here are some examples.

To prevent from firing any transition leading to variable *n320cnt* > *2* (simulation will only be possible for values 0, 1 and 2):

```
filter n320cnt > 2
```

To prevent any output of signal *L_EstabInd*:

```
filter output L_EstabInd
```

To limit to 2 signals the content of the input queue of instance 1 of process *DLC* in block *DLCa* (the part *is_active* is not necessary for a process instance that always exists, such as *dispatch(1)*):

```
filter is_active(DLCa!DLC(1)) and \
  length(DLCa!DLC(1)!queue) > 2
```

To prevent instance 1 of process *DLC* in block *DLCa* from reaching the state *connected*:

```
filter DLCa!DLC(1)!state = connected
```

To block all the transitions contained in the block *DLCa*:

```
filter trans DLCa
```

4.3.7.10 Undo and redo Simulator commands

Undo enables you to go back to any previous simulation step; then you can use *redo* to replay the previously undone steps. *Redo* is also used to replay a loaded scenario.

Two buttons in the Simulator perform *undo* and *redo* of one step at a time: ⏮ and ⏭. To undo or redo a larger number of steps, you can use the textual commands *undo* and *redo*. For example, if you are at Step 27, you can type:

```
undo 16 -- returns to step 27 - 16 = 11
redo 8 -- returns to step 11 + 8 = 19
```

4.3.7.11 Declaring simulation variables

Simulation variables can be declared in the Simulator:

- work variables: using the command *dcl*, like in SDL
- environment variables: using the command *define*

Example of work variable:

```
dcl x integer
let x = 45
print x
let x = x + 1
```

Example of environment variable (extract from the script in Section 4.3.6):

```
define test_failed False
-- etc.
if $test_failed; echo 'TEST FAILED'
else; echo 'TEST PASSED'; fi
```

4.3.7.12 Two important Simulator options

The Simulator behavior can be configured thanks to several options; to see all the options, enter the textual command *define*. For the main options, the *Edit > Configuration* command opens the Simulator configuration window represented in Figure 4.64.

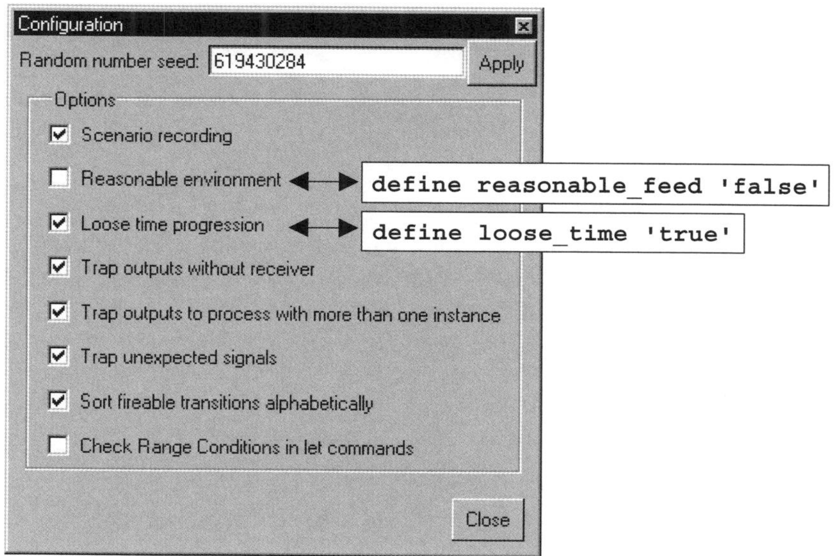

Figure 4.64 The Simulator configuration window

We explain two important options here: *reasonable environment* and *loose time progression*. If you want to set these options automatically, enter the appropriate define command, indicated in Figure 4.64, in one of your Simulator startup files.

Reasonable environment allows the input of an external signal (coming from the feed command) only if the SDL model is in a 'quiet' state. Quiet means that all the input queues in the SDL model are empty and all the start transitions have been executed.

For example, if you start the simulation of V.76 and click on *init* ⏮, after starting process *dispatch* you get nine firable transitions with *reasonable environment* off, instead of three transitions if you turn it on, as illustrated in Figure 4.65. When on, the Simulator wants you to first finish starting all the process instances before enabling the input of external signals.

In practice, the simulation is simpler when *reasonable environment* is on, because there are generally fewer transitions to select in the list. But in certain models, some behaviors can only be simulated when *reasonable environment* is off (for example, here you cannot send any *L_EstabReq* to the model before starting the 4 process instances).

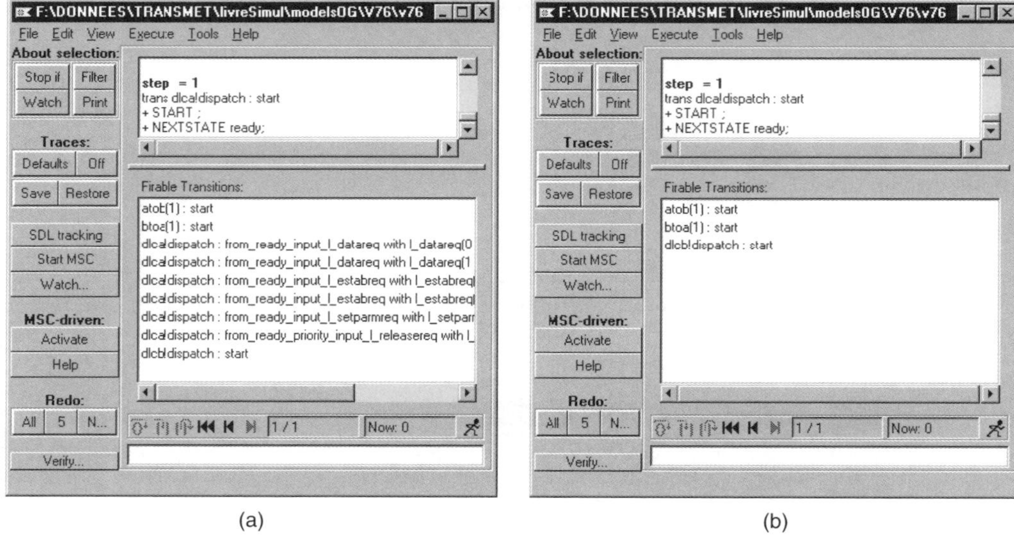

Figure 4.65 *Reasonable environment* off (a) and on (b)

If *loose time progression* is off, the Simulator will not allow any time progression while the simulation is not 'blocked'. We remind you that in the Simulator, the execution duration of the SDL instructions is supposed to be zero. If you simulate the model shown in Figure 4.66, once you have executed the start transition, timer *TIMER1* is started, and process *proc1* is in state *st1*.

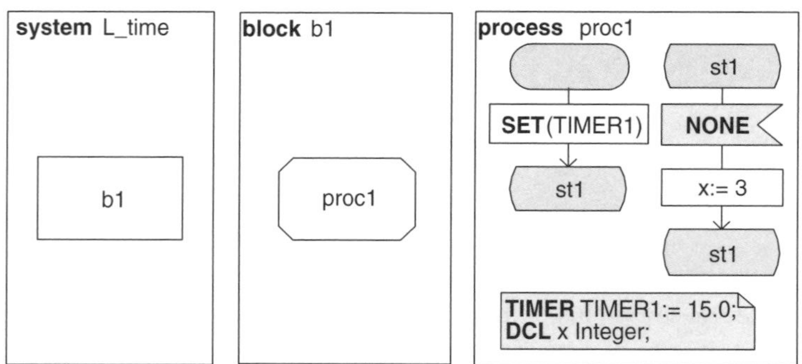

Figure 4.66 Example with timer

If *loose time progression* is off, as shown in Figure 4.67(a), the Simulator will only propose that you execute forever the transition *input none*. Because this transition does not make the time progress (it does not contain any timers), the SDL time, *Now*, is stuck to zero. The only way to get the time progression transition *time(15)* as in Figure 4.67(b) is to set *loose time progression* to on.

In practice, it is easier to set *loose time progression* to on: as soon as you set a timer, you are allowed to time it out. Otherwise you might have to execute first some other transitions, or in the worst cases, it might never time-out, as in the example above.

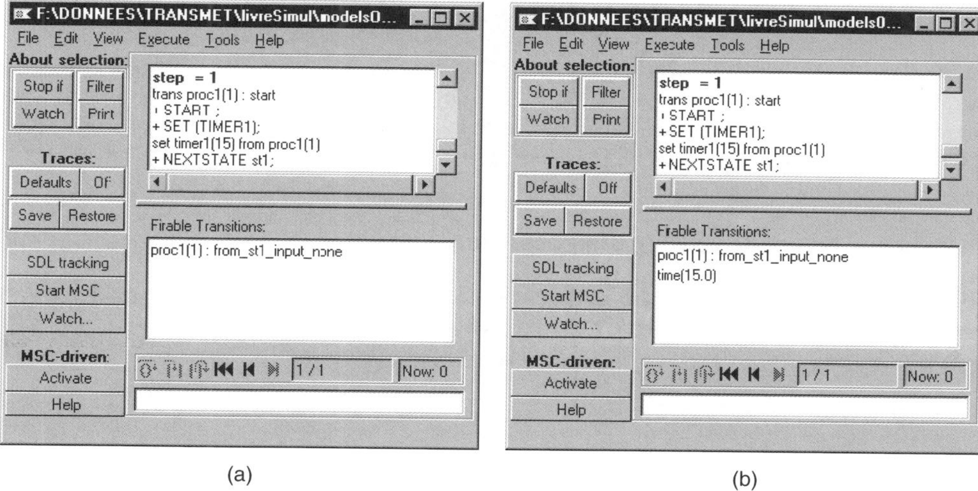

Figure 4.67 Loose time progression off (a) and on (b)

4.3.7.13 Automated generation of feed

To generate automatically the feed commands to transmit external signals to an SDL model, type in the Simulator (feature not available in early versions):

```
feed ?
```

For example, in the V.76 SDL model, the Simulator would answer:

```
-- Feeds skeleton for channel dlcbsu
feed dlcbsu l_estabreq(?)
feed dlcbsu l_estabresp()
feed dlcbsu l_setparmreq()
feed dlcbsu l_setparmresp()
feed dlcbsu l_releasereq(?)
feed dlcbsu l_datareq(?,?)

-- Feeds skeleton for channel dlcasu
feed dlcasu l_estabreq(?)
feed dlcasu l_estabresp()
feed dlcasu l_setparmreq()
feed dlcasu l_setparmresp()
feed dlcasu l_releasereq(?)
feed dlcasu l_datareq(?,?)
```

Then you can copy the feed skeletons and paste them into a file executed at startup. The character ? in the feed skeletons means the default value. For example, for *Integer*, the default value is 0.

4.4 ERRORS DETECTABLE BY INTERACTIVE SIMULATION

In addition to the static errors detected when compiling the SDL model to generate the C code used for simulation (e.g. the ObjectGeode SDL Checker *geodecheck*, called before simulation, detects 541 kinds of errors and warnings), many kinds of dynamic errors are detected by simulation. It enables you to find bugs before writing a line of code.

4.4.1 Dynamic errors detected by Tau SDL suite Simulator

Tau SDL Suite Validator detects the same errors as the Simulator, plus errors detected by state graph analysis, described in Chapter 7 on exhaustive simulation.

4.4.1.1 SDL model errors

- Nonconformance to expected behavior: by examining the simulation textual or MSC trace, you discover that the SDL model does not behave as expected. An example of such error was discovered in Section 4.2.5.2.
- Deadlock: the SDL simulation should continue but no transition is ready.
- Infinite loops: within a transition, detected because the maximum number of symbols *Define-Max-Transition-Length* is reached.
- Discarding an unexpected signal: an example of such error was discovered in Section 4.2.3.1. For example, if signal *sig1* is sent to a process whose current state does not specify any input of *sig1*, the Simulator displays the message: *Signal sig1 caused an immediate null-transition.*

4.4.1.2 SDL dynamic errors

- Decision answer missing: the value of the expression in a decision did not match any of the answers.
- Create process instance: attempt to create more process instances than the maximum number specified.
- Warning in SDL output *to<Pid>* : *Pid* is *Null*, or no path exists between the sender and the receiver that can convey the signal. The signal is discarded.
- Warning in SDL output: no possible receiver found, or signal sent to a stopped process instance. The signal is discarded.
- Error in SDL import: attempt to import from *Null*, from a stopped process instance or from the environment; no process exports this variable or the specified process does not export this variable.
- Errors in SDL view: attempt to view from *Null*, attempt to view from stopped process instance, attempt to view from the environment, the specified process does not reveal this variable, no process reveals this variable.
- Value out of range: for example, in a syntype, in an *Array* operator or in a *String* operator.

- Error when accessing a component: nonactive *choice* or *UNION* component, component is not present, or attempt to access a nonactive optional struct component.
- De-referencing of *Null* pointer: attempt to de-reference a Null pointer defined using the *Ref* generator (Pointers and *Ref* are specific to Tau).
- Attempt to divide by zero in sorts: Integer, Real or Octet.
- Integer overflow using the operator Fix.
- User specified error forced by the SDL *error* expression (e.g. $n := ERROR$ in a task symbol).

4.4.2 Dynamic errors detected by ObjectGeode SDL Simulator

4.4.2.1 SDL model errors

- Nonconformance to expected behavior: by examining the simulation textual or MSC trace, you discover that the SDL model does not behave as expected. An example of such an error was discovered in Section 4.3.5.2.
- Deadlock: the SDL simulation should continue but no transition is ready.
- Infinite loops: inside a transition, detected because the maximum number of symbols *trans_events_limit* is reached.
- Discarding an unexpected signal[47]: an example of such error was discovered in Section 4.3.3.1.

4.4.2.2 SDL dynamic errors

- Decision answer missing: the value of the expression in a decision did not match any of the answers.
- Create process instance: attempt to create more process instances than the maximum number specified (warning).
- Output (without specifying *to<Pid>*) to a process with more than one instance[47] (multiple receivers). If not configured to give an error, the signal is sent to one of the instances.
- Output: no receiver process instance exists for the signal[47] (e.g. the process instance has stopped, the specified *Pid* is *Null*, or no path exists for the signal to the receiver process instance). If not configured to give an error, the signal is discarded.
- View-revealed: revealed statement is missing.
- View-revealed: variable revealed by several process instances.
- Import–Export of variable or procedure: errors similar to output (as exported values or procedures are translated into signal exchanges).
- Value out of range: for example, in a syntype, in an *Array* operator or in a *String* operator.

[47] According to the configuration of the Simulator, this can be an error or not (see *Edit > Configuration*).

- Variant not active: for example, in an ASN.1 *choice* value.
- Optional ASN.1 field not assigned: attempt to access the field.
- Recursive call of procedure with states (this is an ObjectGeode restriction).
- Maximum number of SDL instructions in a transition exceeded: detects infinite loops in a transition. By default set to 1000. Can be changed by typing: *define trans_events_limit '2500'*. Specific to ObjectGeode.
- Maximum number of recursive calls of SDL procedures (without states) exceeded: detects infinite calls. By default set to 100. Can be changed by typing: *define call_depth_limit '60'*. Specific to ObjectGeode.
- User specified error forced by the SDL *error* expression (for example, *n* := *ERROR* in a task symbol).

Remark: division by zero is detected at C code level, not by the Simulator.

4.4.3 Dynamic errors not checked

The following dynamic checks are not performed by Tau or ObjectGeode Simulators:

- Several answers to a decision match the decision expression value: the Simulator will arbitrarily choose one of the existing paths.
- Overflow of Integer and Real values: this is checked at the C level if the actual C system performs such checks.

5

Automatic Observation of Simulations

5.1 PRINCIPLES

5.1.1 Automatic checking of model properties

In the previous chapters, we have seen that the detection of SDL errors such as missing input (signal discarded) is automatically performed by the Simulators. However, we had to watch the simulation traces to check that the SDL model behaves as we expect. The present chapter shows how to formalize the expected behavior in order to automatically monitor the simulation to detect if the SDL model behaves correctly or not.

The idea of automatic observation is not new; before being implemented in SDL tools, it was used in tools based on the Estelle language: VEDA [Jard88] and VESAR [Alga91].

As illustrated in Figure 5.1, the expected property of the system coming from its requirements is formalized into an observer. Then the SDL Simulator permanently monitors the simulation: when the behavior of the SDL model does not comply with the observer, the Simulator signals a violation of the property. If the simulated behavior complies with the observer, the Simulator signals that the property is satisfied.

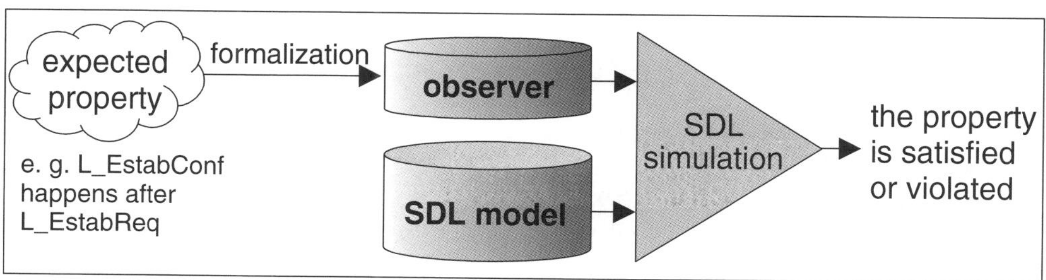

Figure 5.1 Automatic observation of simulations

The observer is executed in parallel with the SDL model: one transition is executed in the SDL model, then the observer is executed, one SDL transition is executed, the observer is executed again and so on. This is called *on-the-fly verification*, as there is no need to finish the computation, for example, of some reachable states graph to start running the verification.

Figure 5.1 refers to an expected property: in such case, we hope that the simulation will prove that the property is satisfied. However, an observer can also represent an unwanted property (example: traffic lights are green at the same time in a crossroad), and we hope that

Validation of Communications Systems with SDL: The Art of SDL Simulation and Reachability Analysis.
Laurent Doldi ISBN: 0-470-85286-0

the simulation will prove that this property cannot be satisfied. More details are provided in Chapter 7.

Table 5.1 shows the observation formalisms that can be used in our two SDL tools. In this book, we call all these formalisms 'observers', but in fact, the name observer, in the tools documentation, is only used for the MSCs or GOAL[1] observers in ObjectGeode and the observer processes in Tau SDL Suite.

Table 5.1 Observation formalisms available in the two SDL tools

Kind of property	Formalization	ObjectGeode Simulator	Tau SDL Validator*
Expression on current model state	easy	stop conditions	rules
Linear signal sequencing	easy	basic MSCs	basic MSCs
Complex signal sequencing	medium	hierarchical MSCs (i.e. with operators)	MSCs with inline operators
—	—	—	HMSCs (High-Level MSCs)
All the above plus memorization and more	medium	GOAL observers	observer processes

*The Simulator can also be used to observe an SDL model with an MSC, using the Organizer command *Simulator Test > New Simulator*.

Examples of observers:

- stop condition (ObjectGeode): detects that instance 1 of process *DLC* in blocks *DLCa* and *DLCb* are in state *connected*;

```
stop if (DLCa!DLC(1)!state = connected) and \
        (DLCb!DLC(1)!state = connected)
```

- MSC (Tau and ObjectGeode), shown in Figure 5.2: checks that in the simulated SDL model, an input of signal *L_EstabReq(0)* by block *DLCa* is followed by the output of signal *L_EstabConf(0)* from the same block. Note that in Tau, exchanges of other signals are considered as a violation, as opposed to that in ObjectGeode. There is no way to memorize the parameter value, thus there must be one MSC per value of parameters (one for 0, one for 1 etc.), or the parameter values must be ignored.

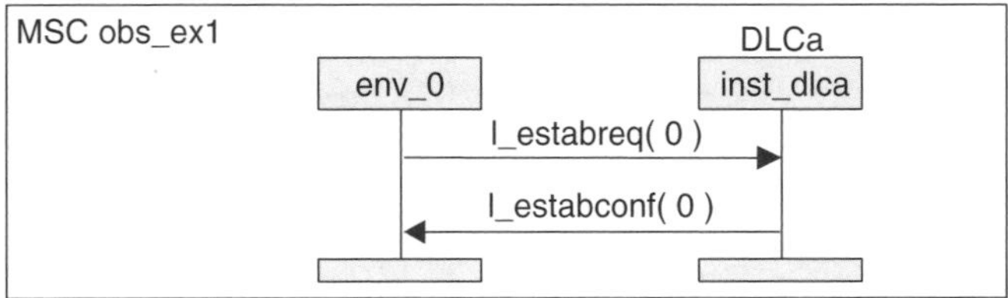

Figure 5.2 Example of observer MSC

[1] Geode Observation Automata Language.

- GOAL observer (ObjectGeode), shown in Figure 5.3, or observer process (Tau): behavior similar to the observer MSC in Figure 5.2, plus memorization of the value of the parameter of signal *L_EstabReq* (the number of the DLC to establish) to check that it is equal to the DLC number transmitted in *L_EstabConf*.

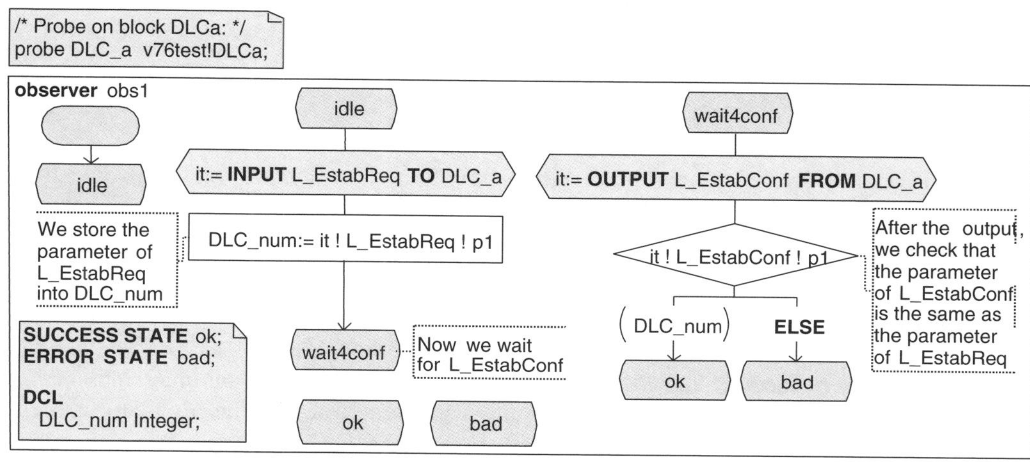

Figure 5.3 Example of GOAL observer

The observation principles described here are available in any simulation mode: interactive, random or exhaustive, plus their variants. In exhaustive simulation, the simulators provide a way to replay interactively the scenarios leading to the satisfaction or to the violation of a property.

5.1.2 Specificity of observation with MSCs in Tau SDL Suite

In Tau SDL Suite Validator[2], the observation of properties using MSCs or HMSC is different compared to ObjectGeode. In Tau SDL Suite Validator:

- Only one MSC (or HMSC) at a time can be used as an observer (i.e. can be loaded by the command *Load-MSC*). To use several MSCs, they can be grouped into an HMSC.
- The MSC (or HMSC) does not have to be compiled with the SDL model.
- As explained in Section 5.2.5.1, Tau observes any signals, as opposed to ObjectGeode that observes only signals present in the observer MSC, by default.
- More important, the MSC (or HMSC) drives the exploration, as a TTCN test case when testing an IUT (Implementation Under Test): considering the MSC in Figure 5.4, Tau Validator will not transmit *sigB* before *sigA* to the SDL model. In ObjectGeode, the same MSC would generate two *feed sigA* and *sigB*, and *sigB* could be transmitted first to the SDL model (if ready to input it). This is why in general Tau Validator checks MSCs faster than ObjectGeode. To speed up the MSC verification in ObjectGeode, you must set the *MSC Simulation Property*

[2] Note that the Simulator can also be used to observe an SDL model with an MSC, using the Organizer command *Simulator Test > New Simulator*.

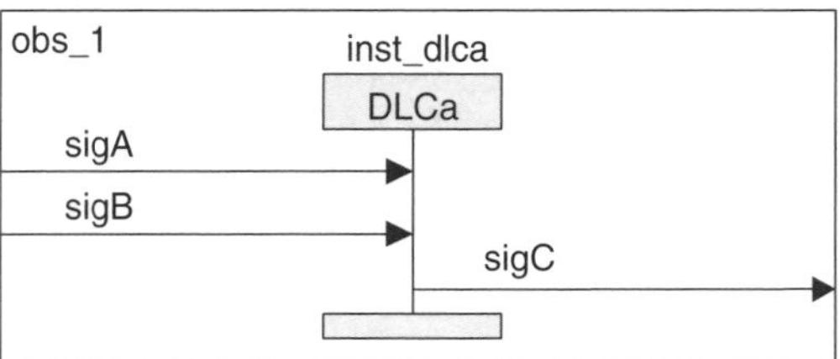

Figure 5.4 The observer MSC *obs_1*

of the MSC to *verify*: in this case, the Simulator will not explore the states leading to an error (a violation) of the MSC, because the default configuration is *error cut* (equivalent to *prune* in Tau).

5.2 CASE STUDY WITH TAU SDL SUITE

In Chapter 4, we have used the Tau SDL Suite Simulator. To benefit from automatic observation features, we will now switch to the Tau SDL Suite Validator. Note that the Simulator can also be used to check the SDL model against an MSC.

5.2.1 Simulate with user-defined rules

In the Validator, only one user-defined rule can be used at a time. To check several conditions, you can use the operator *or* to group them in a single rule.

5.2.1.1 Detect DLC establishment

We want to detect that a DLC is established. This means, in our V.76 SDL model, that:

- instance 1 of process *DLC* in block *DLCa* is in state *connected*, and
- instance 1 of process *DLC* in block *DLCb* is in state *connected*.

It seems that the Validator rules do not accept qualifiers such as <<*Block DLCa*>>. As there are two processes named *DLC*, one in block *DLCa* and the other in block *DLCb*, it is not possible to write a rule to detect that both DLCs are in state *connected*. An observer process could be used instead.

The solution would be to modify the SDL model to have a copy of block *V76_DLC* on each side: transformation of the block type *V76_DLC* into a block named *V76a*, making a copy of it and naming the copy *V76b* and in each block, renaming the *DLC* process, respectively, *DLC_a* and *DLC_b*. Then, the Validator user-defined rule would be:

```
state(DLC_a:1)=connected and state(DLC_b:1)=connected
```

Rather than performing this model modification, you will use rules concerning process *AtoB* and *BtoA*, which do not require the use of qualifiers, as they are unique in the system.

5.2.1.2 *Detect state of processes AtoB and BtoA*

We want to detect that in our V.76 SDL model:

- instance 1 of process *AtoB* is in state *ready* and
- instance 1 of process *BtoA* is in state *ready*.

This condition, translated into a Validator rule, becomes:

```
state(AtoB:1)=ready and state(BtoA:1)=ready
```

Compile the SDL model, start the Validator and test the rule:

A. In the Organizer, select the SDL system *V76test* and do *Generate > Make*: the window represented in Figure 5.5 appears. Select *Microsoft Validation* or *Borland Validation*, and press *Full Make*.

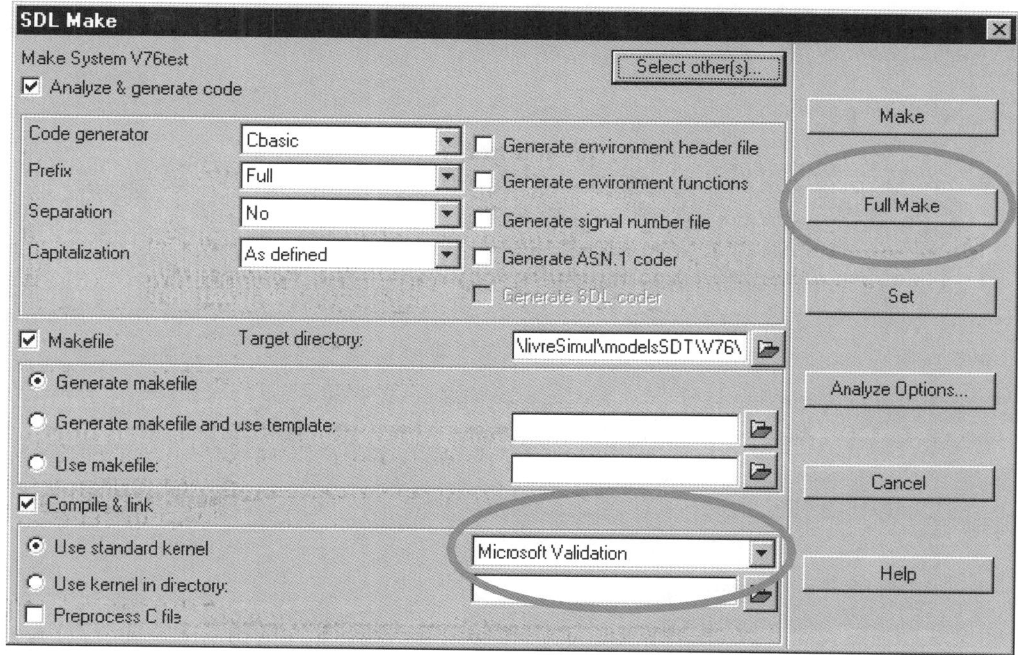

Figure 5.5 The SDL Make window set for validation

B. In the Organizer, press the *Validate* button to start the Validator. The Validator main window appears, as shown in Figure 5.6.

C. In the Validator command line, enter:

```
Define-Rule state(AtoB:1)=ready and state(BtoA:1)=ready
```

D. Then, to test if the rule is satisfied or not for the current SDL model state, enter:

```
Evaluate-Rule state(AtoB:1)=ready and state(BtoA:1)=ready
```

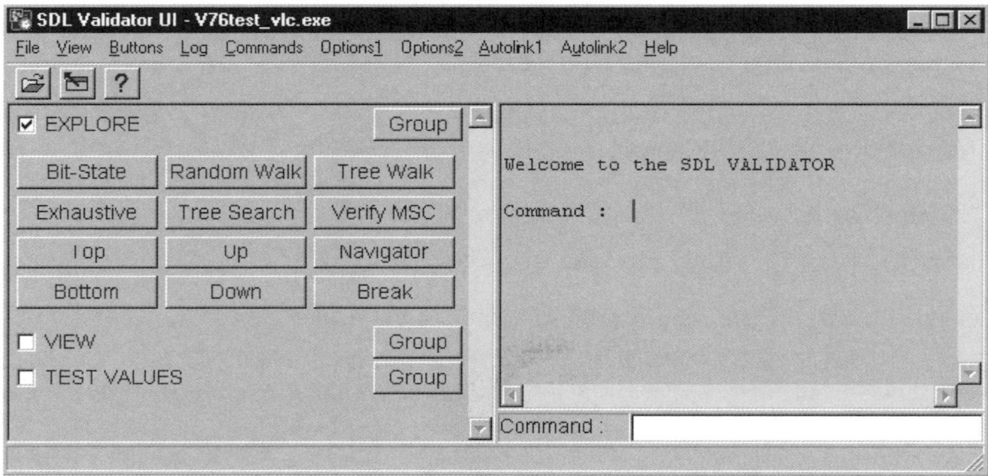

Figure 5.6 The Validator main window

The Validator answers:

```
Evaluating rule:
(( state( AtoB:1 ) = ready) and (state( BtoA:1 ) = ready ))
Rule not satisfied
```

This is normal, as the processes are not yet in state *ready*.

E. Select *View > Command Window*: you see that the processes *AtoB:1* and *BtoA:1* are in state *start*.

F. Press the *Navigator* button. In the Navigator window, double-click two times on *Next 1* to execute the start transitions of *AtoB* and *BtoA*, as shown in Figure 5.7.

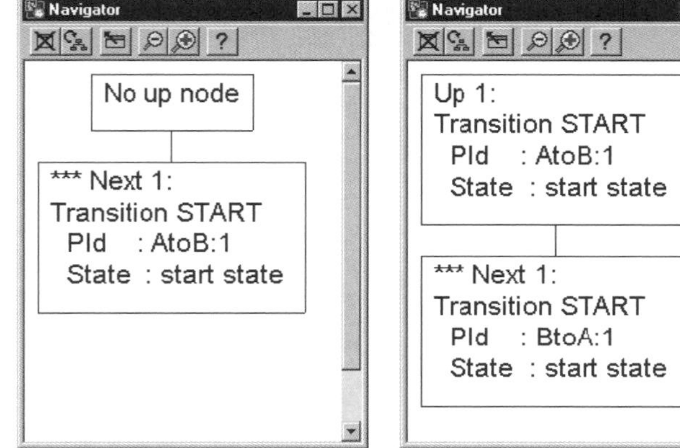

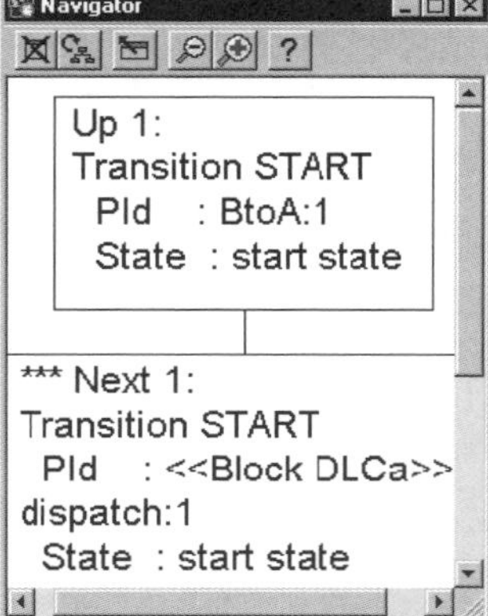

Figure 5.7 Starting processes *AtoB* and *BtoA*

G. In the Validator command line, enter:

```
Evaluate-Rule state(AtoB:1)=ready and state(BtoA:1)=ready
```

As expected, the Validator answers:

```
Evaluating rule:
(( state( AtoB:1 ) = ready) and (state( BtoA:1 ) = ready ))
User-defined rule satisfied.
```

You can check in the *Command* Window that the processes *AtoB* and *BtoA* are in state *ready*.

We have executed the transitions manually using the Navigator; we will see later that the transitions can also be executed automatically by the Validator, using several algorithms such as exhaustive simulation: then reports are automatically generated each time the user-defined rule is satisfied.

5.2.1.3 More user-defined rules

Here are some more user-defined rules that you can enter in the Validator as indicated previously.

To detect that variable *foo1* in process *AtoB* contains 4:

```
Define-Rule AtoB:1->foo1 = 4
```

To detect when the first signal in the input queue of process *AtoB* is *v76frame*:

```
Define-Rule sitype(signal(AtoB:1))=V76frame
```

To detect if process *AtoB* has exactly one instance:

```
Define-Rule instno(AtoB) = 1
```

To detect if a process input queue contains more than one signal:

```
Define-Rule maxlen() > 1
```

To detect that process *AtoB* has not created any process instance (rule always satisfied in our model):

```
Define-Rule offspring(AtoB:1) = Null
```

As you can see, several conditions in the SDL model can be detected by user-defined rules during the validation. However, user-defined rules cannot detect the correct sequencing of signals: for that, you must use MSCs.

5.2.2 Simulate with a basic MSC

You will simulate the V.76 SDL model, observed by a basic MSC. Basic means that the MSC just contains a single scenario, as opposed to MSCs containing inline operators or to High-Level

MSCs (HMSC). To simplify, we reuse the basic MSC *test1.msc*, generated by the Simulator in Chapter 4. Naturally, you could use another MSC drawn from scratch with the Editor or rework an existing MSC.

In Tau SDL Suite Validator[3], an MSC is not only observing but also driving the simulation.

A. Select *File > Restart* in the Validator, and enter the command:

```
load-msc test1.msc
```

Check that in *test1.msc*, there is either a single environment instance named *env_0* or two environment instances named *DLCaSU* and *DLCbSU* (the names of the two external channels in the SDL model); otherwise the simulation would not match the loaded MSC.

When an MSC is loaded, the Validator sends to the SDL model the first environment signal present in the MSC. Then, after receiving the first response from the SDL model, the Validator sends the next environment signal and so on until the MSC is verified or violated.

B. Select *Commands > Toggle MSC Trace*.

C. Press the *Navigator* button in the *Explore* group. The Navigator window appears, as illustrated in Figure 5.8. By double-clicking in this window, you can execute transitions in the SDL model, going forward (down) but also backward (up).

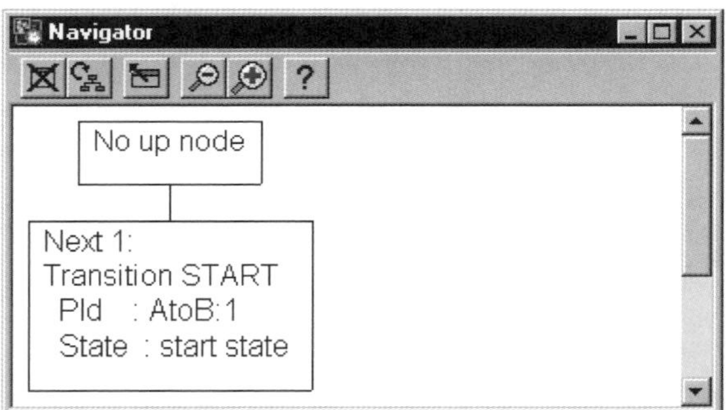

Figure 5.8 The Navigator window

D. In the Navigator window, double-click on the lower rectangle, marked *Next 1*, to execute the corresponding SDL transition.

E. Continue double-clicking; when there are several possibilities, select the left one.

F. After around 45 transitions, the Navigator window displays: *No down node – MSC test1 verified*, as shown in Figure 5.9. It means that the signal sequence present in the loaded MSC has been simulated. You can see that the trace MSC is identical to the loaded MSC.

Note that if the simulated parameter values did not match the MSC expected values, the MSC property would not have been satisfied.

[3] The MSCs can also be verified using the Simulator rather than the Validator.

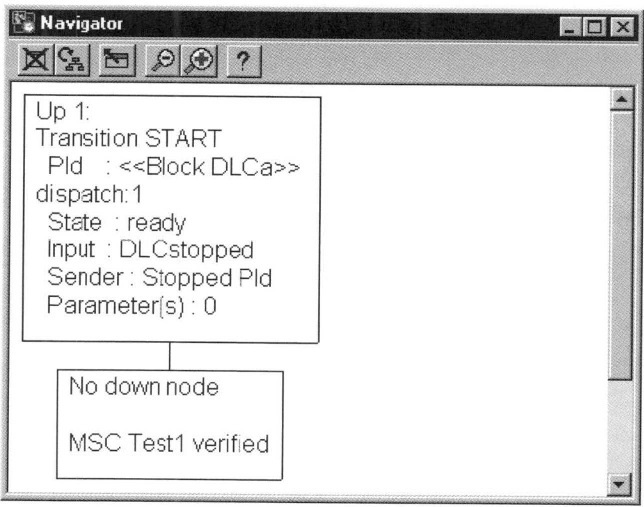

Figure 5.9 The MSC *test1* is verified

5.2.3 Simulate with an MSC containing inline operators

The basic MSC *test1* used in the previous section represents only a single execution sequence. To represent several expected behaviors consistent with the V.76 textual specification without creating too many basic MSCs, we will add inline operators to the MSC *test1*.

5.2.3.1 Create the MSC

With the MSC Editor, you will rework a copy of the basic MSC *test1.msc* generated by the Simulator in Chapter 4, to obtain the MSC *test1inline.msc*, shown in Figure 5.10:

A. In Windows (or Unix), make a copy of the file *test1.msc* into *test1inline.msc*

B. In the Organizer, select *Tools > Editors > MSC Editor*

C. In the MSC Editor, select *File > Open* and choose *test1inline.msc*

D. Enter the MSC name *test1inline*, and remove the XID part (eight signals from *L_SetparmReq* to *L_SetparmConf*).

E. Using copy–paste, duplicate the data transfer part, arrange the pasted signals to reverse the originating side of the second data transfer, and replace 86 by 39, as shown in Figure 5.11.

F. Using copy–paste, duplicate the DLC release part and arrange the pasted signals to reverse the originating side of the second release: release originated by B.

G. Using the Symbol Menu, insert one *loop* and two *alt* inline operators, depicted in Figure 5.10. Resize them and add two expression separators, to obtain the layout shown in Figure 5.11.

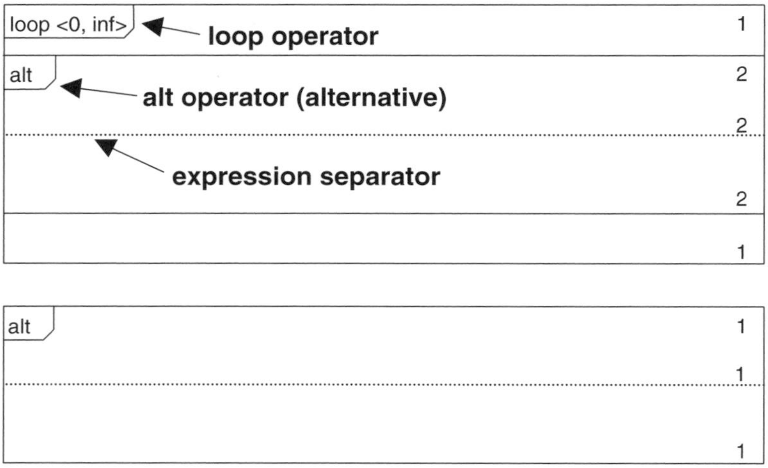

Figure 5.10 The inline MSC operators used in *test1inline.msc*

This MSC represents the following behavior expected from the V.76 SDL model:

1. one connection phase,
2. followed any number of times (between 0 and infinite) by the repetition (operator *loop <0, inf>*) of data transfers from A to B or (operator *alt*) from B to A,
3. followed by the disconnection phase initiated by A or (second *alt* operator) initiated by B.

5.2.3.2 Use the MSC to check the SDL model

A. Start the Validator as indicated in Section 5.2.1.2, and enter the command:

```
load-msc test1inline.msc
```

B. Select *Commands > Toggle MSC Trace*.

C. Press the *Navigator* button in the *Explore* group.

D. Using the Navigator, execute an SDL scenario matching the loaded MSC (when there are two possibilities, select the left one): for example, establish DLC number 0, then transmit data 86, and then release DLC number 0.

E. After around 40 transitions, the Navigator window displays: *No down node – MSC test1inline verified*, as shown in Figure 5.12. It means that one of the signal sequences present in the loaded MSC has been simulated. The Navigator also indicates which MSC operators have been followed and which branch has been taken in each *alt* operator.

You could press the button *Top* in the *Explore* group to go back to the initial SDL model state, and try to verify other paths, for example, no data transfer, or a data transfer from B to A and so on.

F. Exit from the Validator.

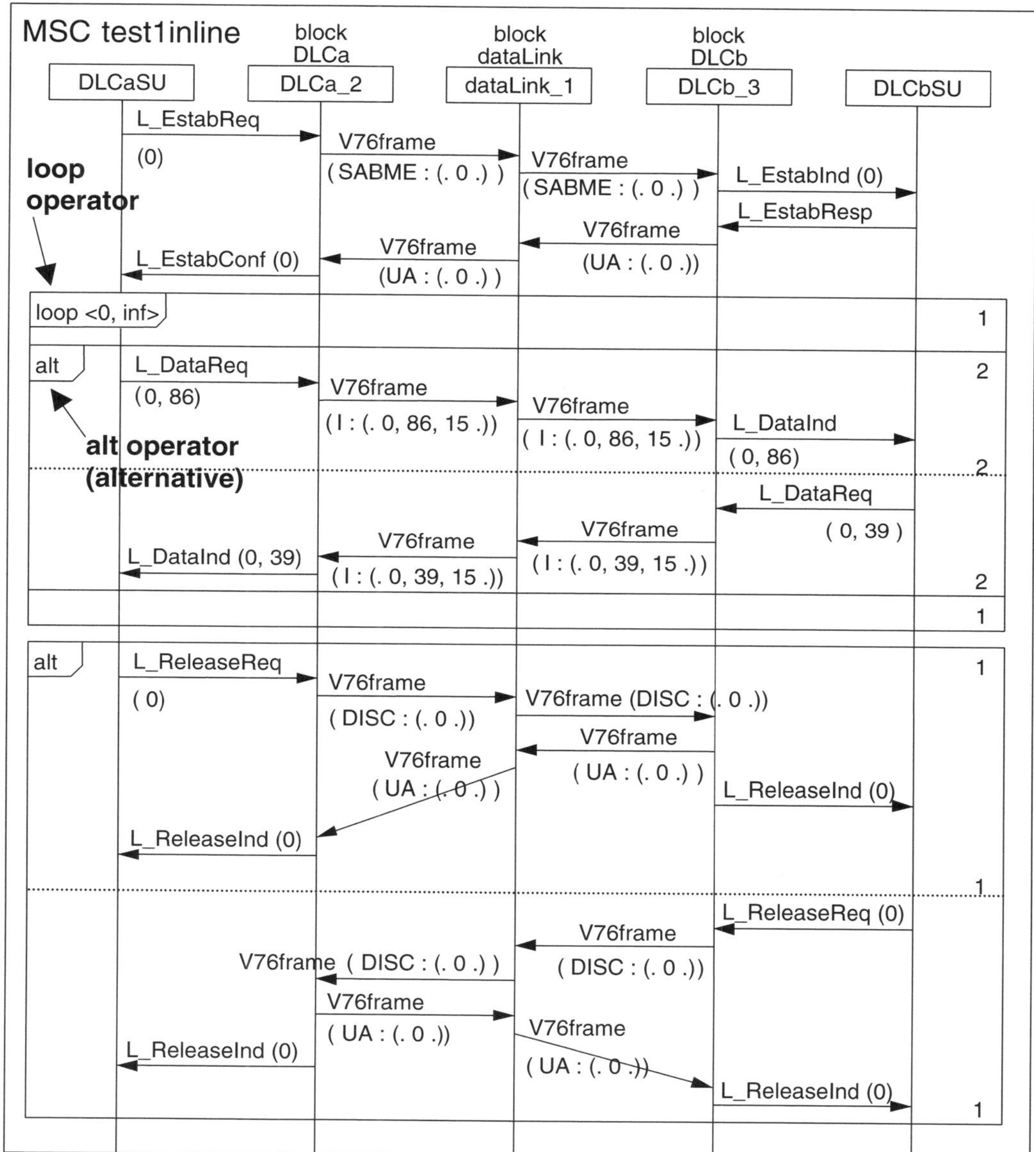

Figure 5.11 The MSC *test1inline.msc*

5.2.4 Simulate with an HMSC

The idea with HMSC [MSC96] is to combine several MSCs (containing or not inline operators) to get a good overview of the expected behaviors.

Remark: to be used in the Validator, all the MSCs used in an HMSC must contain the same entities; for example, it is not allowed to have block *DLCa* in one MSC and *<<block DLCa>> process DLC* in another MSC.

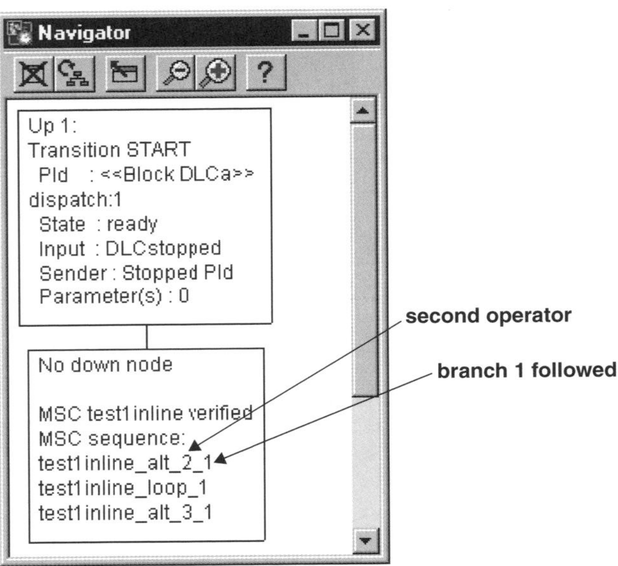

Figure 5.12 The MSC *test1inline* is verified

5.2.4.1 Create the HMSC

With the MSC Editor, you will rework a copy of the basic MSC *test1.msc* generated by the Simulator in Chapter 4, to obtain the hierarchical HMSC *test1ops.msc* shown in Figure 5.13.

This HMSC *test1ops* represents the following behavior expected from the V.76 SDL model[4]:

1. one connection phase (*cnx1h*),
2. followed or not by the repetition of data transfers from A to B (*data_a2b*) or from B to A (*data_b2a*),
3. followed by the disconnection phase initiated by A (*disc_0_by_a*) or initiated by B (*disc_0_by_b*).

A. In the Organizer, select *Edit > Add New*, check *MSC*, choose *HMSC*, enter the document name *test1ops*, as shown in Figure 5.14, and press *OK*.

B. The HMSC Editor appears. Draw the HMSC represented in Figure 5.13, and save it into the file *test1ops.mrm*.

Now create the five MSCs referenced in the HMSC:

C. In Windows (or Unix), make copies of the file *test1inline.msc* into the files: *cnx1h.msc, data_a2b.msc, data_b2a.msc, disc_0_by_a.msc* and *disc_0_by_b.msc*.

D. In the Organizer, select *Edit > Add Existing* and choose *cnx1h.msc*. The MSC is loaded into the MSC Editor.

---

[4] To simplify the example, this HMSC is incomplete: connection initiated by B is missing, DLC number 1 is not tested and so on.

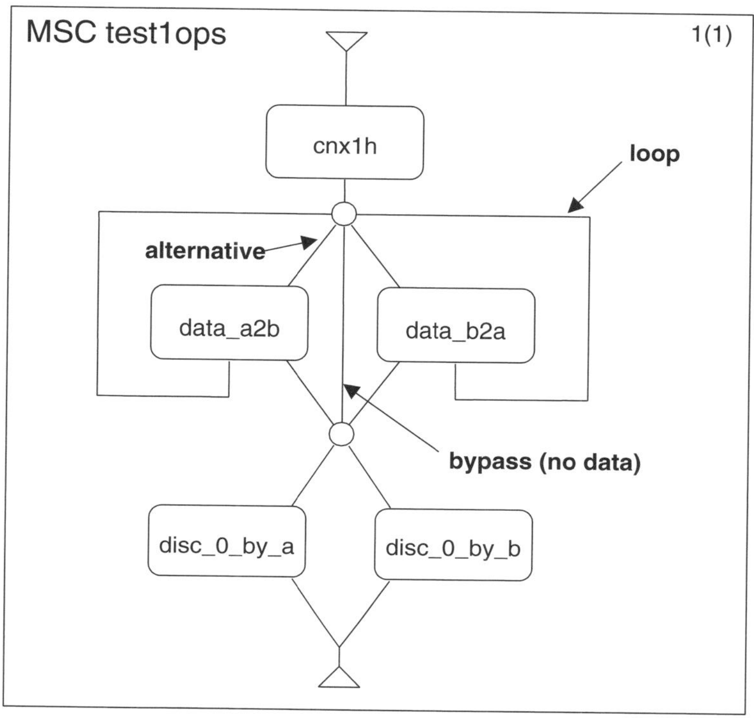

Figure 5.13 The HMSC *test1ops.msc*

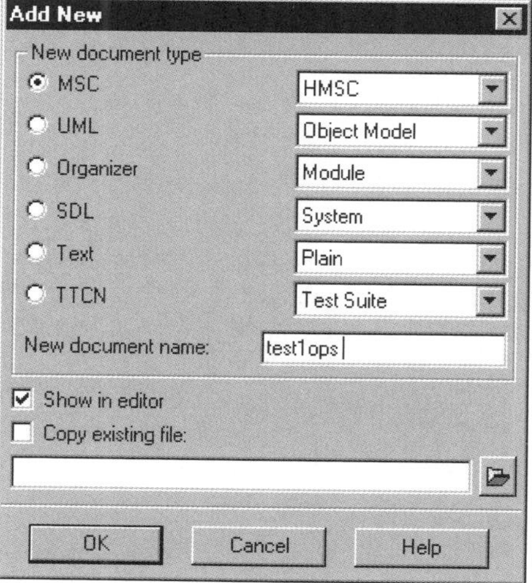

Figure 5.14 Adding HMSC *test1ops*

E. In the MSC Editor, remove all signals or operators except the connection part (the first eight signals), replace (just type directly) *MSC test1inline* by *MSC cnx1h*, and save the MSC into *cnx1h.msc*. The resulting MSC is shown in Figure 5.15.

F. Repeat the two previous steps for the following MSCs: *data_a2b.msc, data_b2a.msc, disc_0_by_a.msc* and *disc_0_by_b.msc*, respectively, shown in Figures 5.16 to 5.19.

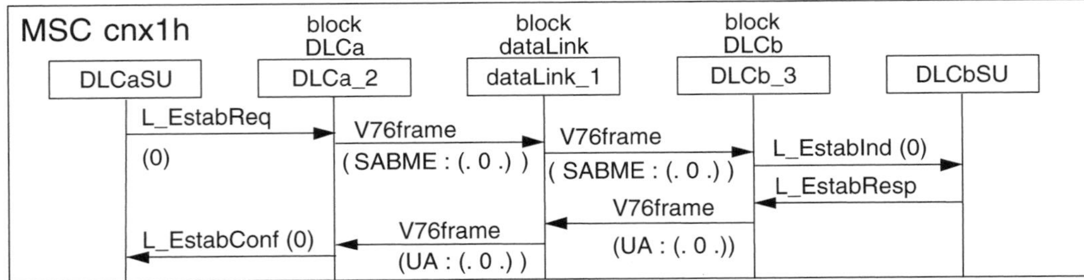

Figure 5.15 The MSC *cnx1h.msc*

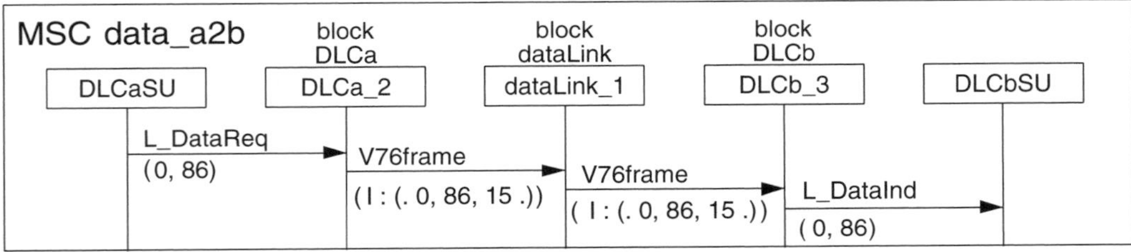

Figure 5.16 The MSC *data_a2b.msc*

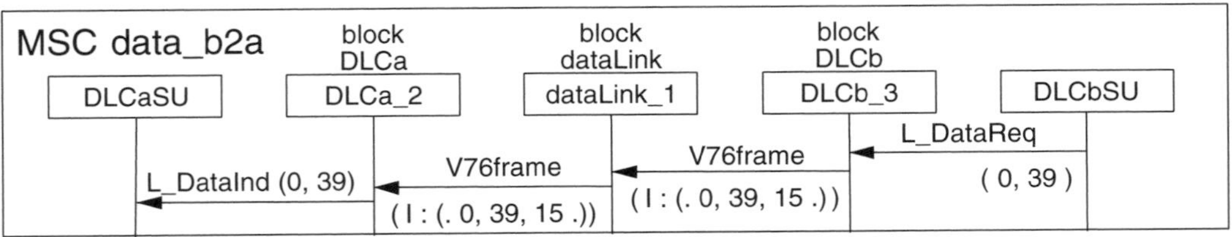

Figure 5.17 The MSC *data_b2a.msc*

G. The Organizer should now look like Figure 5.20. The MSCs have been inserted into the Organizer because it is used to translate the name of the referenced MSCs such as *cnx1h* into their actual file name (which could be different from *cnx1h.msc*).

H. Press *save* in the Organizer to update the . *sdt* file.

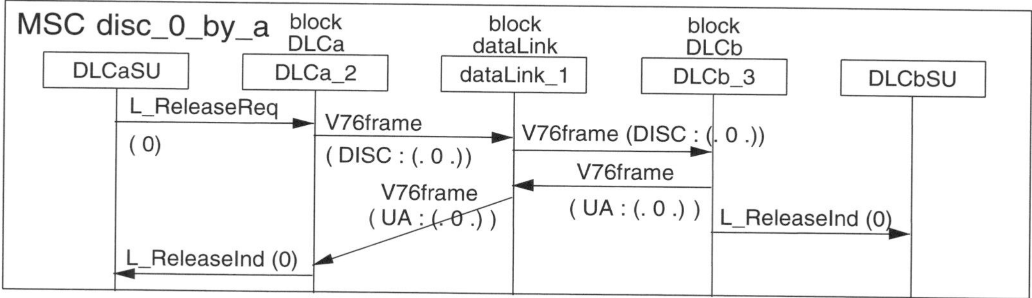

Figure 5.18 The MSC *disc_0_by_a.msc*

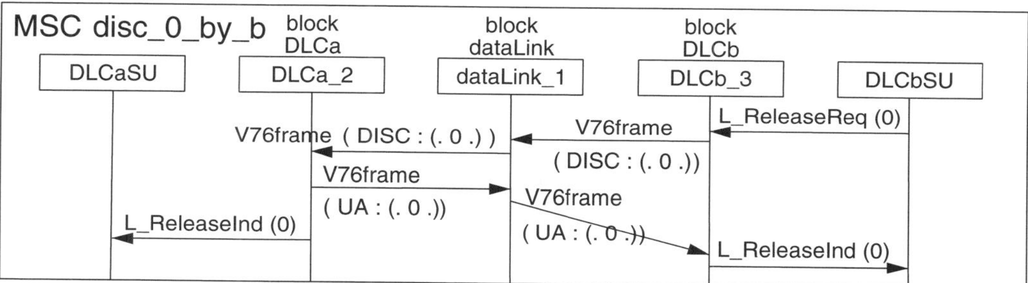

Figure 5.19 The MSC *disc_0_by_b.msc*

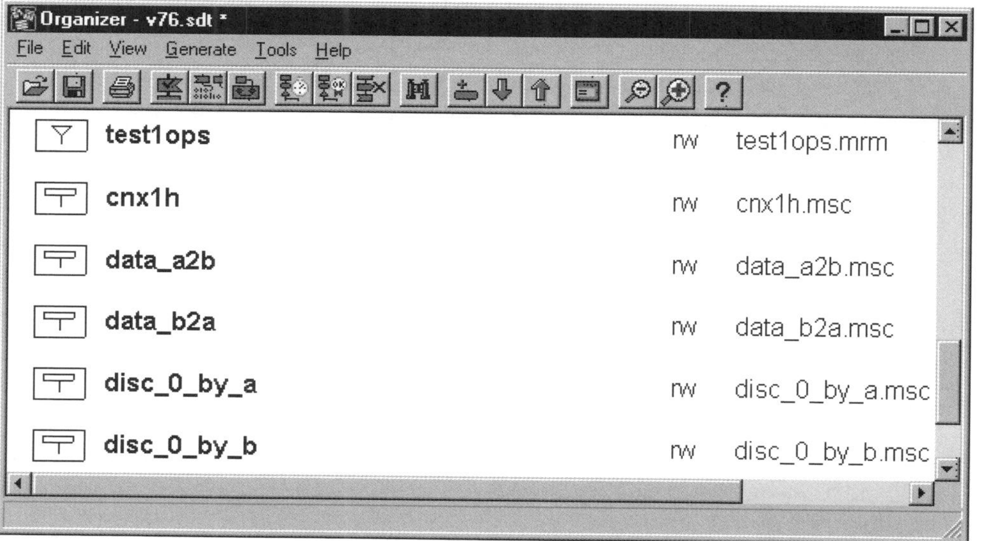

Figure 5.20 The HMSC *test1ops* and the five referenced MSCs in the Organizer

5.2.4.2 *Use the HMSC to check the SDL model*

A. Start the Validator as indicated in Section 5.2.1.2, and enter the command[5]:

```
load-msc test1ops.mrm
```

B. Select *Commands > Toggle MSC Trace*.

C. Press the *Navigator* button in the *Explore* group.

D. Using the Navigator, execute an SDL scenario matching the loaded HMSC (when there are two possibilities, select the left one): for example, establish DLC number 0, then transmit data 86, and then release DLC number 0.

E. After around 37 transitions, the Navigator window displays: *No down node – MSC test1inline verified*, as shown in Figure 5.21. It means that one of the signal sequences present in the loaded HMSC has been simulated. The Navigator also indicates which MSCs have been followed in the HMSC. These MSCs have been manually shown in bold in the figure.

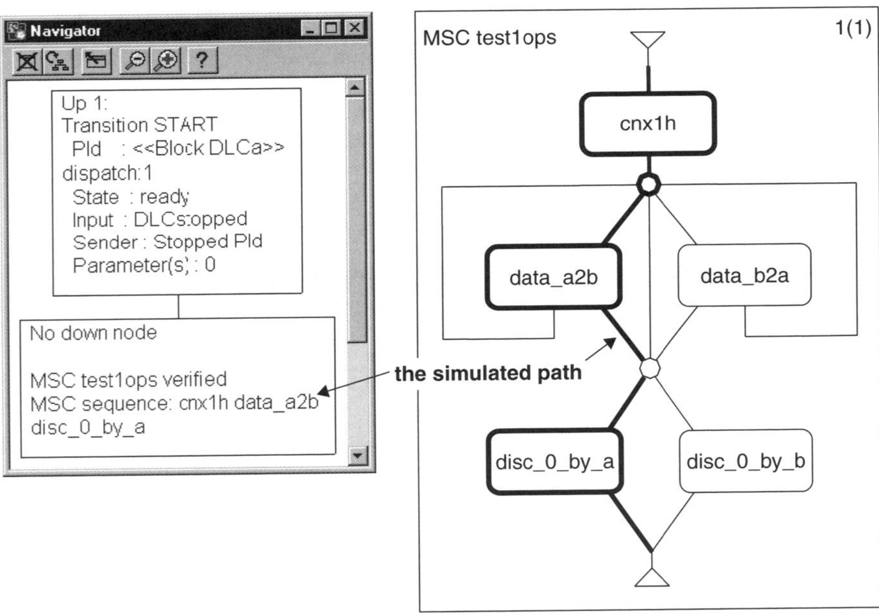

Figure 5.21 The MSC *test1ops* is verified

You could press the buttons *Top* or *Up* in the *Explore* group to go back to the initial or to the previous SDL model states, and try to verify other paths, for example, no data transfer, or a data transfer from B to A and so on.

[5] If the *load-msc* command reports errors, check that inline expression separators (horizontal dashed lines) do not remain in the faulty MSC.

5.2.5 More details on MSCs

5.2.5.1 How the Validator monitors the MSC events

When verifying an MSC (or HMSC), the Validator checks that the events occurring during the simulation are identical to the events specified in the MSC. For example, in Figure 5.22, after the output of *sA*, the output of *sB* is expected. Unfortunately, the output of *sZ* occurs and the Validator detects a violation of the MSC *test_seq*[6].

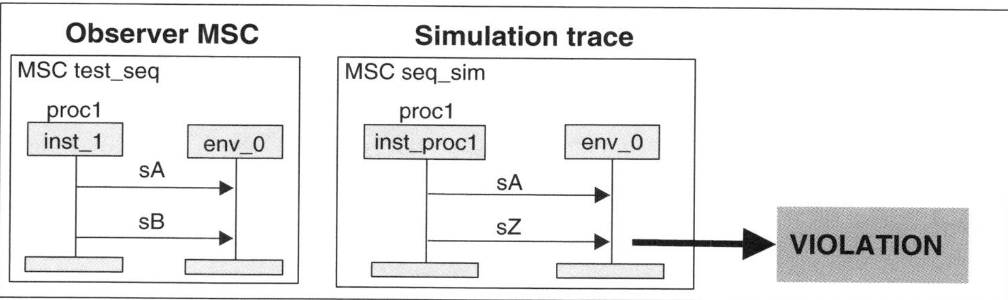

Figure 5.22 MSC violation

5.2.5.2 The MSC symbols used by the Validator

During MSC or HMSC loading (either using the textual command *Load-MSC* or the button *Verify MSC*), the Validator checks that all the elements used in the MSC exist in the SDL model. For example, in Figure 5.23, the following checks are performed during MSC loading:

- The entity named *dataLink* in the MSC *retry1* exists in the SDL model *HDLC*.
- The signal *v76frame* in the MSC also exists in the SDL model.
- The environment instance contains either *env_0* or an external channel name (such as *ch1* here).

The name *dataLink_1* located inside the MSC instance is not supposed to match any SDL name.

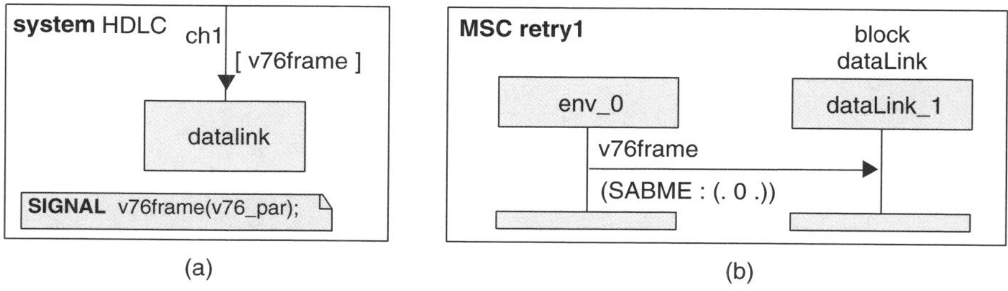

Figure 5.23 An MSC (b) consistent with the SDL model *HDLC* (a)

[6] In ObjectGeode, signal *sZ* would have been ignored because it is not in the observer MSC. To get a violation as in Tau, *sZ* must be declared as unexpected signal in the MSC simulation properties.

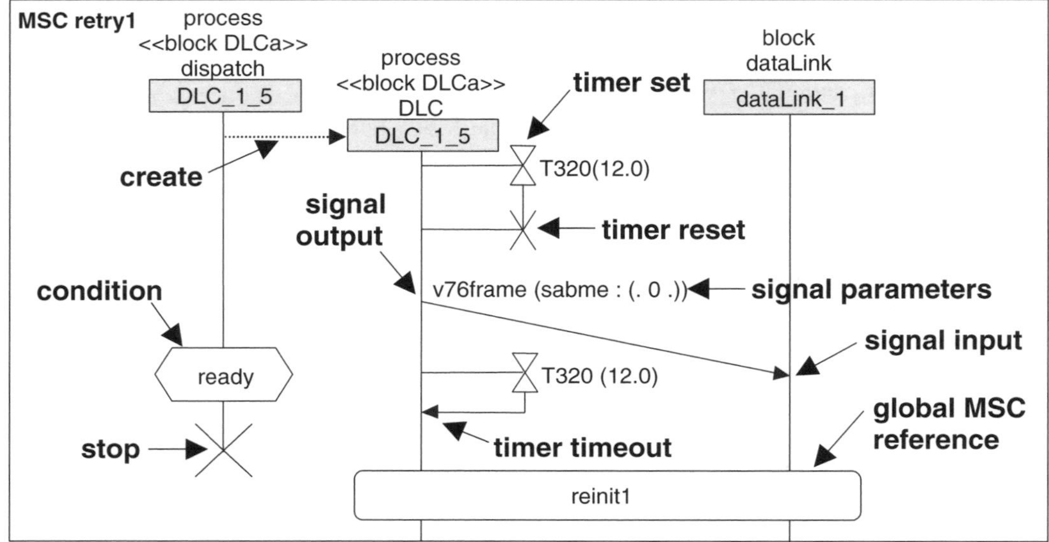

Figure 5.24 The MSC symbols checked dynamically

The following MSC symbols, depicted in Figure 5.24, are dynamically checked by the Validator during simulation:

- Signal input and output and their parameter values,
- Timer set, reset and timeout,
- Process instance create and stop,
- Global MSC reference (without substitution and gates),
- Condition (if *Define-Condition-Check* is on, default is off).

The other symbols present in the MSC, shown in Figure 5.25, are ignored.

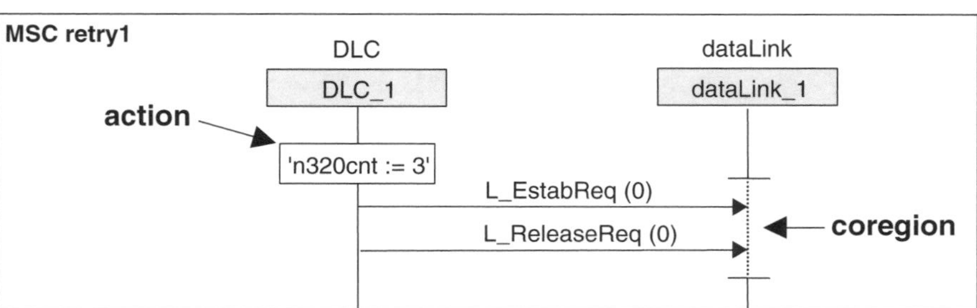

Figure 5.25 The MSC symbols not checked dynamically

5.2.5.3 The MSC signal parameters

The values of the signal parameters in an MSC to be verified can be omitted: in that case, they are not checked.

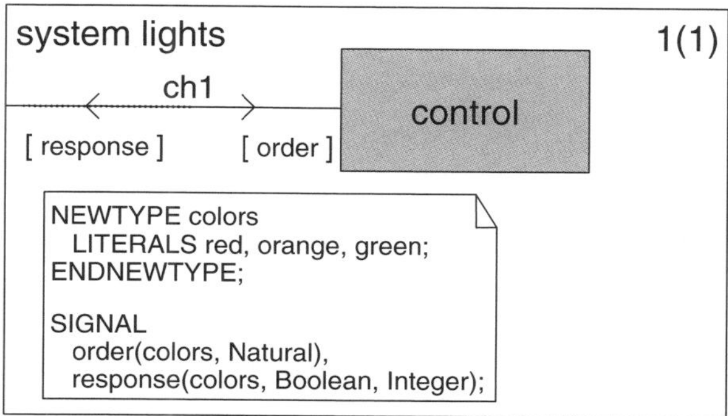

Figure 5.26 The SDL model *lights*

We consider the SDL model *lights* shown in Figure 5.26: it can receive a signal *order*, having one parameter of type *colors* and one parameter of type *Natural*. The signal *response* has one parameter of type *colors*, one parameter of type *Boolean* and one parameter of type *Integer*.

The Validator automatically generates test values for each incoming external signal such as *order*. The default test values are 0 and 55 for *Natural* parameters.

During simulation, the Validator makes a combination of all possible test values, to test all combinations of parameter values in the signals received by the SDL model. For example, in Figure 5.27, the signal *order* is shown with all test value combinations created by the Validator: the first parameter of type *colors* takes three values *red, orange* or *green*, and the second parameter of type *Natural* takes the values 0 or 55. Therefore, we have $3 \times 2 = 6$ possibilities: (*red*, 0), (*red*, 55), (*orange*, 0) and so on.

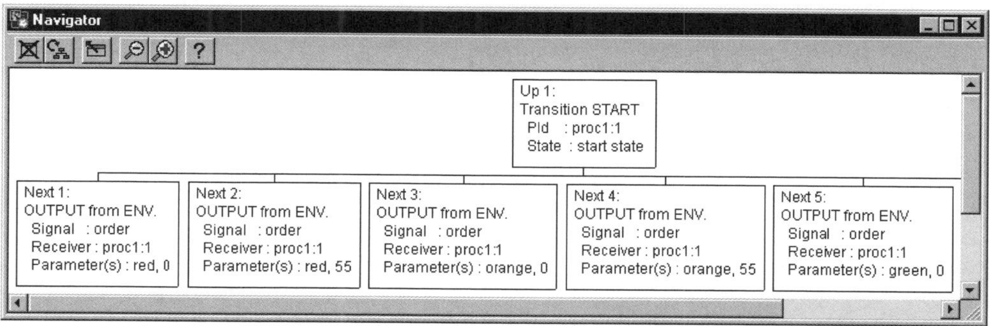

Figure 5.27 Test values generated by the Validator, no MSC loaded

If the MSC *test2* represented in Figure 5.28 is loaded[7] into the Validator:

- parameter values specified in incoming signals are used, such as *red* in *order*,
- nonspecified values in incoming signals, such as the second parameter of *order*, are taken in the Validator test values,

[7] Either using the command *Load-MSC* or the button *Verify MSC*.

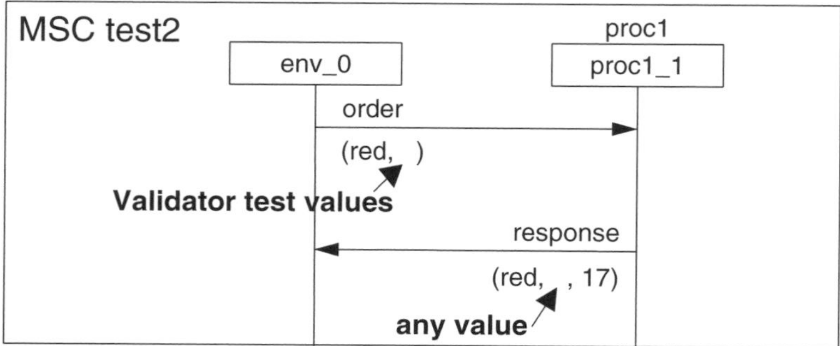

Figure 5.28 MSC loaded to be verified by the Validator

- parameter values specified in outputs must match the simulated values,
- nonspecified values in outputs are ignored by the Validator (i.e. always match).

Figure 5.29 shows the two combinations of parameter values for signal *order* when the MSC *test2* is loaded into the Validator: (*red*, 0) and (*red*, 55).

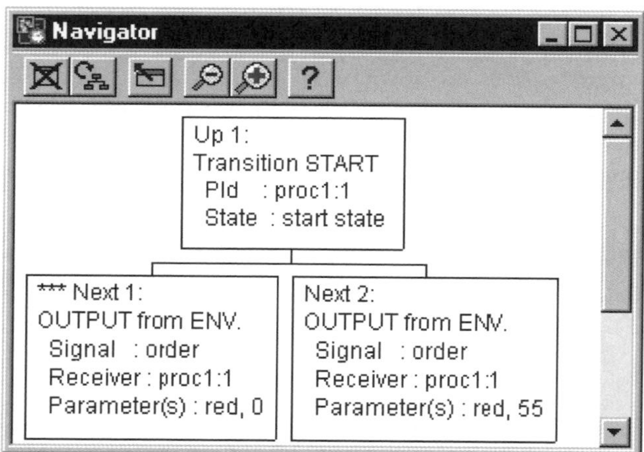

Figure 5.29 Test values generated by the Validator, MSC *test2* loaded

Figure 5.30 shows an MSC simulated in the Validator satisfying the loaded MSC *test2*.

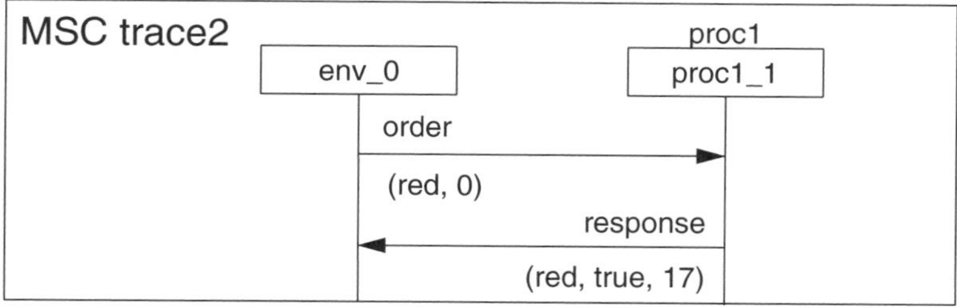

Figure 5.30 Validator trace result of MSC *test2* verification

5.2.5.4 The inline operators accepted in MSCs

The Validator accepts the following inline operators in the MSCs:

- alt: alternative
- loop: repetition
- par: parallel
- seq: sequence
- exc: exception
- opt: optional

These operators are defined in [MSC96].

5.2.5.5 The symbols accepted in HMSCs

The Validator accepts the following symbols in HMSCs, illustrated in Figure 5.31:

- start
- MSC reference (without substitution and gates)
- condition (ignored)
- connection point
- end

These symbols are defined in [MSC96].

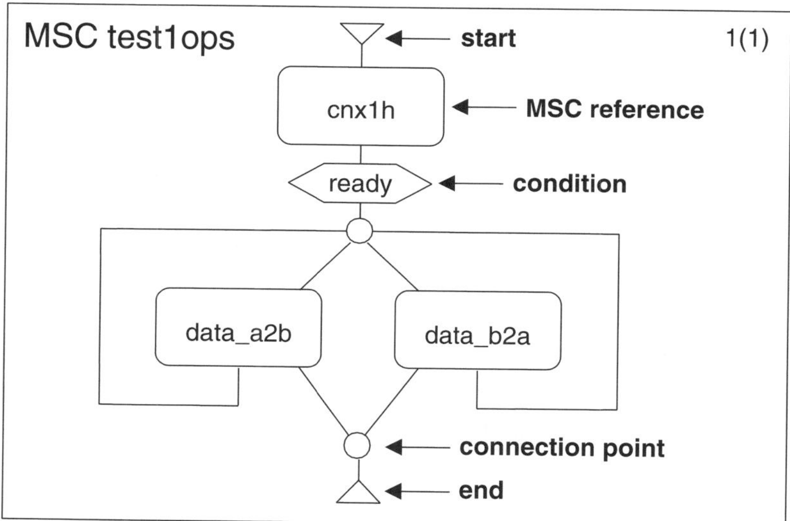

Figure 5.31 HMSCs symbols accepted by the Validator

5.2.5.6 Time in MSCs

When simulating with an MSC or HMSC, the time is considered global to all the instances present in an MSC. For example, if you load the MSC shown in Figure 5.32 into the Validator

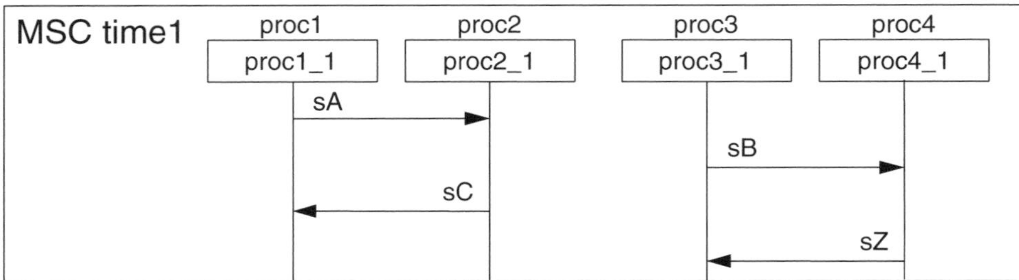

Figure 5.32 MSC with four instances

and simulate the SDL sequence *sA, sC* and *sB*, you get a violation report, because *sB* was expected before *sC*.

5.2.6 Simulate with observer processes

You will add an observer process to detect if the variable *uData* in process *dispatch* in block *DLCa* contains *55*.

A. In Windows (or Unix), make a copy of the file *v76test.ssy* into *v76test_obs.ssy* (to keep a version of the SDL system without observer).

B. In the Organizer, select *V76test*, choose *Edit* > *Connect*, choose *To an existing file*, press the folder-shaped icon and connect to the file *v76test_obs.ssy*.

C. In the Organizer, press the *save* button.

D. In the SDL Editor, check that you are editing the file *v76test_obs.ssy*, and add a block reference named *obs*, as shown in Figure 5.33.

E. Double-click on the block *obs*, confirm the two windows, and add a text symbol containing the include statement shown in Figure 5.34, necessary to call observation operators. Type exactly the same words, respecting the case, and do not add spaces.

F. In the block *obs*, create a process reference named *obs1*. Don't forget to press the *save* button in the Organizer from time to time.
The Organizer should now look like Figure 5.35.

G. In the SDL Editor, double-click the process reference *obs1*: after two confirmations, the content of process *obs1* is displayed.

H. Copy the observer represented in Figure 5.36: in the start transition, *displ* is filled with the Pid of the first instance of process *dispatch*, inside block *DLCa*. From state *testing*, if the variable *uData* in process *dispatch* contains *55*, then a call to the predefined procedure *Report* triggers the generation of a report by the Validator. To avoid a deadlock when the condition is false, we have added a transition always firable, having a lower priority.

   GetPid and *vInteger* are operators predefined in *access.pr*. *Report* is a predefined procedure.

   SIGNALSET; is necessary because the process has no signals.

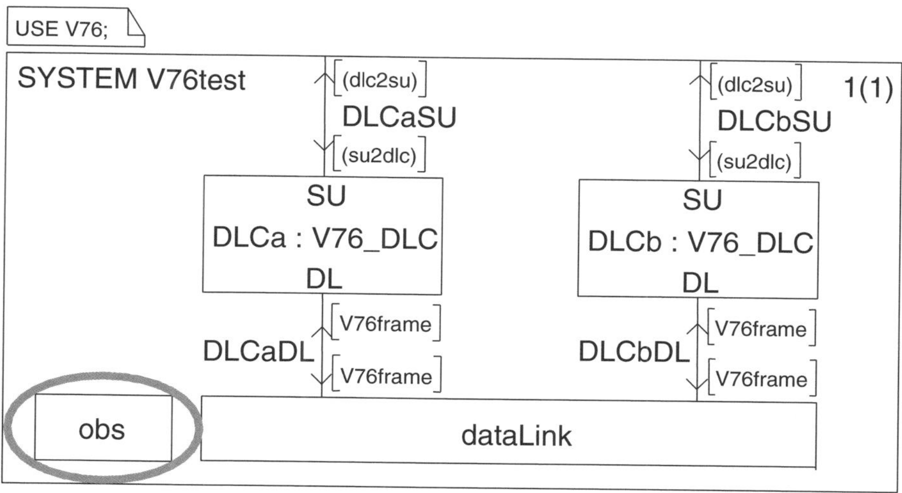

Figure 5.33 Adding the *obs* observer block

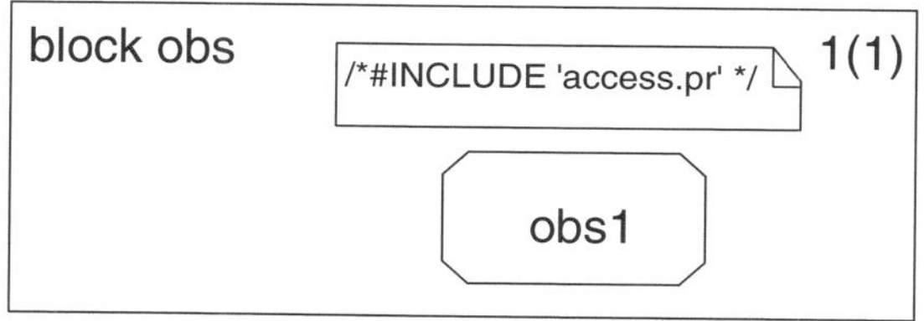

Figure 5.34 The contents of the *obs* observer block

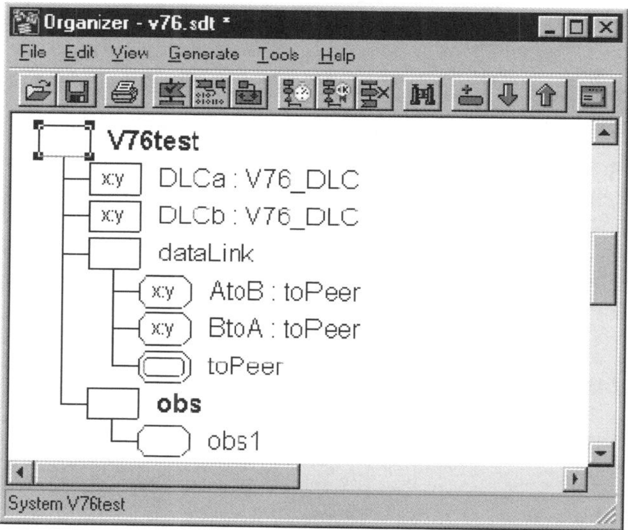

Figure 5.35 The Organizer showing the observer

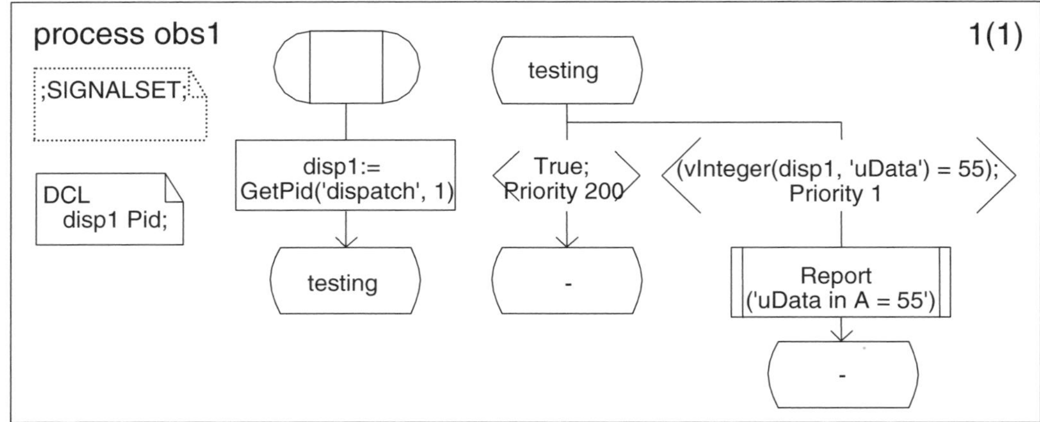

Figure 5.36 The observer process *obs1*

I. Press the *save* button in the Organizer, start the Validator as indicated in Section 5.2.1.2 and enter the command:

```
define-observer obs1
```

This command tells the Validator to execute *obs1* as an observer instead of a regular process: its transitions are no longer proposed in the navigator and so on.

J. Enter the command *list-observers* and check that the Validator answers *obs1*.

K. Select *Commands > Toggle MSC Trace*, and press the *Navigator* button in the *Explore* group.

L. In the Navigator window, double-click on the lower rectangle, marked *Next 1*, to execute the corresponding SDL transition.

M. Continue double-clicking to establish DLC number 0; when there are several possibilities, select the left one.

N. After around 17 transitions, the Navigator window displays many transitions, as shown in Figure 5.37. Scroll right and double-click on the transition transmitting *L_DataReq(0, 55)* to *DLCa*.

O. Double-click on the next transition: the Navigator displays the observer report depicted in Figure 5.38, because *uData* contains 55.

5.2.7 More details on observer processes

Note that several observer processes may be used at the same time in the Validator.

5.2.7.1 Using an observer process to test the state of a process

The observer process *obs2* represented in Figure 5.39 detects when the state of process *dispatch* in block *DLCa* is equal to *waitUA*.

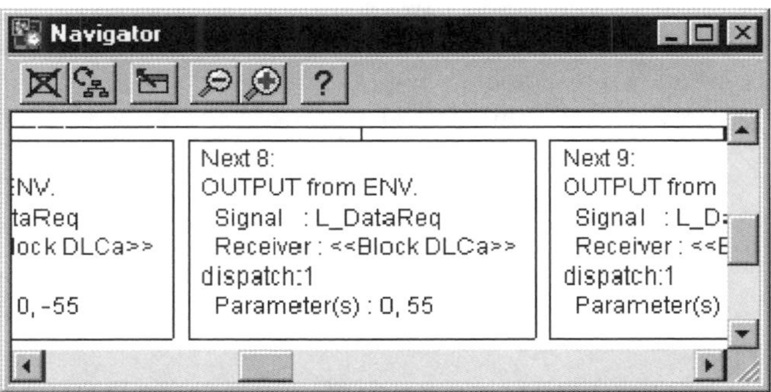

Figure 5.37 The Validator ready to transmit *L_DataReq(0, 55)* to *DLCa*

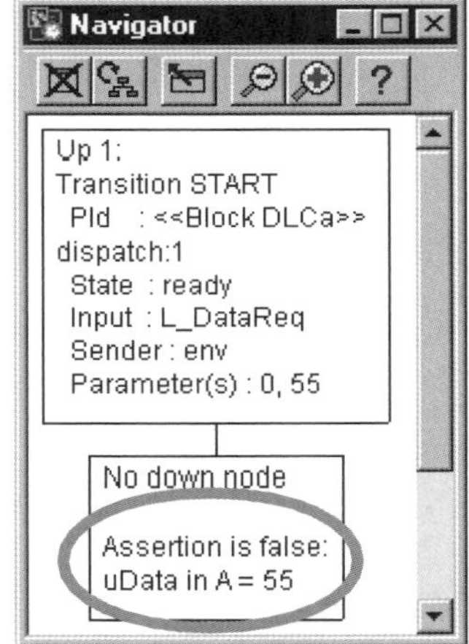

Figure 5.38 The condition has been detected by the observer *obs1*

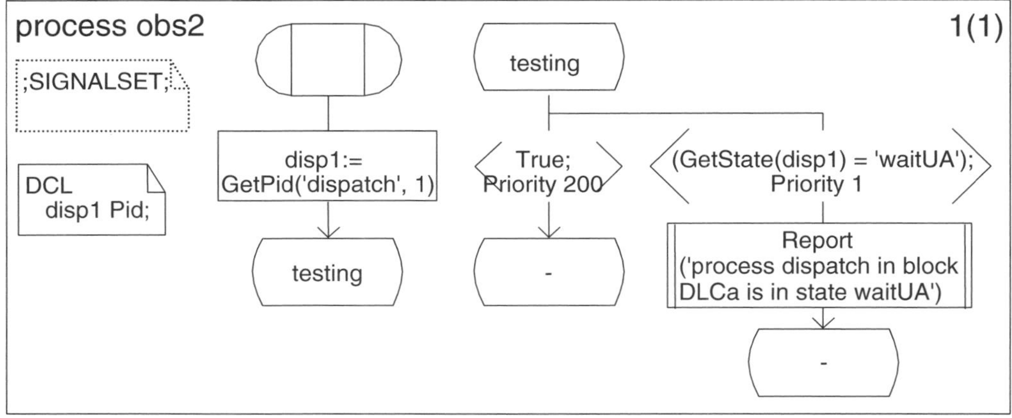

Figure 5.39 Testing if a process is in a certain state

5.2.7.2 Using an observer process to detect process creation and stop

The observer process *obs3* represented in Figure 5.40 detects when the instance of process *dispatch* in block *DLCa* is stopped (this does not occur in the V.76 SDL model).

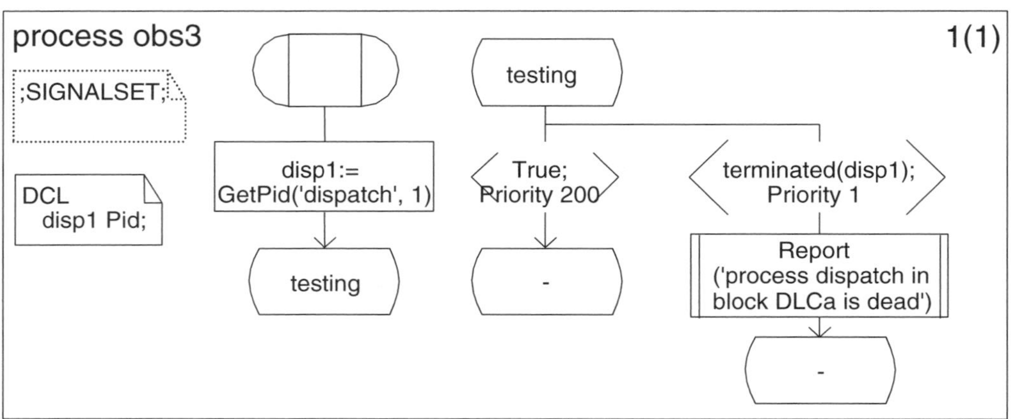

Figure 5.40 Detecting process instance stop

5.3 CASE STUDY WITH OBJECTGEODE

5.3.1 Simulate with stop conditions

5.3.1.1 Detect DLC establishment

We want to detect that a DLC is established. This means, in our V.76 SDL model, that:

- instance 1 of process *DLC* in block *DLCa* is in state *connected* and
- instance 1 of process *DLC* in block *DLCb* is in state *connected.*

A. Start the Simulator to simulate the V.76 model alone.

B. In the Simulator command line, type the stop condition[8] (the \ means continue on next line):

```
stop if (DLCa!DLC(1)!state = connected) and  \
        (DLCb!DLC(1)!state = connected)
```

You can also use the menu command *Edit > Stop Conditions* to enter, delete or modify a stop condition; in such case, do not type *stop if.*

Several stop conditions may be entered: they are considered as being linked by an *or* operator.

[8] To easily get the correct syntax for the name of an entity in the simulator, first use the *print* command; for example, here, if you enter *print state*, you get the pathnames to all the states in the model, which you can copy and paste to build your stop condition.

C. Replay[9] the connection establishment scenario: select *File > Scenario > Load*, choose *cnx1.scn*. Press the button *All*: at the end of the scenario replay[10], the Simulator displays a line showing that the stop condition is satisfied, as depicted in Figure 5.41.

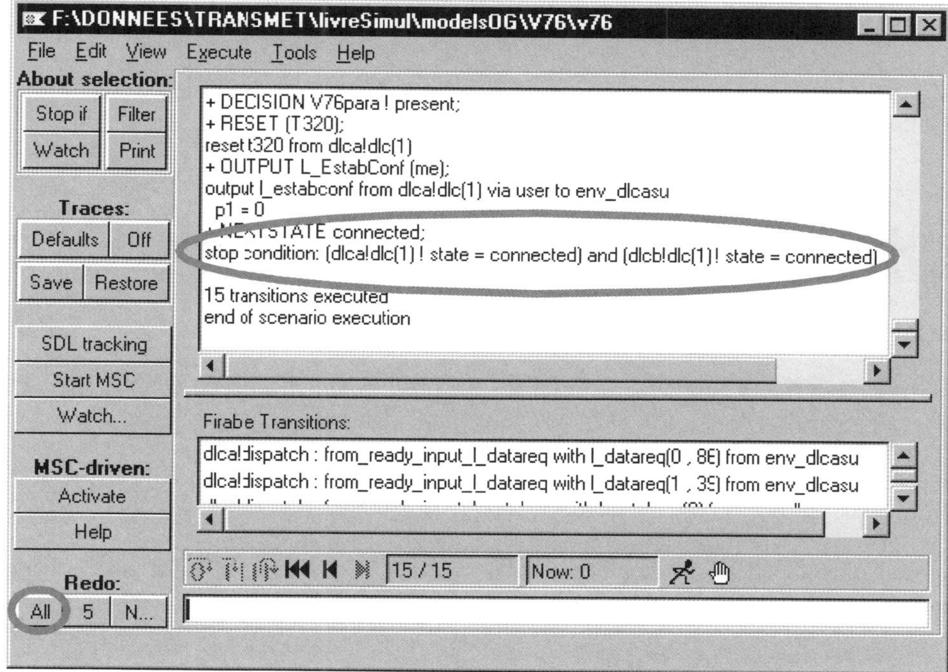

Figure 5.41 The stop condition is satisfied

5.3.1.2 More stop conditions

The syntax of the *stop if* command is identical to the syntax of the *filter* command. Stop conditions can also be inserted into the model startup file (here *v76.startup*), if necessary.

Here are some more stop conditions that you can enter in the Simulator as indicated previously.

To detect when one of the variables *n320cnt* is > *2*:

```
stop if n320cnt > 2
```

To detect when the variable *n320cnt* declared in the first instance of process *DLC* in block *DLCa* is > *2* (replay the scenario *retry1.scn* to get such a value):

```
stop if DLCa!DLC(1)!n320cnt > 2
```

---

[9] Naturally, stop conditions also work when simulating without replaying a stored scenario.

[10] If after executing the four start transitions the Simulator is blocked, it means probably that the feeds are missing. They should be loaded by the model startup file, *v76.startup*. See previous chapters.

To detect if the transitions coverage rate of the SDL model is > 95%:

```
stop if trans_cover_rate(system) > 95
```

To detect any output of signal *L_EstabInd*:

```
stop if output L_EstabInd
```

To detect the output of any signal to the environment:

```
stop if output TO ENV
```

To detect the input of signal *v76frame* by instance 1 of process *AtoB*, coming from instance 1 of process *DLC* in block *DLCa*:

```
stop if input v76frame from dlca!dlc(1) to atob(1)
```

To detect when the input queue of instance 1 of process *DLC* in block *DLCa* contains more than two signals (the part *is_active* is not necessary for a process instance that always exists, such as *dispatch(1)*):

```
stop if is_active(DLCa!DLC(1)) and length(DLCa!DLC(1)!queue) > 2
```

To detect that the first signal in the queue of process *dispatch* in block *DLCb* is named *V76frame* and that process *dispatch* in block *DLCa* is in state *waitUA*:

```
stop if (length(DLCb!dispatch! queue) > 0) and (DLCb!dispatch !
queue(1)!name=v76frame) and (DLCa!dispatch!state = waitUA)
```

To detect the progression of time:

```
stop if trans time
```

To detect if the SDL time is > 100:

```
stop if now > 100
```

To detect if timer *T320* in instance 1 of process *DLC* in block *DLCa* is set (i.e. is armed):

```
stop if dlca!dlc(1)!t320!active
```

To detect the creation of any process instance:

```
stop if create
```

To detect the creation of an instance of process *DLC* in block *DLCa*:

```
stop if create DLCa!DLC
```

To detect the death of any process instance:

```
stop if stop
```

To detect the call of any procedure:

```
stop if call
```

To detect the call of procedure *CRCok* in instance 1 of process *DLC* located in block *DLCb*:

```
stop if is_active(DLCa!DLC(1)) and call DLCb!DLC(1)!CRCok
```

As you can see, any event or combinations of events in the SDL model can be detected by stop conditions during the simulation. However, stop conditions cannot detect the correct sequencing of signals: for that, you must use MSCs.

5.3.2 Simulate with a basic MSC

You will simulate the V.76 SDL model, observed by a basic MSC. *Basic* means that the MSC contains a single scenario, as opposed to a hierarchy of scenarios linked by temporal operators (such as AND, OR, REPEAT...). To simplify, we reuse the basic MSC *test1.msc*, generated by the Simulator in Chapter 4. Naturally, you could use another MSC drawn from scratch with the Editor, or rework an existing MSC.

5.3.2.1 Compile the SDL model plus the MSC

A. In the SDL Editor, load the V.76 SDL model plus the MSC *test1.msc* (generated in Section 4.3.2.4).

B. Unload any other files from the SDL Editor.

C. Quit (do NOT minimize) the ObjectGeode Launcher if running.

D. In the SDL Editor, select *Tools > SDL & MSC Simulator*. The window represented in Figure 5.42 appears.

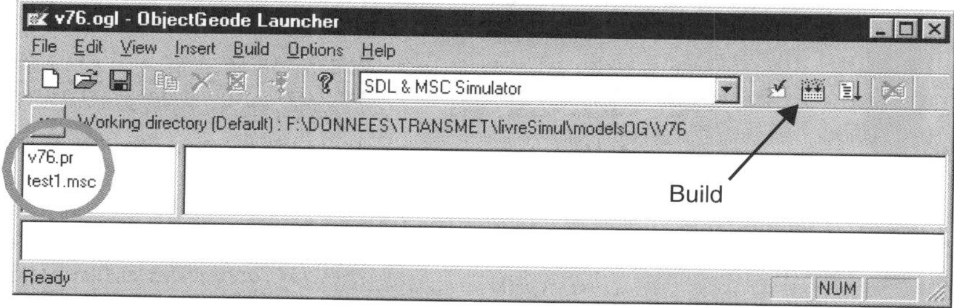

Figure 5.42 The ObjectGeode Launcher with an MSC

E. Check that the left area only contains *v76.pr* and *test1.msc*, as in Figure 5.42, otherwise select any file to remove and do *Edit > Remove*.

F. Select *File > Save* to save the current settings in the file *v76.ogl* in your current directory (feature not available in the Unix version).

G. Press the *Build* button: this checks your SDL model and the MSC, as shown in Figure 5.43. If both are correct and if the MSC is consistent with the SDL model (for example, you would get an error if using a signal in the MSC not declared in the SDL model), it translates the SDL model into C code, and the MSC into an observer (in C). Then the C files are compiled and linked to the simulator library, to produce the executable file *v76.sim*.

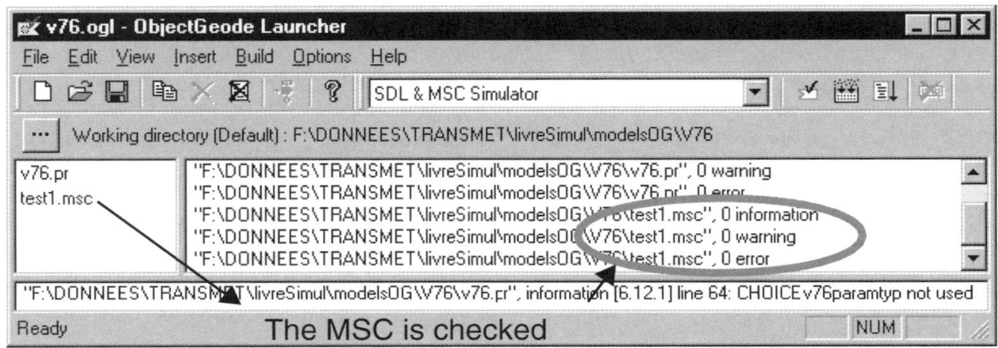

Figure 5.43 After pressing the *Build* button

5.3.2.2 Replay the simulation scenario test1

A. Press the *Execute* button to start the Simulator. The Simulator main window appears.

B. In the Simulator, select *View > Hierarchy*[11]: as depicted in Figure 5.44, you now get a third area containing *observation*. It corresponds to the MSC *test1* you compiled with the SDL model.

C. Unfold the observation part as in Figure 5.44, select *test1* and press the button *Track*: the Editor opens a window showing the hierarchy of MSC *test1*. Double-click the box *test1* to open it: a bold horizontal bar at its top, illustrated in Figure 5.45, shows the next signal input or output expected in the SDL model by the MSC.

D. Replay the Simulation scenario (or fire transitions manually if you prefer): select *File > Scenario > Load*, choose *test1.scn*, and press the *redo* ⏭ Simulator button nine times until Step 9, to get the MSC shown in Figure 5.45.

   In this figure, the lower horizontal bold bar shows you the next expected event in the SDL simulation: input of signal *L_EstabResp* by block *DLCb*.

   To help you, the Simulator displays at each step which events were observed by each observer; here, at Step 9, we have:

[11] To avoid repeating this several times, you can add into the file *v76.startup* the textual equivalent command: *track_msc observation ! test1*

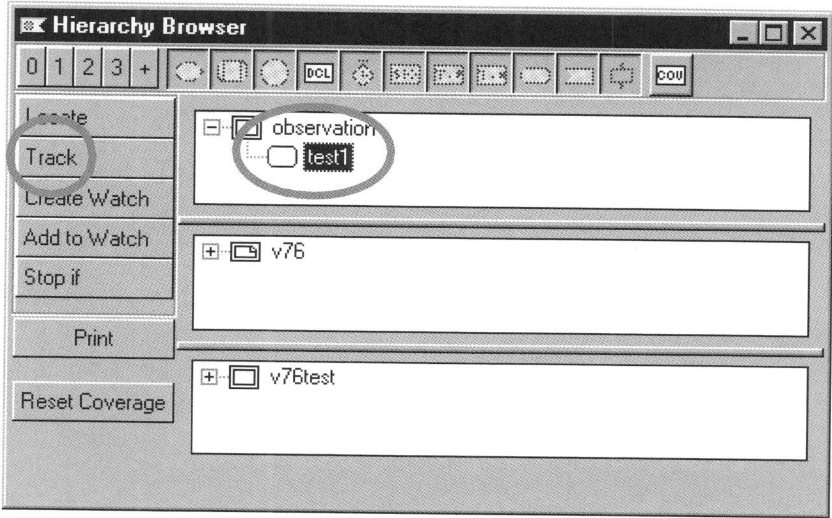

Figure 5.44 The Simulator Hierarchy Browser with the observer *test1*

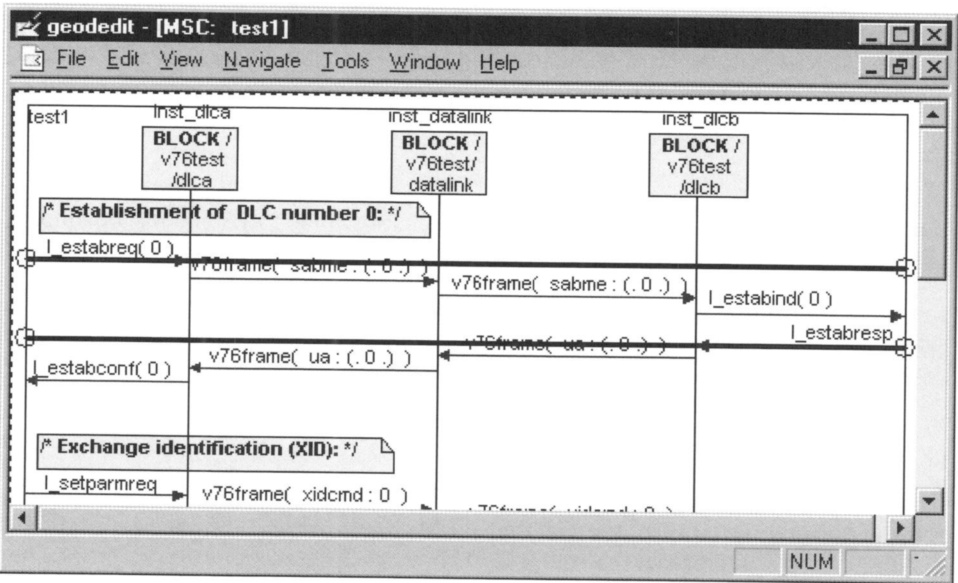

Figure 5.45 Tracking MSC *test1*: waiting for input of signal *L_EstabResp*

```
Events observed in scenario test1 (File F:\V76\test1.msc):
  IN v76frame(sabme,0) FROM datalink TO dlcb
  OUT l_estabind(0) FROM dlcb TO ENV VIA dlcbsu
```

It means, as shown in Figure 5.46, that the last SDL events observed by the MSC were:

- the input by *DLCb* of signal *V76frame* with parameter value *sabme* and *0* and
- the output by *DLCb* of signal *L_EstabInd* with parameter value *0*.

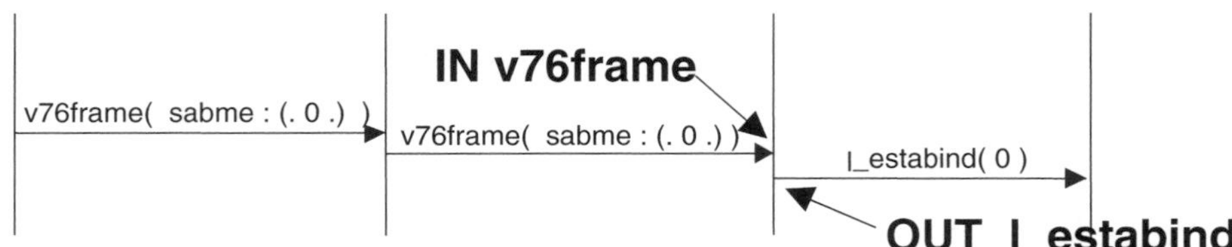

Figure 5.46 Events observed at Step 9

Note that if the simulated parameter values do not match the MSC expected values, the MSC property will not be satisfied.

E. Finish replaying the simulation scenario until its end (Step 41) by pressing the *Redo* or *Redo: All* buttons. You should see the line:

```
SUCCESS state reached in scenario test1
```

It proves that the SDL simulation performed is consistent with the observer MSC.

5.3.3 Simulate with a hierarchical MSC

The basic MSC *test1* represents only a single execution sequence. To represent several expected behaviors consistent with the V.76 textual specification without creating too many basic MSCs, we will create a hierarchical MSC.

You will simulate the V.76 SDL model, observed by two MSCs:

- a basic MSC, representing the XID scenario and
- a hierarchical MSC, containing a hierarchy of scenarios linked by temporal operators (such as AND, OR, REPEAT...), representing the rest of the expected behaviors.

The basic XID MSC is not integrated into the hierarchical MSC because the XID scenario can occur at any moment. Thus, there is no need to sequence it with the other behaviors (connection, data transfer and disconnection).

5.3.3.1 Create the hierarchical MSC

With the MSC Editor, you will rework a copy of the basic MSC *test1.msc* generated by the Simulator in Chapter 4, to obtain the hierarchical MSC *test1ops.msc*, shown in Figure 5.47.

This MSC *test1ops* represents the following behavior expected from the V.76 SDL model:

1. one connection phase (*cnx_0*),
2. followed or not by the repetition of data transfers from A to B (*data_a2b_0*) or from B to A (*data_b2a_0*),
3. followed by the disconnection phase initiated by A (*disc_0_by_a*) or initiated by B (*disc_0_by_b*)[12].

[12] To simplify the example, this MSC is incomplete: connection initiated by B is missing, DLC number 1 is not tested and so on.

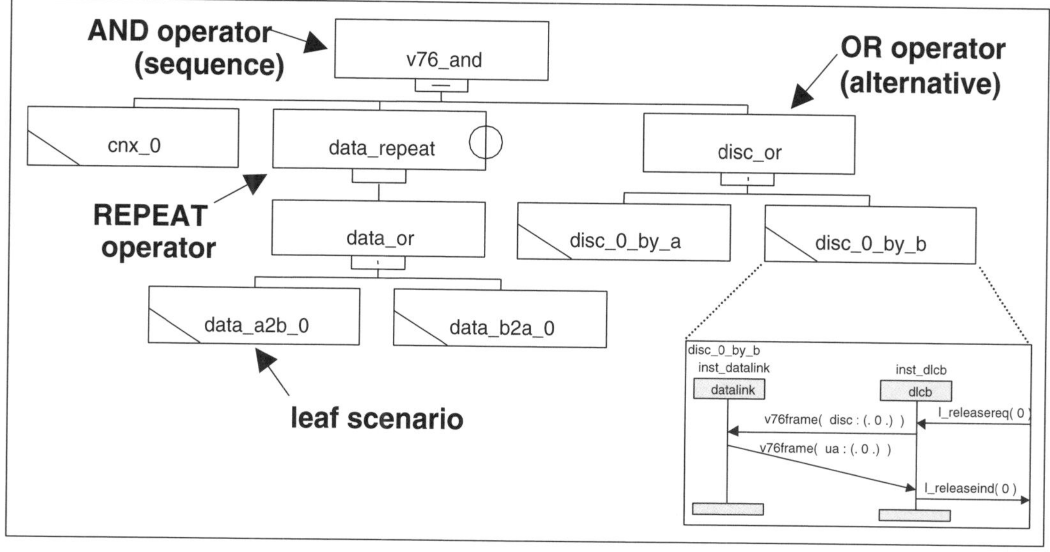

Figure 5.47 The hierarchical MSC *test1ops.msc*

To save time, you will first cut *test1* into four MSCs as shown in Figure 5.50.

A. Unload *test1.msc* from the Editor.

B. In Windows (or Unix), make a copy of the file *test1.msc* into *test1ops.msc*.

C. Open *test1ops.msc* with the Editor[13].

D. In the Editor Framework window, select *test1* as in Figure 5.48, copy it and paste it three times.

E. Rename the four scenarios as in Figure 5.49.

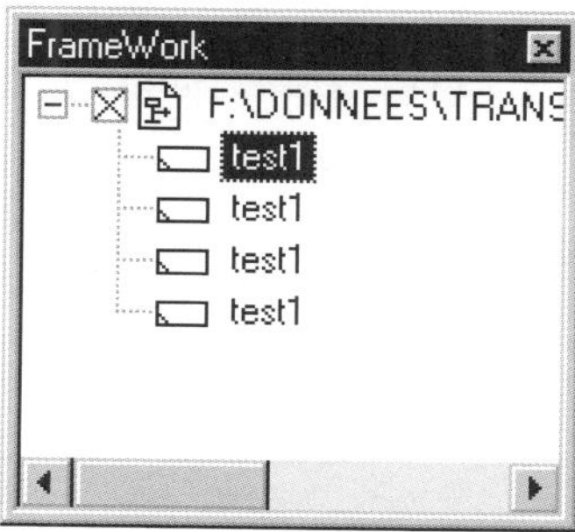

Figure 5.48 The Framework window after duplicating scenario *test1*

[13] The MSC Editor is integrated into the SDL Editor.

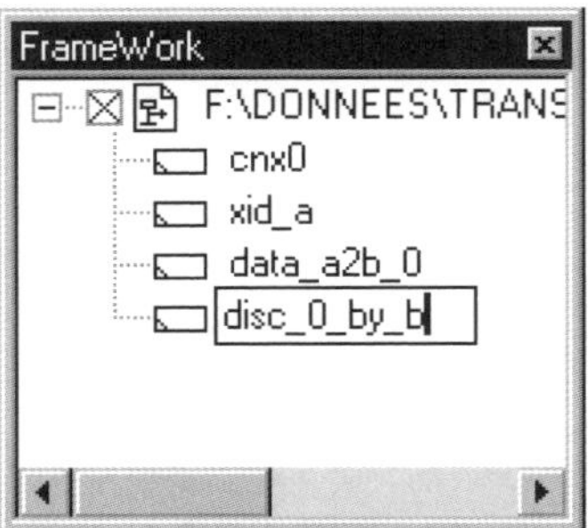

Figure 5.49 The Framework window after renaming the scenarios

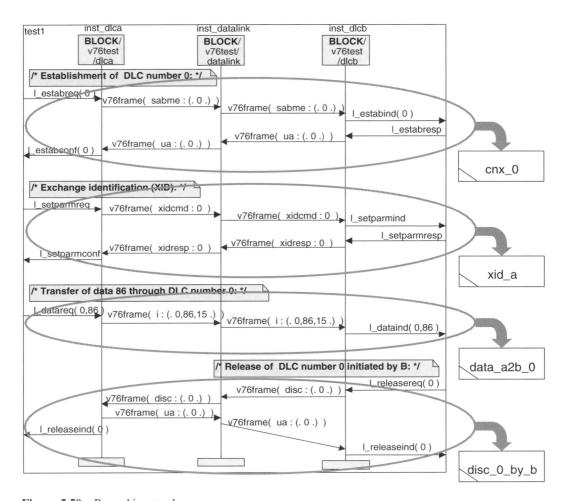

Figure 5.50 Reworking *test1.msc*

F. Double-click on *cnx_0*: the Editor opens a window containing the MSC shown in Figure 5.50. Remove all the arrows except the upper part representing the connection phase (you can drag the mouse to select several elements at a time, but inside the frame only).

G. Perform the same operation for the scenarios *xid_a*, *data_a2b_0* and *disc_0_by_b*, as shown in Figure 5.50. In *xid_a*, you can remove the block *datalink* and the signals exchanged with it, as shown in Figure 5.51.

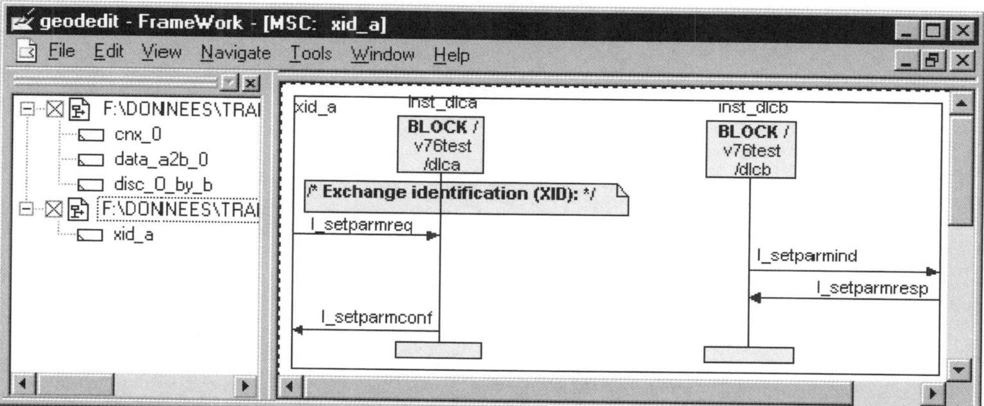

Figure 5.51 The scenario *xid_a* moved into a new file

H. In the Framework window, select *File > New > MSC* and drag the scenario *xid_a* into the new file, as depicted in Figure 5.51. Select the new file, do *File > Save as* and enter *xid1.msc*.

Now you will add operators to organize our basic MSCs into a hierarchy.

I. In the Framework window, select the file *test1ops.msc*, press the AND palette button, and type *v76_and*, to obtain the configuration shown in Figure 5.52.

J. Double-click on the AND operator named *v76_and* (click the symbol, not its name): the MSC hierarchy view appears, as illustrated in Figure 5.53.

K. In the Framework window, drag the three scenarios *cnx0*, *data_a2b_0* and *disc_0_by_a* under the AND operator *v76_and*, to obtain the configuration shown in Figure 5.54.

L. In the Framework window, insert a REPEAT and two OR operators, rename them, respectively, *data_repeat* and *data_or* or *disc_or*.

M. Drag them to obtain the MSC represented in Figure 5.55.

N. In the Framework window, copy and paste the scenario *data_a2b_0*; rename the pasted scenario *data_b2a_0*.

O. In the Framework window, copy and paste the scenario *disc_0_by_b*; rename the pasted scenario *disc_0_by_a*, to obtain the hierarchy shown in Figure 5.47.

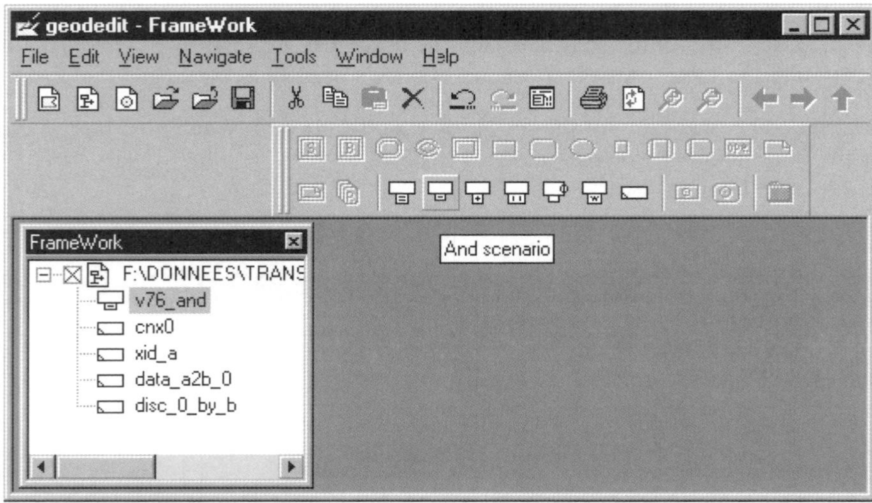

Figure 5.52 After inserting the AND scenario

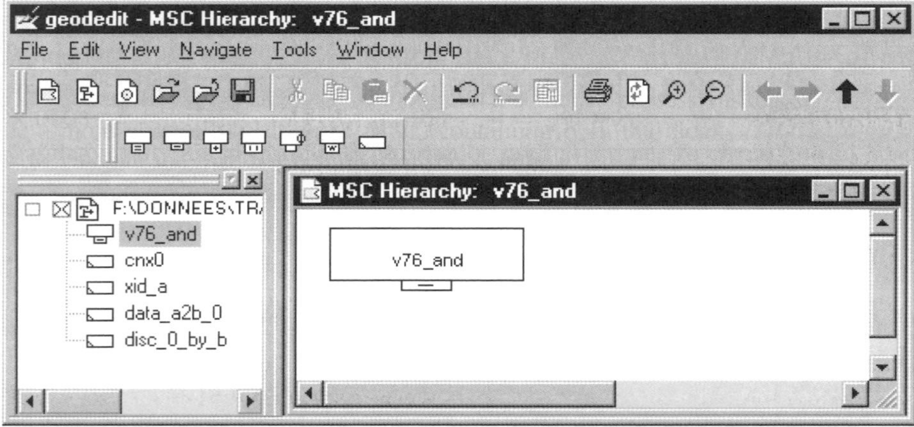

Figure 5.53 The MSC hierarchy window

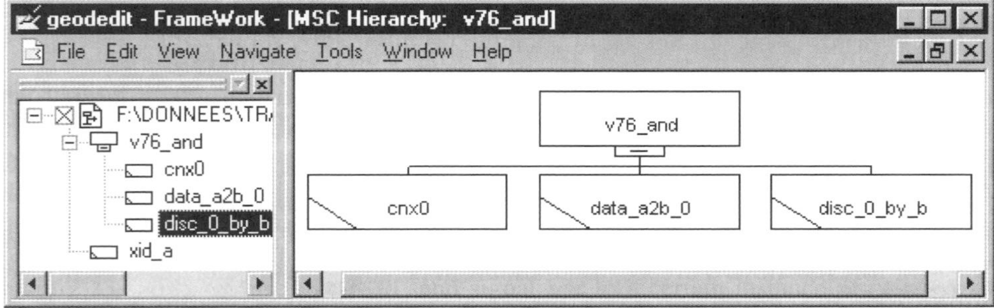

Figure 5.54 After moving three scenarios under the AND

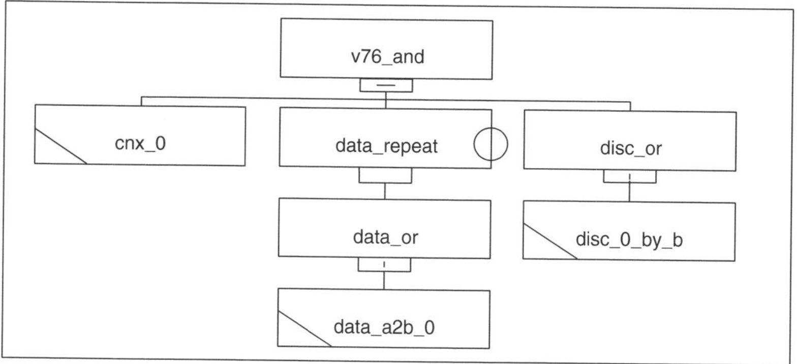

Figure 5.55 Building the MSC

P. Double-click the scenario *data_b2a_0*, create the four new signals[14] (message button), copy the text from the old signals to the new signals, delete the four old signals, and finally replace 86 and 15 by 97 and 23 to obtain the scenario shown in Figure 5.56.

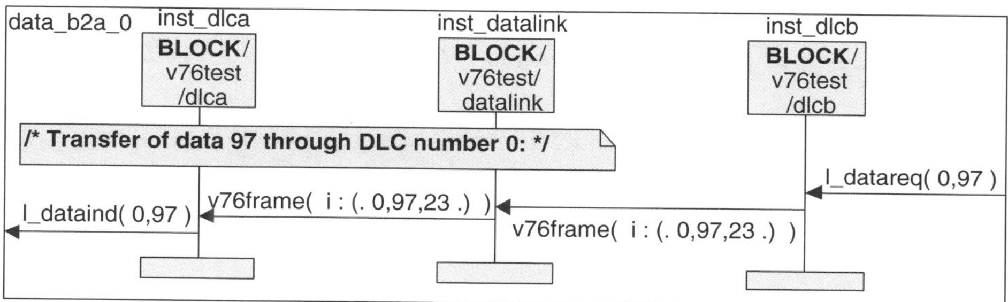

Figure 5.56 The scenario *data_b2a_0*

Q. Double-click the scenario *disc_0_by_a*, create the seven new signals, copy the text from the old signals to the new signals, and delete the seven old signals to obtain the scenario shown in Figure 5.57, taking care to have exactly the same ordering between signal outputs and inputs.

R. In the Framework window, select the file *test1ops.msc* and do *File > Save*.

The hierarchical MSC *test1ops* is now ready for tracking.

5.3.3.2 Compile the SDL model plus the two MSCs

A. In the SDL Editor, check that the V.76 SDL model plus the two MSCs contained in *test1ops.msc* and *xid1.msc* are loaded.

[14] Do not move the existing signals because the Editor stores some *VIA <channel_name>* in the *.msc* file, which may provoke compilation errors (for example, when moving a signal from *DLCa* to *DLCb*).

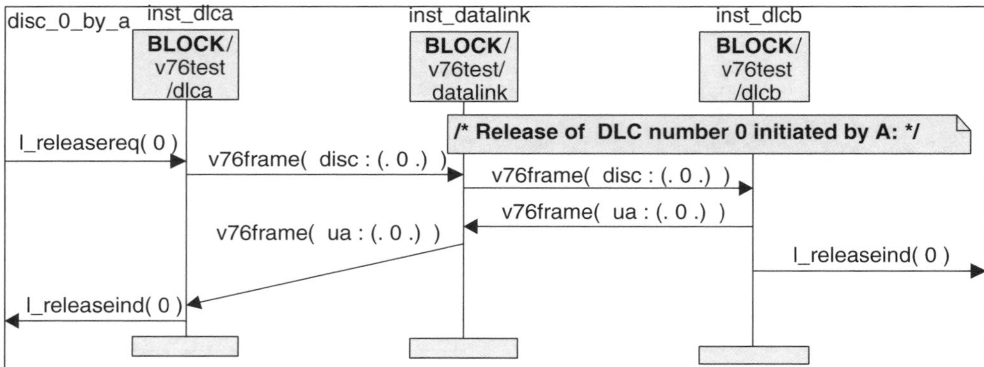

Figure 5.57 The scenario *disc_0_by_a*

B. Unload any other files from the SDL Editor, and close all the diagrams opened in the right area.

C. Quit (do NOT minimize) the ObjectGeode Launcher if running.

D. In the SDL Editor, select *Tools > SDL & MSC Simulator*. The ObjectGeode Launcher appears: check that its left area only contains *v76.pr, test1ops.msc* and *xid1.msc*.

E. Press the *Build* button: this checks your SDL model and the two MSCs and translates the SDL model and the MSCs into C code, to produce the executable file *v76.sim*.

5.3.3.3 Replay the simulation scenario test1

A. Press the *Execute* button to start the Simulator. The Simulator main window appears.

B. In the Simulator, select *View > Hierarchy*: as depicted in Figure 5.58, you now get a third area containing *observation*. It corresponds to the MSCs *xid_a* and *v76_and* contained in the files *xid1.msc* and *test1ops.msc*.

C. Unfold the observation part as in Figure 5.58, select *xid_a* and press the button *Track*: the Editor opens a window showing the MSC *xid_a*. Double-click if necessary in the Editor to display the content of *xid_a*.

D. Do the same for the MSC *v76_and*, double-click on *data_a2b* in the Framework area, and select *Window > Tile Horizontally* to obtain the disposition shown in Figure 5.59.

E. Replay a Simulation scenario: select *File > Scenario > Load*, choose *test1.scn*, and press the *Redo: All* Simulator button[15]: at Step 23, the simulation should stop, and display:

```
SUCCESS state reached in scenario xid_a
```

It proves that the SDL simulation performed is identical to the behavior described in the observer MSC *xid_a*.

[15] In case the step number and so on are no longer displayed in the Simulator trace area, press on *Traces: Defaults*.

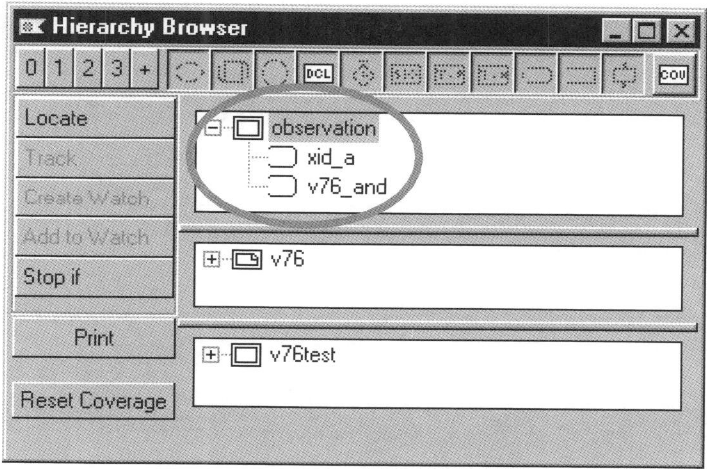

Figure 5.58 The Simulator Hierarchy Browser with two observers

Important: if the MSC tracking does not work or no success is reached, unload the MSC from the Editor, open the *.msc* file with a text editor and remove all the occurrences of *VIA <some_name>*.

F. Continue replaying the scenario: press the *redo* ⏭ Simulator button to reach Step 26, where you should see the same situation as in Figure 5.59.

In this figure, the bold horizontal bar shows that the next SDL event expected by the MSC *data_a2b_0* is an output of signal *v76frame* by block *datalink*.

Also the window named *MSC Hierarchy v76_and* shows in bold which scenarios can accept events: as we are in *data_a2b_0*, the other scenarios located after it such as *disc_or* are not yet ready; *data_a2b_0* must be finished first.

G. Finish replaying the simulation scenario until its end (Step 41) by pressing the *Redo* or *Redo: All* buttons. You should see the line:

```
SUCCESS state reached in scenario v76_and
```

It proves that the SDL simulated behavior is included in the expected behaviors described in the observer MSC *v76_and*.

You can now simulate manually other scenarios and try reaching again the success state in the MSC observers. Chapter 7 will show you how the Simulator can discover automatically such scenarios.

5.3.4 More details on MSCs

5.3.4.1 The operators used in MSCs

To represent a more complex behavior than a single basic scenario, operators specific to Object-Geode can be used to create a hierarchy of MSCs. This is similar to the HMSCs used in Tau. The operators are:

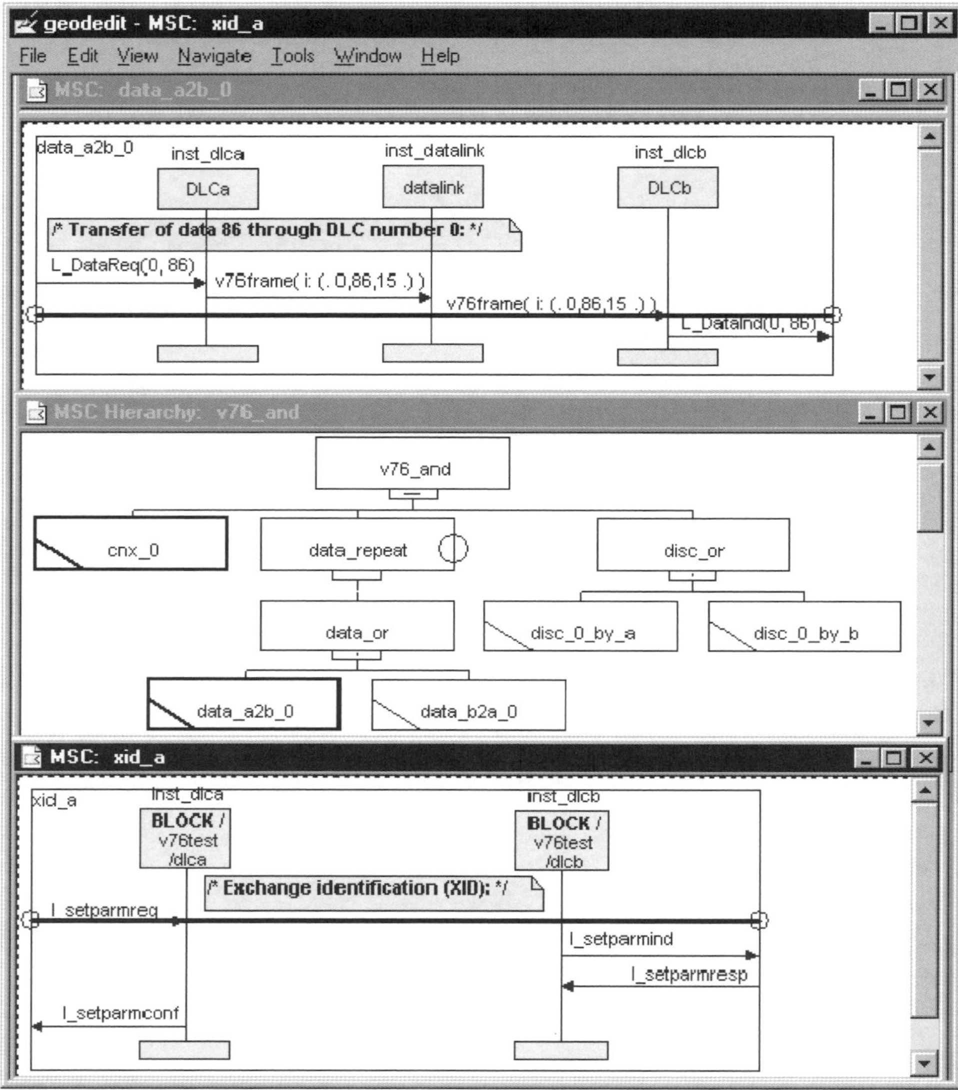

Figure 5.59 Tracking the two MSCs *v76_and* and *xid_a* during simulation

- AND: sequence
- OR: alternative
- REPEAT: repetition 0 to *n* times
- EXCEPTION: description of behavior expected after an error
- PARALLEL: independent execution of several MSCs.

We will only detail the most used operators: AND, OR and REPEAT.

The operator AND is used to split one long scenario into two, three or more shorter ones. For example, in Figure 5.60, the sequence *sA*, *sB*, *sC*, *sD* is split into *sA*, *sB* and *sC*, *sD*.

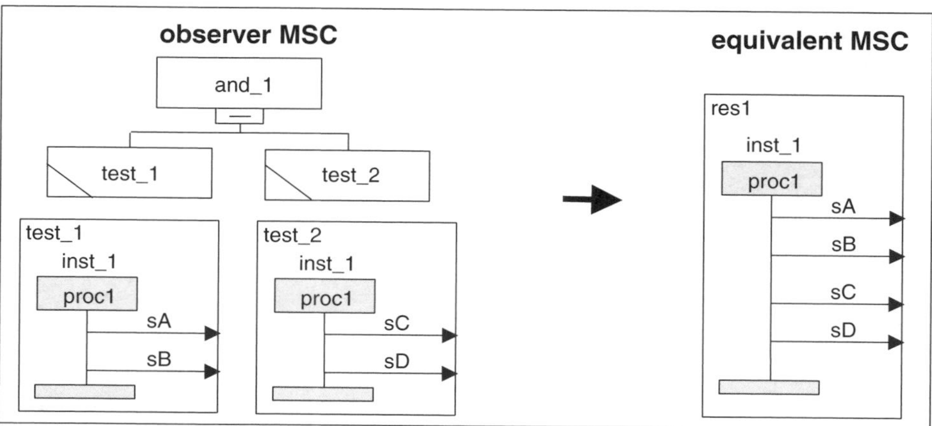

Figure 5.60 The operator AND

The operator OR means that the expected behavior is an alternative between two or more MSCs. For example, in Figure 5.61, the expected behavior is:

- either the sequence *sA*, *sB*
- or the sequence *sA*, *sC*.

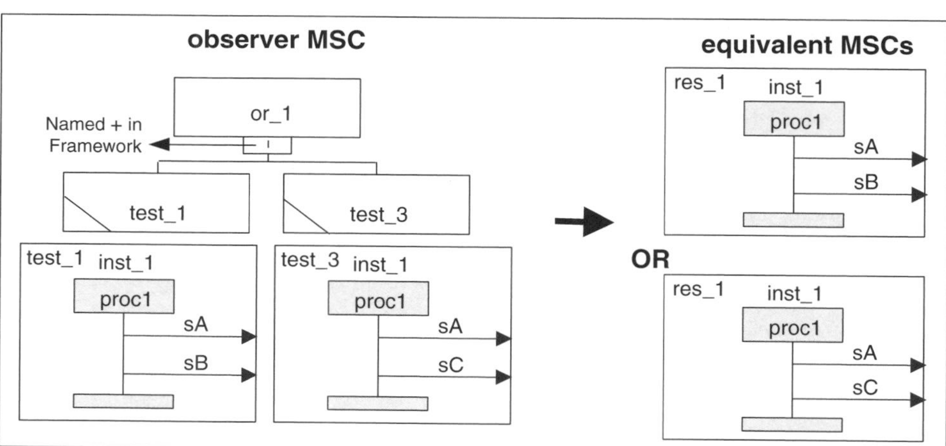

Figure 5.61 The operator OR

This example also illustrates the fact that there can be a common part between several MSCs under an OR operator (here, *sA* is common to *test_1* and *test_3*).

The operator REPEAT means that the expected behavior is the repetition from 0 to *n* times of an MSC. For example, in Figure 5.62, the expected behavior is:

- either nothing,
- or the sequence *sA*, *sB*, *sC*,
- or the sequence *sA*, *sB*, *sC*, *sA*, *sB*, *sC*,

and so on.

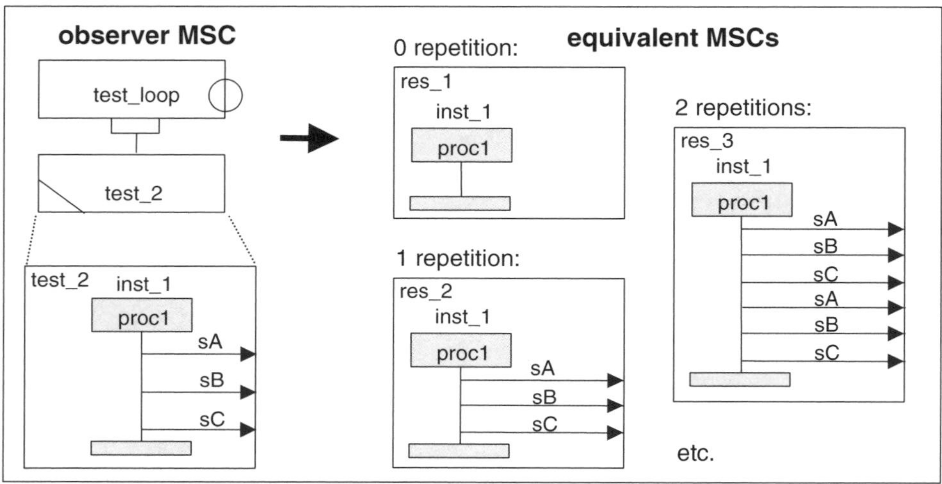

Figure 5.62 The operator REPEAT

5.3.4.2 The MSC symbols used for observation

When an MSC is compiled with an SDL model, ObjectGeode checks that:

- All the elements used in the MSC exist in the SDL model.
- The scope in the MSC is compatible with the SDL model.

For example, in Figure 5.63, the following checks are performed during compilation:

- The entity named *dataLink* in the MSC *retry1* exists in the SDL model *HDLC*.
- The signal *v76frame* in the MSC also exists in the SDL model.

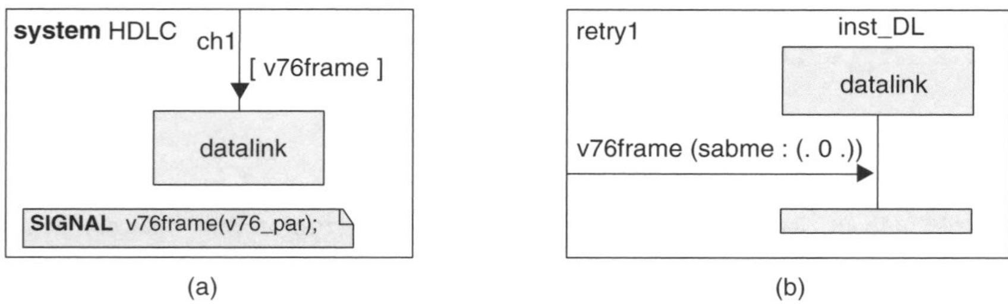

Figure 5.63 An MSC (b) consistent with the SDL model (a)

- The signal *v76frame* being received by *dataLink* in the MSC, a channel carrying *v76frame* and reaching *dataLink* exists in the SDL model.

The name *inst_DL* located on top of *datalink* in the MSC is not supposed to match any SDL name.

Then, when using an MSC as an observer, only the following symbols, depicted in Figure 5.64, are dynamically checked during simulation:

- Signal input and output and their parameter values,
- Timer set, reset and timeout.

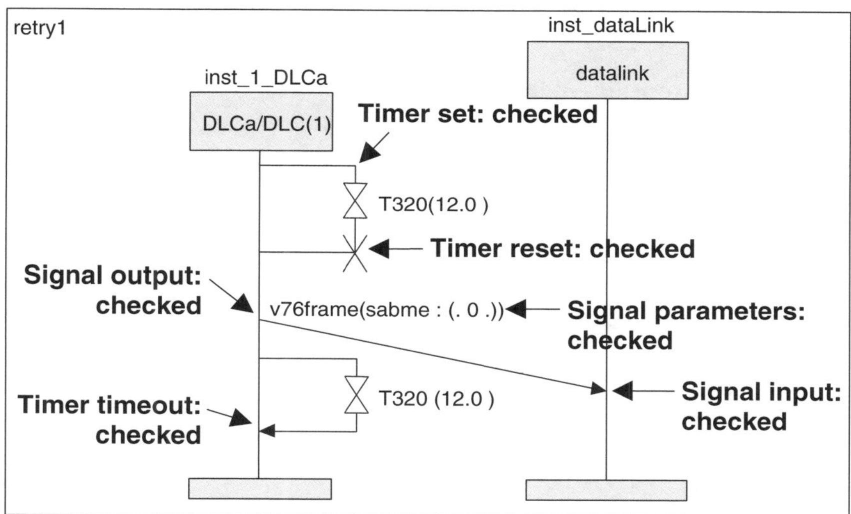

Figure 5.64 The MSC symbols checked during observation

The other symbols present in the MSC, shown in Figure 5.65, are ignored.

5.3.4.3 The syntax for MSC signal parameters

A special notation has been introduced for signal parameters in observer MSCs. An example is shown in Figure 5.66.

In this example:

- The first occurrence of *sig* means that the output of *sig* is expected from *BTS* with value *17* as first parameter, *cyan* as second parameter and *True* as third parameter.
- In the second occurrence, *sig* is expected with any value as first parameter, *cyan* as second parameter and any value as third parameter.
- In the third occurrence, *sig* is expected with *17* as first parameter, *cyan*, *black* or *yellow* as second parameter and *True* as third parameter.
- In the fourth occurrence, *sig* is expected with a value between *4* and *16* as first parameter, *cyan* as second parameter and *False* as third parameter.

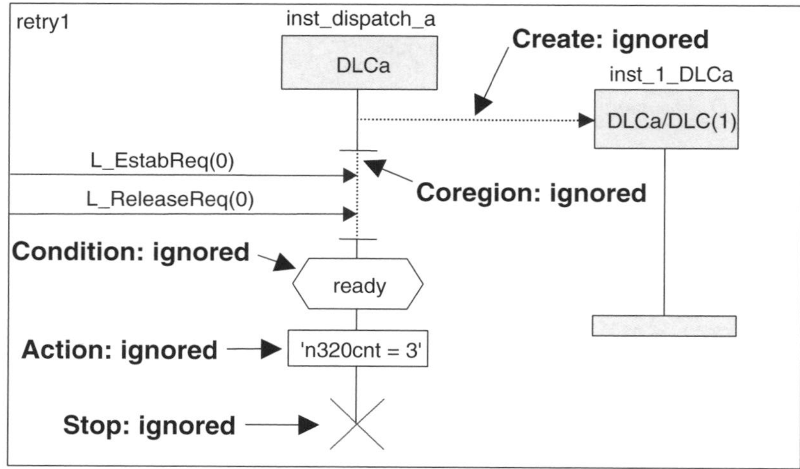

Figure 5.65 The MSC symbols not checked during observation

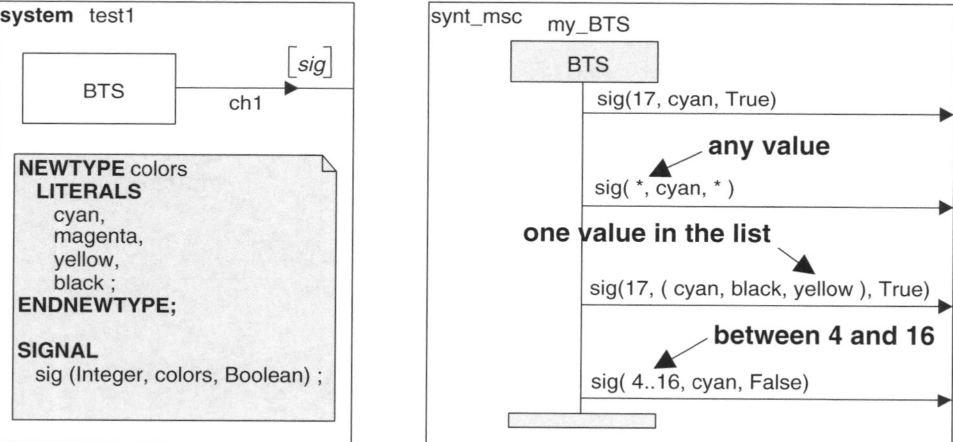

Figure 5.66 Example of observer MSC with special parameters

5.3.4.4 Observer MSCs generate feed commands

When an MSC is compiled with an SDL model, the inputs coming from the environment in the MSC are automatically translated into feed commands in the simulation. For example, in Figure 5.67, if the MSC *test4* is compiled with the SDL model *test2*, after launching the Simulator you will see the following feed already set:

```
> list feed
feed ch1 sig2( -8, true)
feed ch1 sig2( 45, false)
```

This feature saves time when using MSCs to test an SDL model.

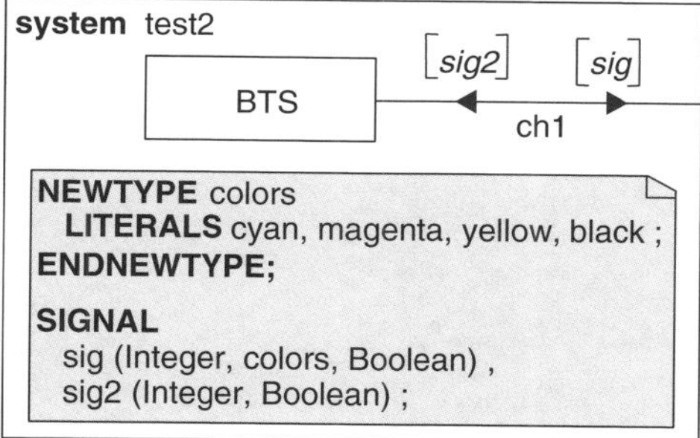

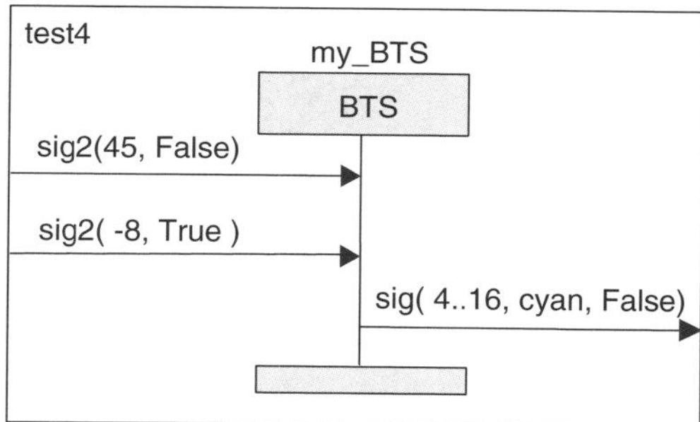

Figure 5.67 MSC *test4* observing the SDL model *test2*

5.3.4.5 The MSC attributes search and verify

According to the kind of properties you want to check and the kind of SDL model you use, you can select one of the following MSC attributes for the simulation (which are an MSC extension specific to ObjectGeode):

- *search*: detects behaviors conforming to the MSC
- *verify*[16]: like *search*, plus detects behaviors nonconforming to the MSC.

Generally, it is better to use *verify* because the Simulator will not only detect if your model conforms to the MSC but will also indicate when it encounters errors[17]. However, there are some subtle differences that we will illustrate in the following lines.

To change the attributes of an MSC, load it into the Editor, select it and do *Edit > MSC Simulation Properties*: the window shown in Figure 5.68 appears. The last three choices in the *Goal* part (*Complete test purpose...*) are reserved for Test Composer, the test case generator.

Figure 5.69 shows the simulation results with the basic observer MSC *test_seq*. It expects output of signals *sA*, *sB* and *sC*.

If we set the simulation attribute of the MSC observer to *search*:

- the first occurrence sequence of *sA*, *sB*, *sC* is detected as a success;
- the two consecutive *sA*, *sA* do not provoke any error;
- the second occurrence of the sequence *sA*, *sB*, *sC* is also detected as a success.

When using *search*, the simulator performs a loopback in the observer MSC to detect several occurrences of the MSC behavior in the simulation.

If we set the simulation attribute of the MSC observer to *verify*:

- the first occurrence of the sequence *sA*, *sB*, *sC* is detected as a success;
- the two consecutive *sA*, *sA* provoke an error;
- then, at the end of the last sequence *sA*, *sB*, *sC*, no other success is detected.

---

[16] Do not confuse this MSC attribute with the name of the command used to run the exhaustive simulation.
[17] In previous versions of ObjectGeode, *verify* MSCs only detected errors.

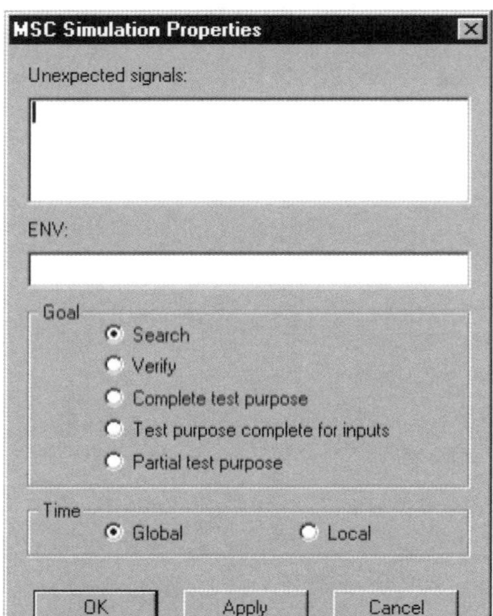

Figure 5.68 The Editor window showing the MSC simulation properties

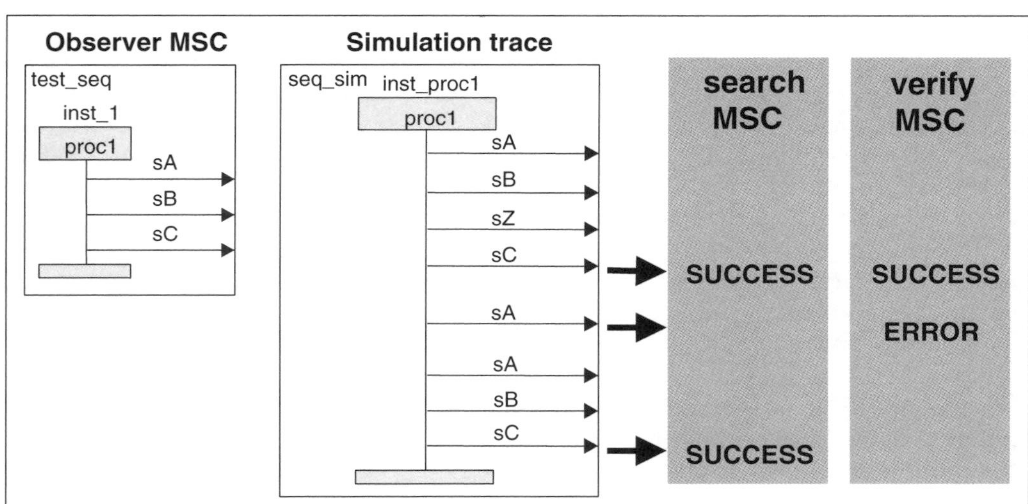

Figure 5.69 Simulation results with the basic MSC *test_seq*

Figure 5.70 shows the simulation results with the hierarchical observer MSC *test_loop*. The observer is the same as in Figure 5.69 (it expects output of *sA*, *sB* and *sC*), but it has been inserted under the *repeat* operator *test_loop*: the sequence *sA*, *sB* and *sC* can now occur, and be repeated 0 to *n* times.

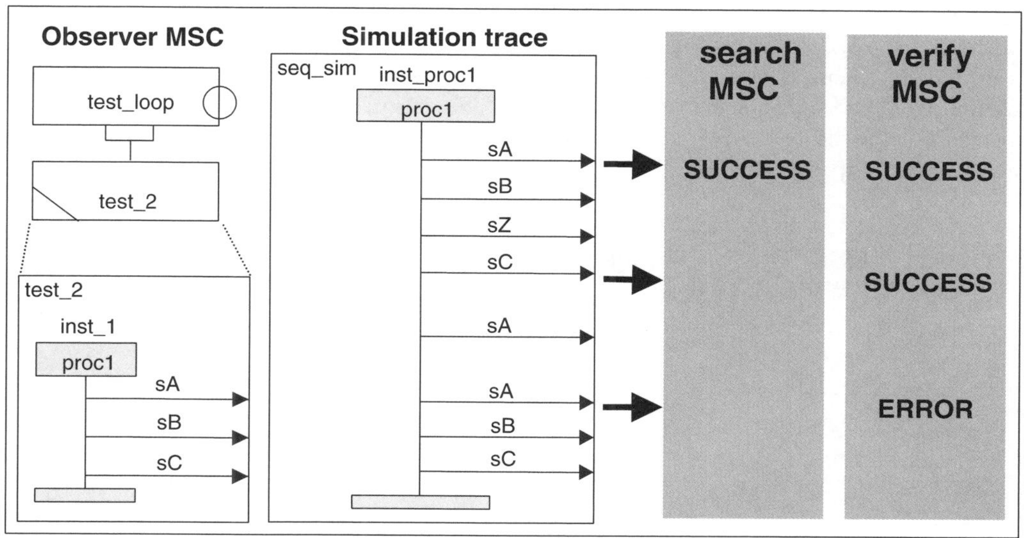

Figure 5.70 Simulation results with the hierarchical MSC *test_loop*

If we set the simulation attribute of the MSC observer to *search*:

- the first output of signal *sA* is detected as a success: this corresponds to the execution of the repeat operator 0 times;
- then, no other success is observed.

It means that the *repeat* operator must be avoided on top of a hierarchy of MSCs when using *search*; but the *repeat* is not necessary here because the simulator performs a loopback in case of *search*, to detect several occurrences of the MSC behavior in the simulation.

If we set the simulation attribute of the observer MSC to *verify*:

- the first output of signal *sA* is detected as a success: this corresponds to the execution of the repeat operator 0 times;
- then at the end of the sequence *sA*, *sB*, *sC*, another success is detected (repeat one time).
- if a third *sA* is simulated after the second, an error is detected (because *sA*, *sA* has been observed, instead of *sA*, *sB*).

To simplify, we have used examples of observation with MSC using interactive simulation, but errors and successes are also detected during random or exhaustive simulation.

5.3.4.6 Unexpected signals

In Figure 5.69, you see in the simulation trace that the output of signal *sZ* is ignored: this is because *sZ* is not present in the observer MSC *test_seq*. The only signals monitored in this case are *sA*, *sB*, *sC*[18].

[18] In Tau SDL Suite, the Validator considers by default that signals not present in the MSC must also be monitored: in the present example, if an output of *sZ* occurs during the simulation, the Validator will report an MSC violation.

To monitor signals that are not in the observer MSC, you can add them into the field *Unexpected signals* shown in Figure 5.68.

For example, here, if you type *sZ* in the field *Unexpected signals* and recompile the model plus the MSC, you get the behavior depicted in Figure 5.71: after *sA* and *sB*, the observer expected *sC*, but *sZ* has occurred, which is not ignored here, and provokes an error; then no other success happens.

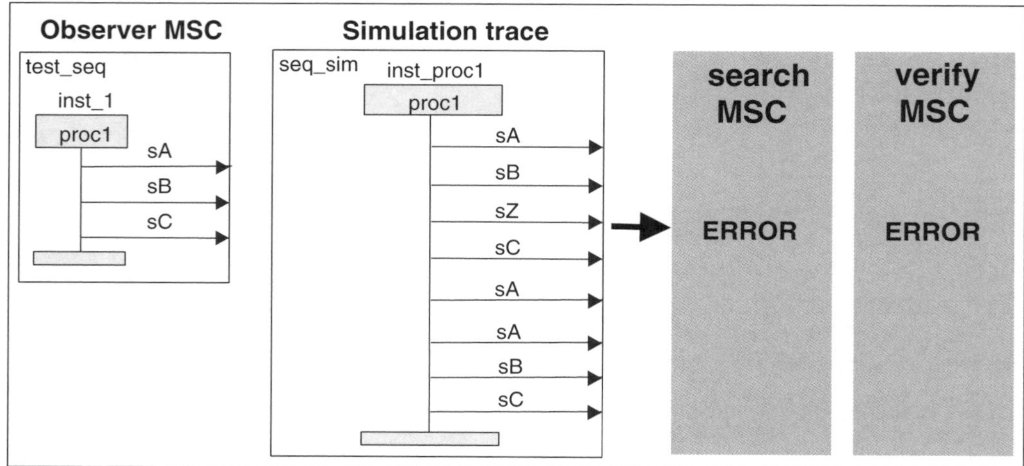

Figure 5.71 Simulation results with *sZ* as *unexpected signal*

5.3.4.7 Time local or global

In MSC simulation properties shown in Figure 5.68, you see two options concerning time: *global* or *local*.

If set to *global*, the ordering of events is global to all the instances present in the observer MSC. For example, in Figure 5.72, if you simulate the SDL sequence *sA*, *sC* and *sB*, you get an error (or you do not get a success), because *sB* was expected before *sC*.

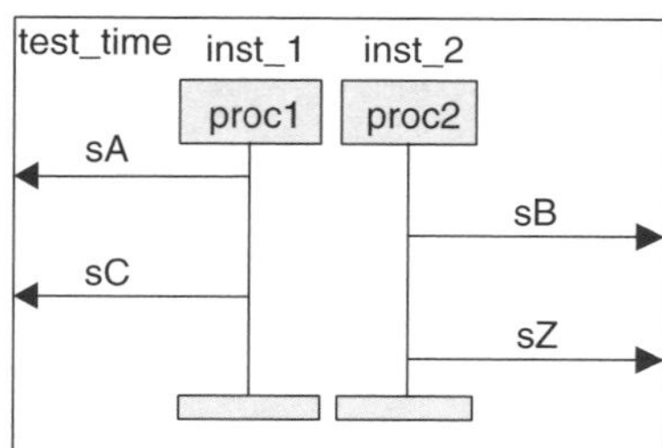

Figure 5.72 Observer MSC with two instances

If set to *local*, the ordering of events is local to each instance present in the observer MSC. In the same example, if you simulate the SDL sequence *sA*, *sC* and *sB*, you do not get an error (or you get a success), because each instance *proc1* and *proc2* are observed separately. The only things checked are that *sC* occurs after *sA* (you get a success) and that *sZ* occurs after *sB* (you get another success).

5.3.5 Simulate with GOAL observers

You will create and simulate a GOAL observer checking that the DLC establishment works, that is, after an establishment request an establishment confirmation occurs. In other words, the observer must check that:

- block *DLCa* in our model receives signal *L_EstabReq*,
- signal *L_EstabConf* is transmitted by block *DLCa* and
- the parameter of *L_EstabConf* (the number of the created DLC) is equal to the parameter of *L_EstabReq* (the number of the DLC to create).

This could have been checked by the MSC shown in Figure 5.73, except that it only works for one parameter value: one MSC might be used for 0, another for 1 and so on.

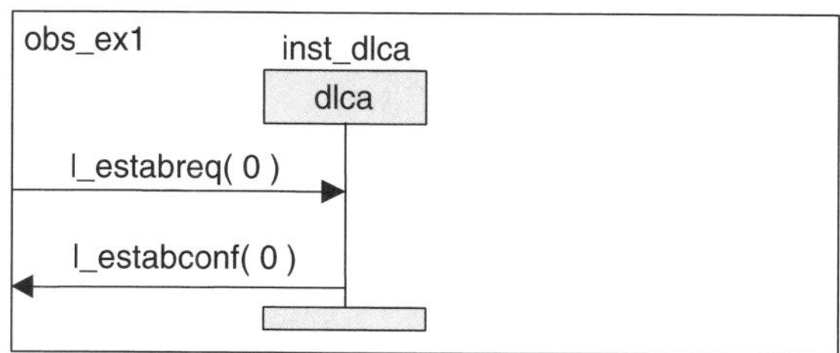

Figure 5.73 An observer MSC cannot memorize values

5.3.5.1 Create the GOAL observer

A. In the Editor, select *File > New > GOAL Observer*. Select *untilted1.obs*, do *File > Save As* and enter *obs_ex2.obs*.

B. Press the *Observation* palette button and name the observation *v76test*.

C. Press the *Observer* palette button to create an observer and name it *obs1*.

D. Select the observation *v76test* and press the *Text* palette button: this creates a *PR Declaration*. Your Framework view should look like Figure 5.74.

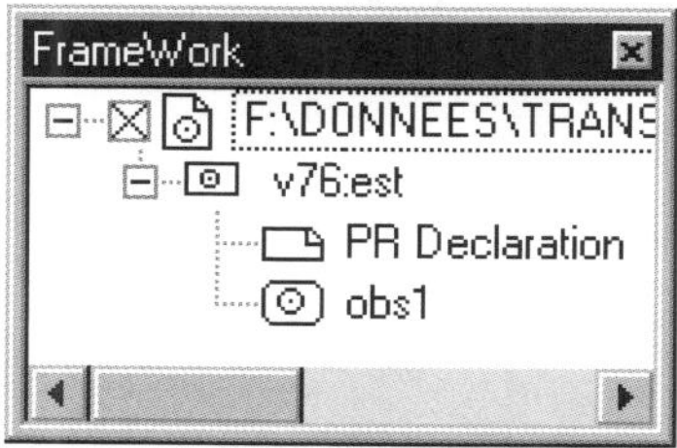

Figure 5.74 The Editor Framework view

E. Double-click the *PR Declaration* and enter:

```
probe DLC_a v76test!DLCa;
```

This creates the probe *DLC_a* pointing toward the block *DLCa* in the SDL system *v76test*. The observer *obs1* will use this probe to access the contents of the block.

F. Double-click on *obs1*: the Editor opens the observer *obs1*, empty.

G. Enter the observer shown in Figure 5.75, and save it.

An observer is similar to an SDL process, except that:

- the transitions are not triggered by incoming signals but by the events occurring in the SDL model (WHEN),
- the ERROR and SUCCESS states must be defined to indicate to the Simulator which states are expected and which are unexpected.

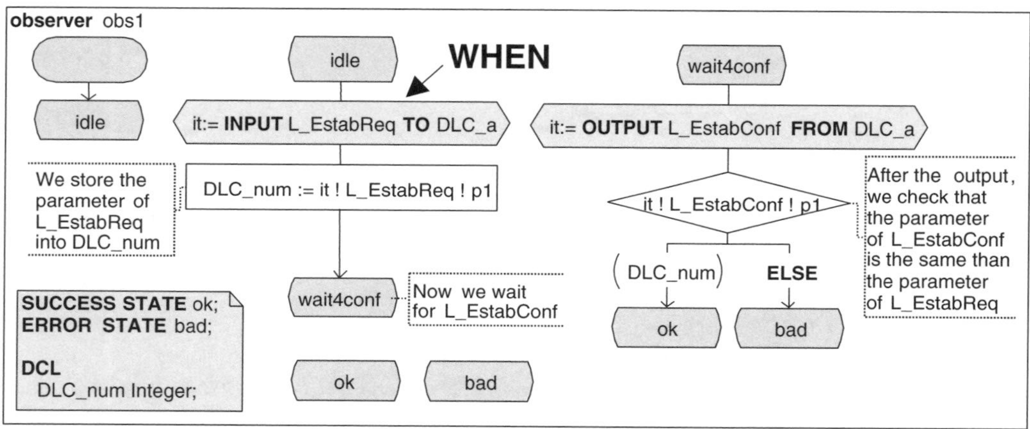

Figure 5.75 The GOAL observer *obs1*

In Figure 5.75, the self-declared variable *it* is used like a hook to memorize the INPUT or the OUTPUT in order to access the parameter of the signals.

The first WHEN could be translated as: when an input of signal *L_EstabReq* to probe *DLC_a* is observed, execute the transition until state *wait4conf*.

5.3.5.2 Compile the SDL model plus the GOAL observer

A. In the SDL Editor, load the V.76 SDL model, plus the GOAL observer contained in *obs_ex2.obs*.

B. Unload any other files from the SDL Editor, and quit (do NOT minimize) the ObjectGeode Launcher if running.

C. In the SDL Editor, select *Tools > SDL & MSC Simulator*. The ObjectGeode Launcher appears: check that its left area only contains *v76.pr* and *obs_ex2.obs*.

D. Press the *Build* button: this checks your SDL model and the GOAL observer and translates them into C code.

5.3.5.3 Replay the simulation scenario test1

A. Press the *Execute* button to start the Simulator. The Simulator main window appears. As opposed to observer MSCs, GOAL observer cannot be tracked.

B. In the Simulator, press the *Watch* button, then press *States*: as depicted in Figure 5.76, the state of the observer *obs1* is displayed.

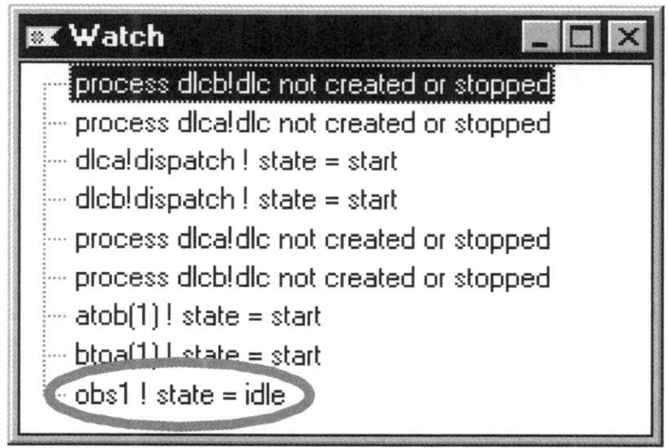

Figure 5.76 The state of observer *obs1*

C. Press the *Start MSC* button.

D. Replay a Simulation scenario: select *File > Scenario > Load*, choose *test1.scn*, and press the *Redo: All* Simulator button: at Step 15, the simulation should stop, and display:

```
stop condition:  not verifying() and success(obs1)
```

This means that observer *obs1* has reached a success state: in the watch, you see that the state of *obs1* is *ok*. The expression *not verifying()* prevents the Simulator from stopping during an exhaustive simulation.

It proves that the SDL simulation performed matches the behavior expected in the observer MSC *obs1*: you can check in the MSC trace that block *DLCa* has received *L_EstabReq*, then that signal *L_EstabConf* has been transmitted by block *DLCa*, and that the parameter of *L_EstabConf*, *0*, is equal to the parameter of *L_EstabReq*.

5.3.6 More details on GOAL observers

GOAL observers can read and change any SDL variable, including the contents of process queues, and have also access to constants and types defined in the observed SDL model. One *.obs* file can contain several GOAL observers.

5.3.6.1 Ordinary and event GOAL observers

By default, only one transition in each GOAL observer is executed after each SDL model transition. For example, the observer in Figure 5.78 will not detect any success in the SDL model represented in Figure 5.77, because the SDL model outputs the signals *orange* and *red* in the same transition; then the Simulator executes one transition only in the observer *obs1*: it goes from state *waitOrange* to state *waitRed*. State *good* is not reached, signal *red* is not detected by the observer.

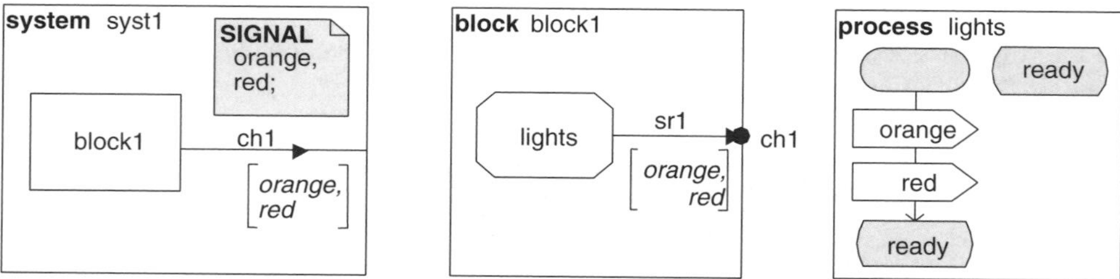

Figure 5.77 Model with two consecutive outputs

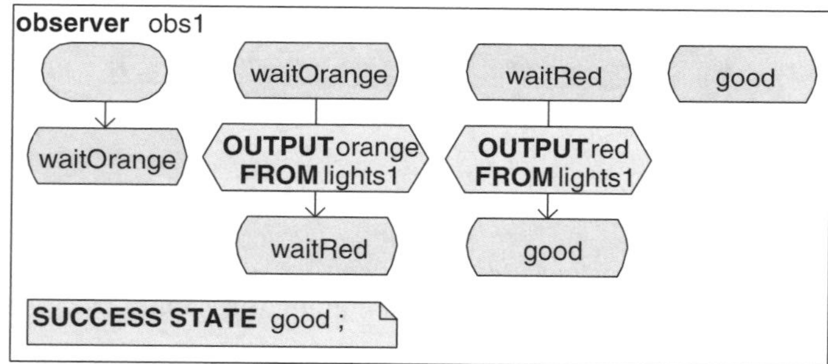

Figure 5.78 Ordinary GOAL observer: no success

The following probe is declared to access instance 1 of process *lights*:

```
PROBE lights1 syst1 ! block1 ! lights(1);
```

If we transform the ordinary observer into an event observer, after each SDL model transition several transitions in each observer are executed. Now the observer in Figure 5.79 will detect a success in the SDL model represented in Figure 5.77, because the Simulator executes two transitions in the observer *obs1*: it goes from state *waitOrange* to state *waitRed* and then to state *good*.

To enter the keyword *EVENT*, type it before *obs1* in the Framework view.

Remark: the observers generated when compiling observer MSCs are event observers.

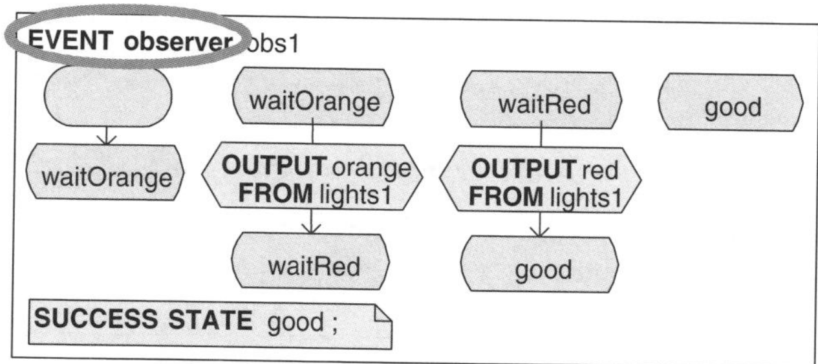

Figure 5.79 Event GOAL observer: success reached

5.3.6.2 *Using a GOAL observer to detect the execution of a transition*

The second transition in process *lights*, as shown in Figure 5.80, is labeled *tr1*. The GOAL observer *obsTrans* represented in Figure 5.81 detects the execution of transition *tr1*. Then, if *x* is not equal to 0, the observer goes to state *good*.

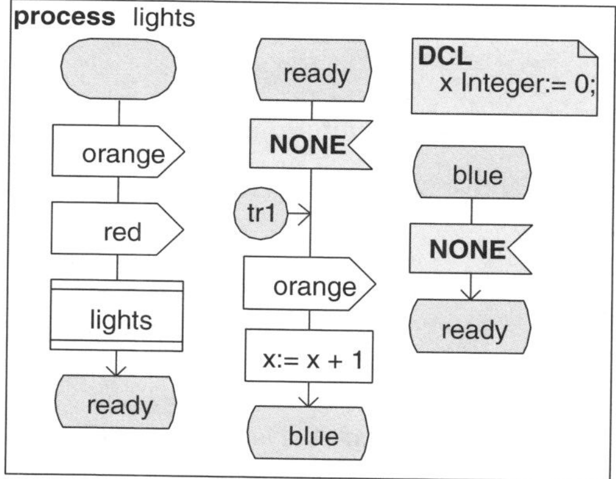

Figure 5.80 Process *lights* modified

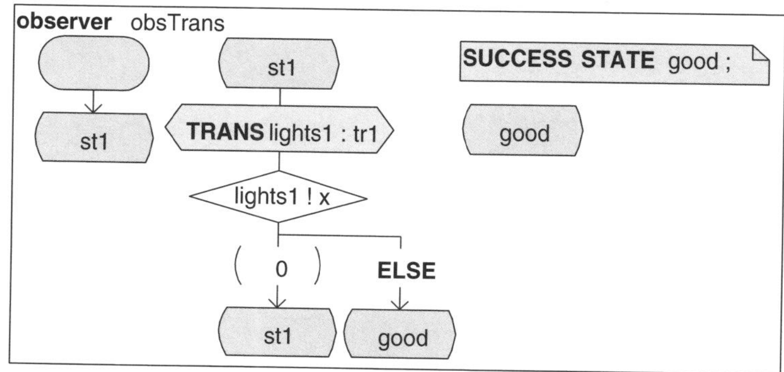

Figure 5.81 Detecting execution of transition *tr1*

Remember that the following probe is declared to access instance 1 of process *lights*:

```
PROBE lights1 syst1 ! block1 ! lights(1);
```

5.3.6.3 Using a GOAL observer to detect a value and modify an SDL variable

The GOAL observer *obsCond* represented in Figure 5.82 detects if the variable x in process *lights* (see Figure 5.80) is > 0, and puts -99 into x.

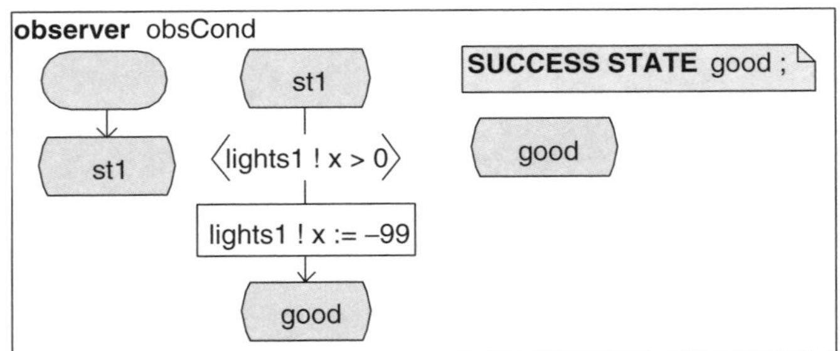

Figure 5.82 Detecting $x > 0$ and changing x

5.3.6.4 Using a GOAL observer to test the state of a process

The GOAL observer *obsState* represented in Figure 5.83 detects if the state of process *lights* (see Figure 5.80) is equal to *blue*.

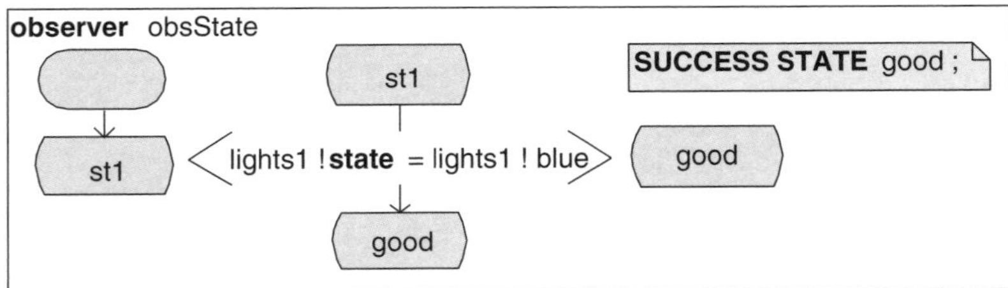

Figure 5.83 Testing the state of a process

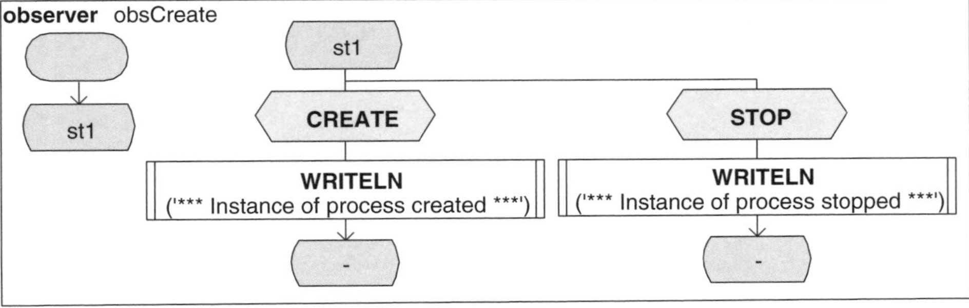

Figure 5.84 Detecting process creation and stop

Remember that the following probe is declared to access instance 1 of process *lights*:

```
PROBE lights1 syst1 ! block1 ! lights(1);
```

5.3.6.5 Using a GOAL observer to detect process creation and stop

The GOAL observer *obsCreate* represented in Figure 5.84 writes a message if there are any creation or stop of process instances in the SDL model. It is also possible to know which process has been created.

6

Random Simulation

6.1 PRINCIPLES

In the previous chapters, we have simulated interactively: when there was a choice between several SDL transitions to execute, the user selected manually the desired transition. For example, the user could send either an *L_DataReq* to transfer more data or an *L_ReleaseReq* to release a connection. In order to perform automatic simulations to test the maximum of model behaviors, the simulators provide an automatic mode, which we generically call random, where the choices between several firable transitions are made randomly. The simulation runs, executing thousand or millions of transitions, to test the SDL model quickly. When an error is encountered, such as a range overflow or a bad behavior detected by an observer, the simulation stops, and the user gets the error message and can use the debugging features of the simulator to understand it.

6.2 CASE STUDY WITH TAU SDL SUITE

In Tau SDL Suite, random simulation mode is available in the Validator, not in the Simulator.

6.2.1 Random simulation without observers

You will run the random simulation on the V.76 SDL model to discover errors automatically.

6.2.1.1 Run the random simulation

A. If you added an observer process to the model as specified in Chapter 5, go back to the version without observer process: in the Organizer, select *V76test*, choose *Edit > Connect*, select *To an existing file*, press the folder-shaped icon and connect to the file *v76test.ssy*. Also remove block *obs* if it remains in the Organizer (but do not delete its file).

B. In the Organizer, press the *save* button, select the SDL system *V76test* and do *Generate > Make*.

C. In the SDL Make window, select *Microsoft Validation* or *Borland Validation*, and press *Full Make*.

D. In the Organizer, press the *Validate* button to start the Validator.

E. In the Validator, press the *Navigator* button.

Validation of Communications Systems with SDL: The Art of SDL Simulation and Reachability Analysis.
Laurent Doldi © 2003 John Wiley & Sons, Ltd ISBN: 0-470-85286-0

F. In the Validator command line, enter:

```
Random-Down 500
```

The Validator performs a random simulation (limited to 500 transitions), and answers (you may get a different number):

```
No down node after 31 steps
```

If you look at the Navigator, as shown in Figure 6.1, you see why the simulation stopped so early:

```
Warning: implicit signal consumption of V76frame etc.
```

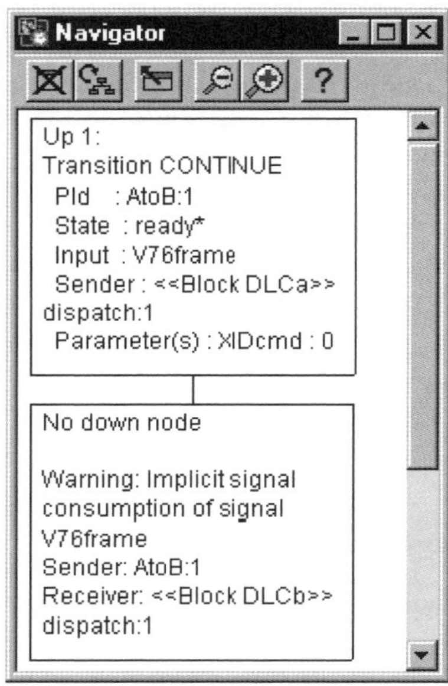

Figure 6.1 After 31 random simulation steps

If you did not get the same error, repeat the sequence *Top, Random-Down 500*. This error means that when signal *v76frame* has been transmitted to process *dispatch* in block *DLCb, dispatch* was not in a state where it could input (or save) the signal.

G. In the Validator, select *Commands > Toggle MSC Trace*: the MSC trace shows the simulation executed by the Validator.

6.2.1.2 Analyze the error

To understand the bug, you will search in which state process *dispatch* (in block *DLCb*) was when process *DLC* transmitted to it the signal *DLCstopped*.

A. In the Validator, select *View > Command Window*: as shown in Figure 6.2, the state of each process instance is displayed. You see that process *dispatch* in block *DLCb* is in state *waitParmResp*.

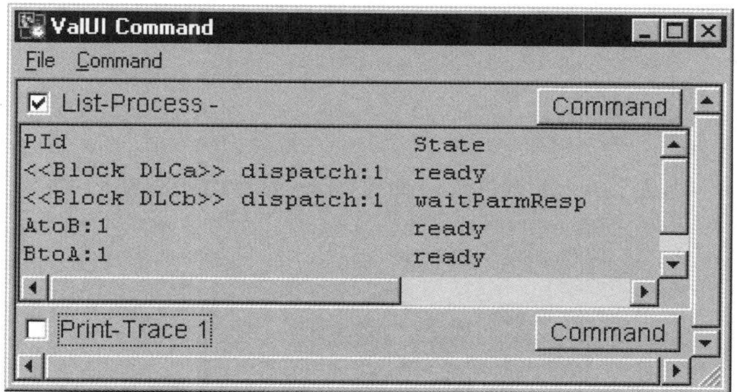

Figure 6.2 Process states in the *Command* window

B. In the Organizer, double-click on process *dispatch*: in the page *part1*, you see that in state *waitParmResp*, the only input is *L_SetParmResp*; therefore, when a *V76frame* is first in the process queue, it is discarded.

The error is also easy to understand by looking at the MSC trace. We will not correct this bug, because we will learn how to find it with exhaustive simulation.

6.2.2 Multiple random simulations

The random simulation algorithm used in the Validator is based on a pseudorandom number called *seed*. The initial default *seed* value, 1, can be changed using *Define-Random-Seed*.

At each random simulation step, the Validator executes a transition selected among the firable transitions according to a random number. For example, if there are two firable transitions, depending on the random number, the first or the second transition is executed.

In the previous section, we have used the textual command *Random-Down*, which performs a single-shot random simulation. To simulate different scenarios in order to hit more errors, you could repeat this simulation several times by returning to the model's initial state (button *Top*) and entering again the command *Random-Down*.

To simplify your work, the Validator command *Random-Walk* performs this repetition automatically. To illustrate this:

A. Restart the Validator.

B. Select *Options1 > Show Options*. The Validator displays:

```
...
Random walk options
Search depth : 100
Repetitions  : 100
...
```

It means that maximum 100 transitions will be executed from the starting step and that the random simulation will be repeated 100 times. It is similar to 100 times the sequence:

```
Random-Down 100
Top
```

C. To stop the simulation after the first error, enter in the Validator command line (can be abbreviated to *d-r-a -*):

```
Define-Report-Abort -
```

D. Press on *Random Walk* to start the random simulation: the Report Viewer appears. Double-click on *1 ImplSigCons* to expand the report, as shown in Figure 6.3. You may hit a different error.

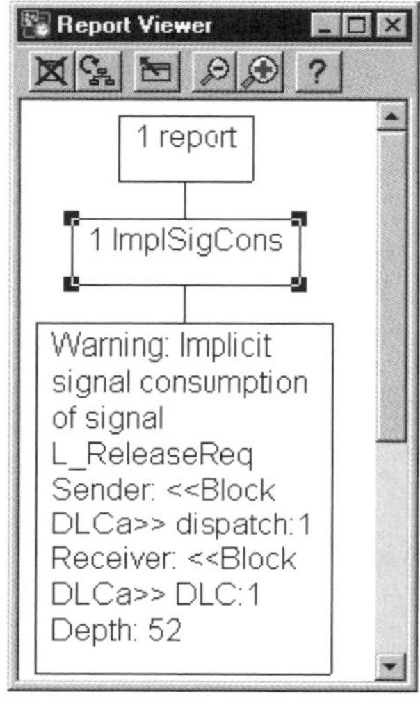

Figure 6.3 The Report Viewer after the random walk

E. Double-click on the warning in the Report Viewer to generate the MSC trace corresponding to the error: signal *L_ReleaseReq* has been transmitted to process *DLC* in block *DLCa*, which is unfortunately in state *waitUAdisc* expecting another signal.

The Validator had also displayed the symbols coverage rate at the end of the random walk:

```
** Random walk statistics **
No of reports: 1
Gen states    : 1093
Max depth     : 100
Min depth     : 100
Symbol coverage :  81.54
```

More details on coverage can be found in the Coverage Viewer (*Commands > Show Coverage Viewer*).

6.2.3 Random simulation with observers

You will run the random simulation on the V.76 SDL model monitored by an observer MSC, to automatically check that the simulated behavior complies with the expected behavior.

A. Restart the Validator, without saving the options.

B. Enter in the Validator command line:

```
Load-MSC test1.msc
```

Now the simulation is driven by the MSC *test1*: the signals transmitted to the SDL model and the value of their parameters are those present in the MSC rather than the Validator default test values.

C. Press on *Random Walk* to start the random simulation: after a few seconds, the Report Viewer appears. The Validator has discovered 1 deadlock, 2 MSC violations and 1 MSC verification scenarios. If you do not get any MSC Verification, you can press again on *Random Walk*. Double-click on *2 MSCViolation* and on *1 MSCVerification* to expand the reports, as shown in Figure 6.4.

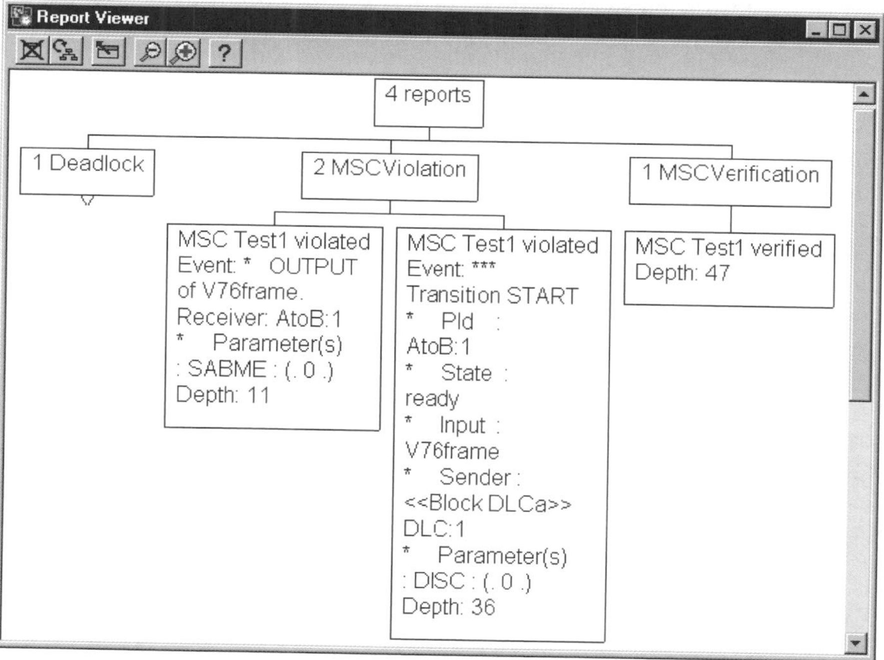

Figure 6.4 The reports after the random walk

D. Double-click on the left MSC violation in the Report Viewer to generate the corresponding MSC trace: timer *T320* has timed-out, therefore signal *V76frame* has been retransmitted; this scenario does not match the MSC *test1*.

E. Double-click on the MSC verification in the Report Viewer to generate the corresponding MSC trace: the MSC Editor displays a scenario identical to *test1*.

Remark: the button *Verify MSC* loads an MSC and starts a bit-state exploration (or a tree-search) rather than a random walk. *Verify MSC* is explained in Chapter 7.

6.3 CASE STUDY WITH OBJECTGEODE

6.3.1 Random simulation without observers

You will run the random simulation on the V.76 SDL model to discover errors automatically.

6.3.1.1 Run the random simulation

A. In the SDL Editor, unload all files except *v76.pr*, and select *Tools > SDL & MSC Simulator*.

B. In the ObjectGeode Launcher, remove any file other than *v76.pr*, press the *Build* button, then if you do not get any SDL errors, press the *Execute* button. The Simulator starts.

C. If the Simulator has not executed automatically the four start transitions (step should be equal to 4), the file *v76.startup* is missing (see Chapter 4).

D. Check that the feed commands (loaded by the file *v76.startup*) have been executed. See Chapter 4 if typing the Simulator command *list feed* does not give the following result:

```
> list feed
feed dlcbsu l_releasereq(0)
feed dlcbsu l_setparmresp()
feed dlcbsu l_estabresp()
feed dlcasu l_datareq(1 , 39)
feed dlcasu l_datareq(0 , 86)
feed dlcasu l_releasereq(1)
feed dlcasu l_setparmreq()
feed dlcasu l_estabreq(1)
feed dlcasu l_estabreq(0)
```

E. Select *Edit > Stop conditions*, enter *rstep* = 500 as shown in Figure 6.5, then press *Add* and *Close*. *r* in *rstep* stands for relative. It asks the Simulator to stop after the execution of 500 transitions from the current step number; here, stop at 4 + 500 = 504. The condition *step = 500* would have meant stop at step = 500: when starting from Step 650, for example, the simulation would never stop.

F. Press on the *go* button to start the random simulation (or type the command *go*): the textual trace displays the step number and the executed SDL statements. The simulation runs much faster if the trace is removed (button *Traces: Off*).

G. At Step 261, a message indicates (you may not hit the same error):

```
Unexpected signal v76frame in dlcb!dispatch, line 312 of v76.pr
```

It means that when signal *v76frame* had been transmitted to process *dispatch* in block *DLCb*, *dispatch* was not in a state where it could input (or save) the signal.

H. Save the Simulator scenario leading to the bug: in the Simulator, select *File > Scenario > Save As*, enter *random1* and press *save*.

Do not exit from the Simulator.

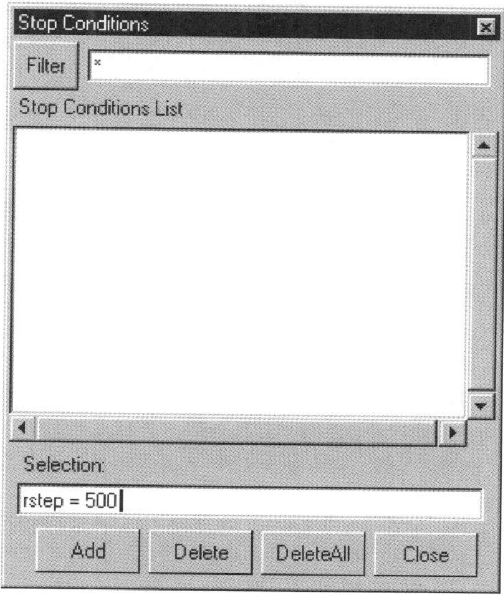

Figure 6.5 Adding the stop condition

6.3.1.2 Analyze the error

To understand the bug, you will search in which state process *dispatch* was (in block *dlcb*) when process *DLC* transmitted to it the signal *dlcstopped*.

A. In the Simulator, press the button *Watch* and select *States*: a watch appears, shown in Figure 6.6, displaying the state of each process instance. You see that process *dispatch* in block *dlcb* is in state *waitparmresp*.

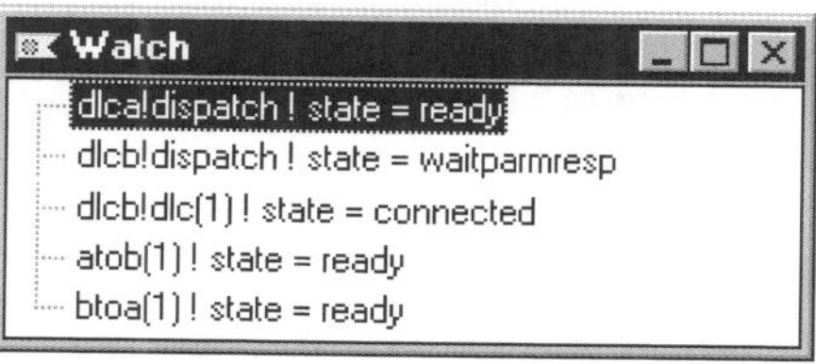

Figure 6.6 The watch window

B. In the Editor, open the partition *part1* of process *dispatch*: you see that in state *waitparm-resp*, the only input is *L_SetParmResp*; therefore, when a *v76frame* is first in the process queue, it is discarded.

   If required, the MSC trace corresponding to the bug can be generated by pressing the buttons *Start MSC, undo* ⏮ and then *redo* ⏭.

   We will not correct this bug, because we will learn how to find it with exhaustive simulation.

6.3.2 Multiple random simulations

The random simulation algorithm used in the Simulator is based on a pseudorandom number called *seed*. The initial default *seed* value (619430284 in the version used) can be changed using *Edit > Configuration*. At each random simulation step, the Simulator:

- executes a transition selected among the firable transitions according to the current *seed* value, (for example, if there are two firable transitions, depending on the *seed* value, the first or the second transition is executed),
- computes the new *seed* value function of the current *seed* value.

It means that the *seed* changes at every simulation step. Also, if you go back (*undo* or *init* commands) to previous simulation steps, *seed* does not return to its previous value: this is to simulate different branches easily. Otherwise, using the same *seed* from the same model's state would simulate the same scenario.

To illustrate this:

A. Quit the Simulator, start the Simulator as indicated in Section 6.3.1.1 and enter the stop condition *rstep = 500*.

B. Press on [icon] to start the random simulation: at Step 261 (you may get another number), the simulation stops (unexpected signal).

C. Press on *init* ⏮, four times on *redo* ⏭ and then on [icon]: this time, the simulation stops at Step 264 (you may get another number).

D. Repeat the same sequence: this time, the simulation stops at Step 48 (you may get another number).

The fact that the random simulation starting from the same SDL model state stops at Step 261, 264 or 48 means that different scenarios have been simulated, because the *seed* was always different (computed at every step, then never reinitialized during init).

To automatically perform the previous random simulations, we are going to write a script.

E. With a text editor, type the following lines and save the file as *rand.wri*:

```
init  -- return to step 0
source start.scn -- execute the 4 start
untrace all -- remove textual traces
define trace_stmt 'false' -- remove PR traces
-- Protection against double declaration of sim_ok:
if '$vars_declared' = ''
  define vars_declared 'Yes'
  dcl sim_ok Boolean
fi

let sim_ok = true
while sim_ok
```

```
    print seed
    go until step = 500 -- random simulation
    if step /= 500
      let sim_ok = false
    fi
    print step
    init
    redo 4 -- play again the 4 starts
  endwhile
```

F. In the Simulator, select *Edit > Configuration* and uncheck the box *Trap unexpected signals*; otherwise our script would stop during the first iteration.

G. In the Simulator, type the command:

```
source rand.wri
```

The Simulator performs four times the while loop (you may get different results), as shown in Figure 6.7: three times without discovering any error before Step 500, then one more iteration where an error is encountered at Step 60.

The error encountered is: *No receiver for output l_releasereq from dlcb!dispatch via dlcs to dlcb!dlc(1).* It means that instance 1 of process *DLC* in block *DLCb* was stopped before the signal *L_ReleaseReq* was transmitted to it by process *dispatch.*

As performed previously, you could press the *Traces: Defaults* button to turn the trace on (it was removed by the script to speed up the simulation), and use the *undo* command to get the MSC trace to understand the error and fix it.

If you launch again the script *rand.wri* without exiting from the Simulator, the seed values will be different, and the number of while loops performed will change. If the simulation never terminates (no error is encountered), you can stop it by pressing the *halt* button.

To see the states, transitions and basic blocks coverage obtained by the script execution, you could type, respectively:

```
print state_cover_rate(system)
print trans_cover_rate(system)
print bb_cover_rate(system)
```

The Simulator answers the following percentages:

```
state_cover_rate(v76test) = 100.00
trans_cover_rate(v76test) = 93.10
bb_cover_rate(v76test) = 89.28
```

More details on coverage can be found in the Hierarchy Browser (*View > Hierarchy*).

6.3.3 Random simulation with observers

You will run the random simulation on the V.76 SDL model monitored by an observer MSC, to automatically check that the simulated behavior complies with the expected behavior.

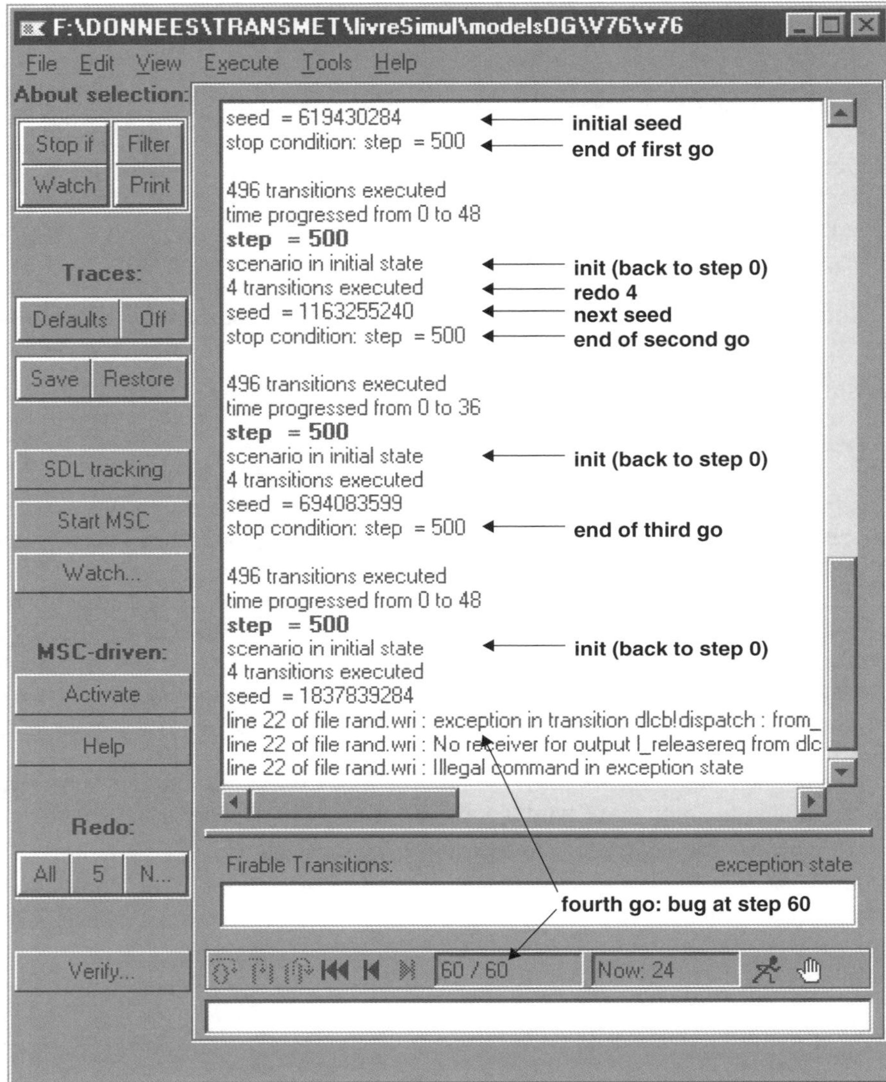

Figure 6.7 The simulation trace of the script *rand.wri*

6.3.3.1 Random simulation not guided

A. Quit the Simulator and the ObjectGeode Launcher.

B. With a text editor, open the file *v76.startup* and add the comment delimiter -- as below to prevent the feed commands execution (because when an MSC is compiled with the SDL model, feeds are automatically created, making duplicates):

```
-- source v76_feed.wri
```

C. In the SDL Editor, unload all files except *v76.pr*, and load the file *test1.msc*.

D. Select *Tools > SDL & MSC Simulator*.

E. In the ObjectGeode Launcher, remove any file other than *v76.pr* and *test1.msc*, press the *Build* button, then, if you do not get any errors, press the *Execute* button. The Simulator starts.

F. In the Simulator enter the stop condition *rstep = 1000*.

G. Press the button *Traces: Off* to speed up the simulation.

H. Press on [go button] to start the random simulation: at Step 46 (you may get a different number), the simulation stops (unexpected signal).

I. Press on *init* ⏮, four times on *redo* ⏭ (to skip the start transitions) and then on the *go* [go button] button: this time, the simulation stops at Step 59 (no receiver for output).

J. Repeat the same sequence several times: the simulation stops, respectively, at Steps 41, 56, 46, 69 and 319 (you may get different numbers).

K. You could continue the sequence *init; redo 4; go* many times, without reaching the observer MSC's success state.

If you look at the MSC *test1*, this is normal: the probability to randomly simulate the successful connection, followed by an XID, then a data transfer and finally a connection release are extremely low.

In our V.76 SDL model, there are several possible cascading simulation choices, as in a complex maze. Let's count the probability to verify the MSC:

- Step 4: choice between sending *L_DataReq*, *L_EstabReq*, *L_SetparmReq* or *L_ReleaseReq* to the SDL model; therefore, four possibilities,
- Step 5: choice between sending *L_ReleaseReq* or starting *DLC* process instance,
- Step 6: choice between sending *L_ReleaseReq*, timer *T320* or input of *v76frame*,
- etc.

If we multiply all the number of possibilities offered at each step, we get:

- $4 \times 2 \times 3 \times 2 \times 3 \times 2 \times 3 \times 3 \times 2 \times 3 \times 5 = 77\,760$ to establish the connection (reception of *L_EstabConf*),
- followed by $4 \times 4 \times 2 \times 5 \times 4 \times 5 \times 2 \times 5 = 32\,000$ for the XID exchange,
- followed by $4 \times 4 \times 4 \times 2 \times 5 = 640$ for the data transfer,
- followed by $4 \times 4 \times 5 \times 2 \times 5 \times 4 \times 5 \times 2 \times 6 \times 4 \times 3 \times 4 = 9\,216\,000$ for the connection release.

Therefore, the probability to simulate the expected behavior *test1* in this configuration can be estimated at: $1/77\,760 \times 32\,000 \times 640 \times 9\,216\,000 = 1/14.677.000.000.000.000.000$.

As certain choices are equivalent, the probability is not so low, but certainly too small to verify the MSC, even performing years of random simulation.

This is why we must guide the simulation to reduce the number of possibilities at each simulation step, as explained in the next section.

6.3.3.2 MSC-driven random simulation

In a way similar to Tau SDL Suite Validator, we will perform a simulation guided by the MSC *test1*.

A. Quit the Simulator and the ObjectGeode Launcher.

B. With a text editor, check that the file *v76.startup* contains a comment delimiter -- as below in front of the feed source line:

```
-- source v76_feed.wri
```

C. In the SDL Editor, check that the only files loaded are *v76.pr* and *test1.msc*.

D. In the SDL Editor, select the MSC *test1* and choose *Edit > MSC Simulation Properties*: in the *Goal* part, select the option *Verify*. Press *OK* and save the MSC.

E. Select *Tools > SDL & MSC Simulator*.

F. In the ObjectGeode Launcher, remove any file other than *v76.pr* and *test1.msc*, press the *Build* button, then press the *Execute* button.

G. Check that the current simulation step is 4 (otherwise your startup file has not been executed). If not, execute the 4 start transitions.

H. Press the button *Traces: Off.*

I. Press the button *MSC-driven: Activate*: this adds the filter condition *filter error(observation)*, which removes the transitions leading to a violation of the observer MSC.

J. Choose *Edit > Configuration*, select the box *Reasonable environment* and deselect the box *Loose time progression.*

K. Select *Edit > Filter Conditions* and add successively:

```
trans atob(1) : decision_lose_the_frame('Yes')
trans btoa(1) : decision_lose_the_frame('Yes')
```

This prevents the Simulator from losing the frame in block *dataLink*.

L. Press on to start the random simulation: at Step 12 (you may get a different number), the simulation stops (deadlock).

M. Press on *init* ⏮, press four times on *redo* ⏭ (to skip the start) and then on : the simulation stops at Step 41, indicating *success(test1)*. It means that this time, the Simulator has selected the correct transitions, simulating the behavior expected by the MSC *test1*.

If you simulate manually with the same Simulator settings (filter etc.), you will see that most of the time there is only one transition to fire; therefore the probability to randomly play the expected scenario is high. If you remove *filter error(observation)*, the success is extremely difficult to obtain randomly.

6.3.4 Details on random simulation

6.3.4.1 Testing the simulator randomness

To know exactly the repartition of random choices performed by the Simulator, we have simulated the model shown in Figure 6.8. Each spontaneous transition increments a variable, to count the number of time it is executed by the Simulator.

After executing the first transition (start), the Simulator, as expected, proposes five firable transitions, as illustrated in Figure 6.9.

Then we remove the traces to speed up the simulation and enter the following command to run 1000 random simulation steps:

```
go until rstep = 1000
```

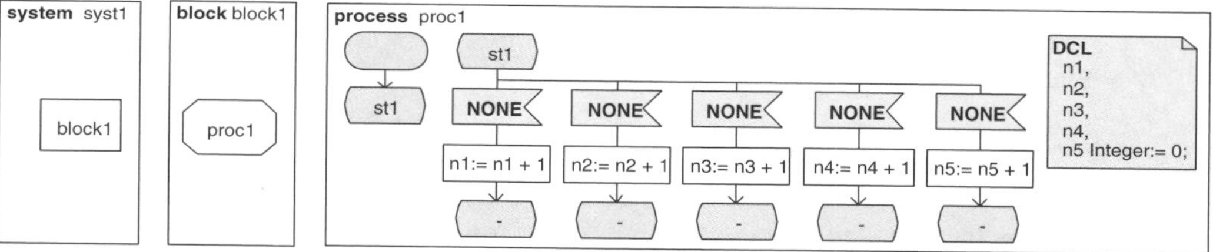

Figure 6.8 SDL model to test random choices

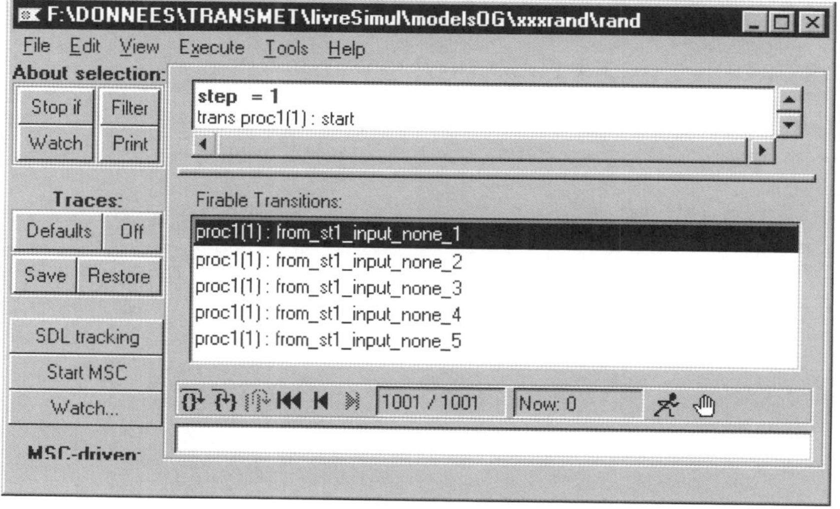

Figure 6.9 The five firable transitions

After the 1000 steps, the following results are obtained:

```
> print n1
proc1(1) ! n1 = 131
> print n2
proc1(1) ! n2 = 240
> print n3
proc1(1) ! n3 = 244
> print n4
proc1(1) ! n4 = 256
> print n5
proc1(1) ! n5 = 129
```

It shows that the random function used by the Simulator is not perfect as the first and last transitions have been executed half as frequently as the others.

6.3.4.2 Running very long random simulations

During simulation (except in exhaustive modes), the following information is stored:

- the scenario, to be able to go back to any step using undo: one line per transition executed,
- the textual traces: several lines per transition executed, stored into the file *model_name.log* and stored into the Simulator window in order to scroll up.

If you plan to run a very long random simulation (many million steps), you can select *Edit > Configuration* and uncheck the box *Scenario recording*. However, in such case, no *undo* or *redo* is possible. Concerning the textual traces, it is wise to turn them off; otherwise the log file will be huge and many lines will be stored in the window.

6.4 ERRORS DETECTABLE BY RANDOM SIMULATION

The errors that random simulation can detect are the same as those that are detectable by interactive simulation. The only difference is that random simulation runs automatically and reports the error.

7

Exhaustive Simulation

In this chapter, after a presentation of the exhaustive simulation algorithms and of the two simple examples, you will learn how to validate the V.76 SDL model using exhaustive simulation: automatically detecting bugs in a few seconds, detecting nonsimulated symbols, exploring millions of states in a few minutes, and using observers (stop conditions, rules, MSCs, processes or GOAL). Then more simulation algorithms are described (supertrace, liveness etc.), and a strategy is given to simulate SDL models with infinite or very large states graphs. Finally, the list of errors that can be detected by exhaustive simulation is presented, for the two Simulators used.

7.1 INTRODUCTION

This chapter concerns exhaustive simulation, where all the reachable states of an SDL model are computed, and also nonexhaustive algorithms such as bit-state, where some of the reachable states can remain unexplored. The reader easily understands that computing all the reachable states of an SDL model is not always possible, especially using exhaustive algorithms, because exhaustive simulation needs to store the global states of the model (or at least their hash-code) in RAM memory. The SDL model and the simulator must be tuned, to limit the memory used. For example, more priority can be given to internal SDL events.

When I was using exhaustive simulation on a Unix server around 1993, equipped with 256 MB of memory (RAM), some people did not believe that such a huge memory could exist. At the time of writing this book, 256 MB of memory on a PC cost only 30 euros or US$30: anybody can run exhaustive simulation on a cheap PC!

Exhaustive simulation in systems engineering can be compared to ultrasound scan in medicine: it reveals exactly what is inside a system, finding harmful behaviors early that would otherwise be detected too late. Exhaustive simulation can find security failures in network software and protocols before they are discovered and used by hackers to attack a system: for example, a simplified model of a PC connected to the Internet could reveal dangerous scenarios that could be fixed during the software design, instead of after several intrusions or virus attacks after product delivery.

7.1.1 Exhaustive simulation

The aim of exhaustive simulation is not only to execute all the SDL symbols at least once (static coverage) but also to execute all the behaviors of an SDL model (dynamic coverage): that means executing all the SDL transitions from all the global states. The global states are also

Validation of Communications Systems with SDL: The Art of SDL Simulation and Reachability Analysis.
Laurent Doldi ISBN: 0-470-85286-0

called the *reachable states*. The global states and the transitions between them are an oriented graph, called the *states graph*.

7.1.1.1 All the states must be stored

Figure 7.1 illustrates how exhaustive exploration works. Note that not all the required data structures are described. To execute all the behaviors of an SDL model, an SDL transition from a certain global state of the model is executed (1). The obtained global state of the SDL model is compared with each previously stored state (2). If none are identical to it, the new state *sD* is stored (3). From *sD*, an SDL transition is executed (4). Again, the obtained global state is compared with each previously stored state (5). If it is identical to *sC* (6), an edge is stored from *sD* to *sC* (7). The exploration stops when all the transitions from all the states have been executed.

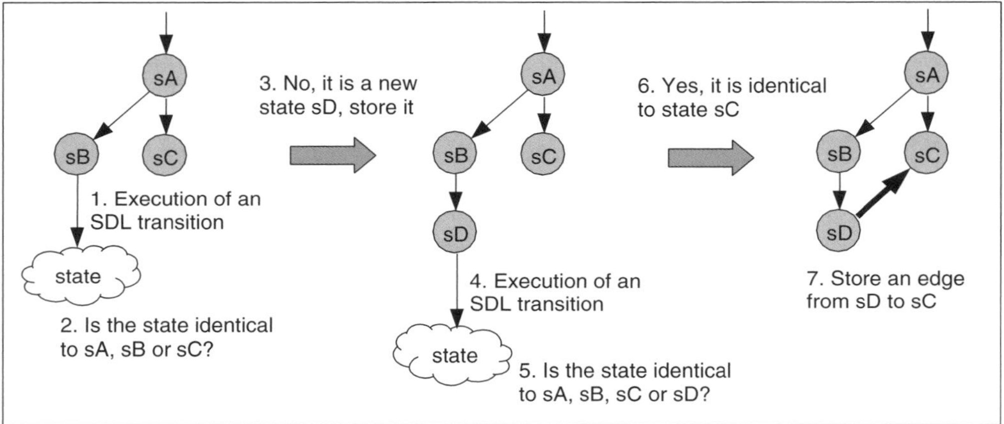

Figure 7.1 Exhaustive exploration of the states graph

Because all the global states of the SDL model must be stored in RAM (otherwise the comparison would be too slow), exhaustive exploration could require a huge amount of memory.

Fortunately, exhaustive algorithms do not need to store each whole SDL state, but, for example, cut each state in slices and store only the slices that have changed after a transition. In an SDL system, when a transition is executed, most of the times the only parts that change are the sender process and the receiver process instances; for a system with, for example, 10 process instances, only 2 instances will change between 2 states.

7.1.1.2 What is a global state

As shown in Figure 7.2, each global state is a snapshot of the SDL model. It contains the state of each process instance and of its input queue. The state of each process instance is composed of the state of the state machine, the value of each variable and timer and the value of the predefined expressions *self*, *offspring*, *sender* and *parent*. The state of each input queue is composed of the name of each signal present and of the value of each parameter of each signal.

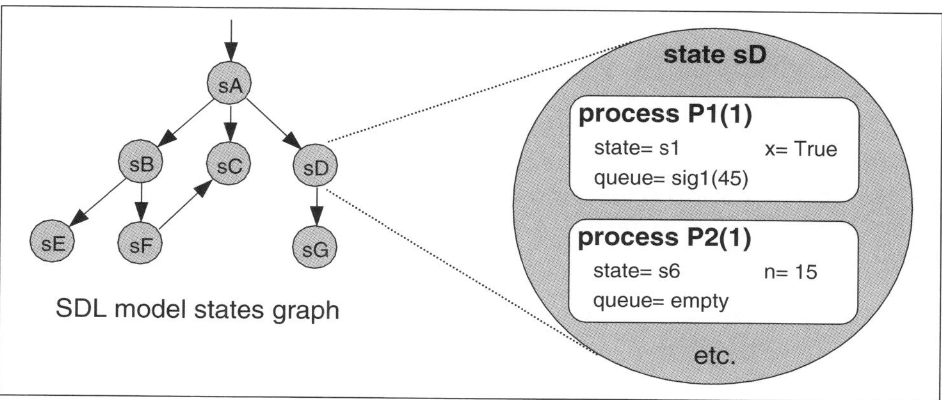

Figure 7.2 Contents of each state of an SDL model

To avoid infinite state graphs, the value of *Now*, the SDL time, is not stored, but special algorithms allow timers to work.

7.1.1.3 Breadth-first and depth-first exploration

To explore a graph, there are two main possibilities: either to explore one level completely before exploring the next level or to go down through all the levels and then return to finish the unexecuted transitions.

Figure 7.3 shows the two exploration modes, with numbers indicating the exploration order: on the left part, from state *sA*, breadth-first explores *sB*, then returns to *sA*, explores *sC*, returns to *sA* and finally explores *sD*. Now that all transitions from *sA* have been executed, the exploration goes one level down. From *sB*, *sE* and *sF* are reached; from *sC*, there are no transitions; from *sD*, *sG* is reached. The exploration goes one level down: the only transition is from *sF* back to *sC*. The exploration is finished.

On the right part of Figure 7.3, from state *sA*, depth-first explores *sB* and *sE*. No transition exists from *sE*, the exploration backtracks to *sB*, explores *sF* and *sC*. No transition exists from *sC*, the exploration backtracks to *sA*. From *sA*, *sC* is reached, the exploration backtracks to *sA*, and explores *sD* and *sG*. The exploration is finished.

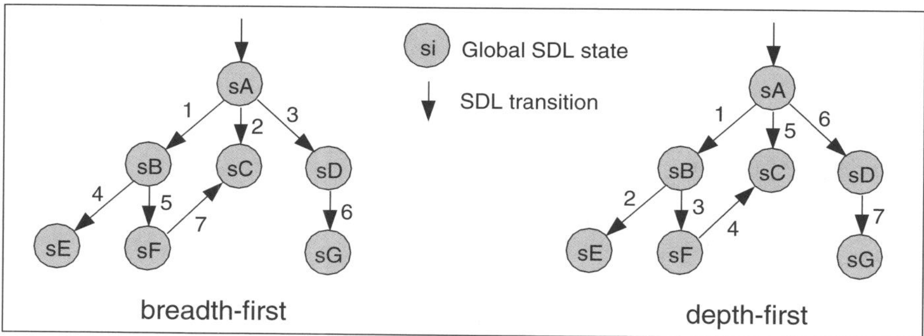

Figure 7.3 Breadth-first and depth-first exploration

7.1.2 Bit-state simulation

To reduce memory usage during the graph exploration, the bit-state algorithm [Holz91] stores the hash-code of each model state instead of the whole state itself.

Figure 7.4 illustrates (partially) the idea of this algorithm. After executing an SDL transition (1), rather than comparing the obtained state and storing it if it is new, bit-state computes h, the hash-code of the state (2) (the hash-code is a kind of checksum). For example, if h = 4, the fourth element in an array of bits is examined: as it contains 0, it means that the state has never been explored (3). We set the fourth element to 1 in the array (4). After execution of another transition (5), the hash-code is computed (6): as the corresponding bit in the array already contains 1, it means that the state has already been explored (or that two different states have the same hash-code) (7).

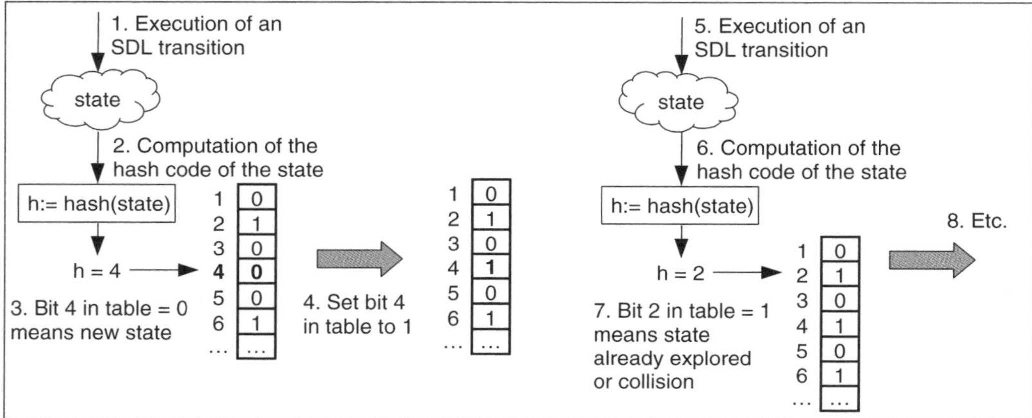

Figure 7.4 Bit-state exploration of the states graph

This algorithm is implemented in ObjectGeode Simulator (named *Supertrace*) and in Tau SDL Suite Validator.

To reduce the collision risk, the number of states to explore must be around 100 times smaller than the size of the array of bits. For example, for a system with 16 millions of states, the memory usage is only: 16 millions $\times$ 100/8 = 200 MB. In addition, two hash-coding functions can be used instead of one.

Bit-state is very efficient, but note that it does not perform a true exhaustive simulation, because it is impossible to guarantee that a hash-coding function will not sometimes give the same result for two different states.

7.1.3 On-the-fly validation

During the exploration of a state graph, after each execution of an SDL transition, the simulators check that no error occurred (receiver dead, no deadlock etc.) and that the observers, if any, do not detect any special event (error or success).

It means that even if the exploration is not finished because the states graph is too large, the simulation can produce results. There is no need to wait till the end of the computation of a state graph before you start checking properties on it.

In addition, as the simulators can be tuned to stop after discovering one error, the simulation campaigns are very fast.

7.2 SIMPLE EXAMPLES

To illustrate how the exhaustive simulation actually works, we have first run it on two very simple SDL examples: the ping TCP/IP command and a model with counters.

7.2.1 Exhaustive simulation of the ping TCP/IP command

We have created an SDL model with very few global states: the *ping* model has only eight global states; therefore, its states graph fits in less than one page.

7.2.1.1 The ping TCP/IP command model

We have created a simplified SDL model of the *ping* command. Ping is generally part of TCP/IP implementations, available, for example, in Unix or Windows. This command is generally used to test if a certain host is present on a network and responding. The *ping* command, executed on a client computer, transmits an echo request to the server computer. The server answers with an echo reply.

The following example shows the use of *ping* on a Windows NT client to test if the server named *nepal* is responding (similar results can be obtained in Unix):

```
C:\>ping nepal

Pinging nepal [196.200.100.99] with 32 data octets:

Reply from 196.200.100.99 : octets=32 time<10ms TTL=128
```

In the following example, some firewalls certainly prevent the *ping* command from reaching the server *www.airfrance.com*:

```
C:\>ping www.airfrance.com

Pinging double6.airfrance.com [193.57.244.15] with 32 data
octets:

Waiting delay exceeded.
```

The SDL model of *ping* contains two blocks, *client* and *server*, as depicted in Figure 7.5, connected through the channel *IP*. The signal names are self-explanatory.

Each block contains one process, named, respectively, *client* and *server*, as shown in Figure 7.6.

Figure 7.7 shows the contents of the processes *client* and *server*. When a *PING* signal is received, process *client* transmits an *echo_req* to the *server*, and starts timer *T1*. When *echo_req* is received by process *server*, an informal decision allows the simulator to either respond with *echo_reply* or not. If the answer *'Yes'* is selected, signal *echo_reply* is received by process *client* before the time-out of timer *T1*, and signal *REPLY* with parameter *'Host is alive'* is

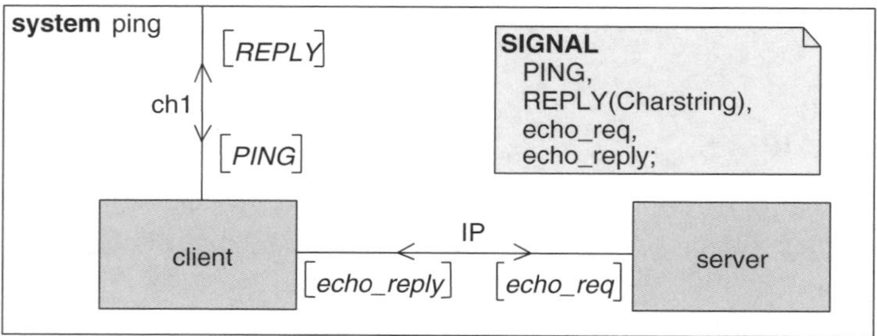

Figure 7.5 The *ping* SDL model

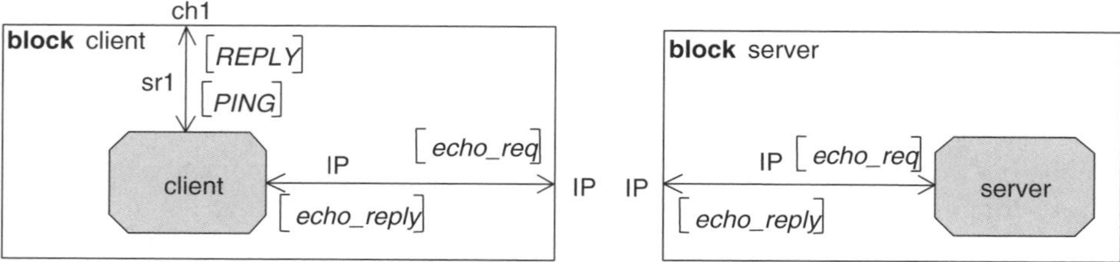

Figure 7.6 The contents of blocks *client* and *server*

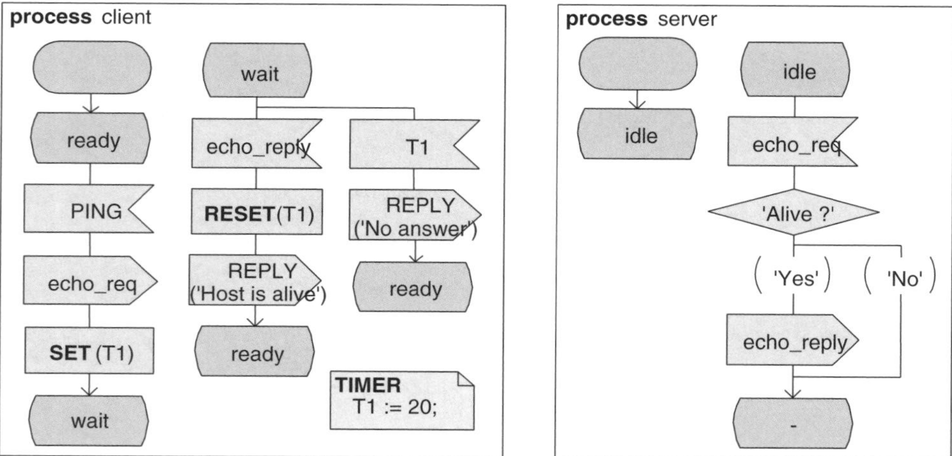

Figure 7.7 The states machines of *client* and *server*

transmitted. If *'No'* is selected, timer *T1* times-out, and signal *REPLY* with parameter *'No answer'* is transmitted.

This behavior is illustrated in Figure 7.8, which is the interactive simulation MSC trace of the model *ping*.

7.2.1.2 Exhaustive simulation of model ping

Running the exhaustive simulation (in the default mode *breadth-first*) on model *ping* generates eight global states. The tool used here is the ObjectGeode Simulator, but similar results are

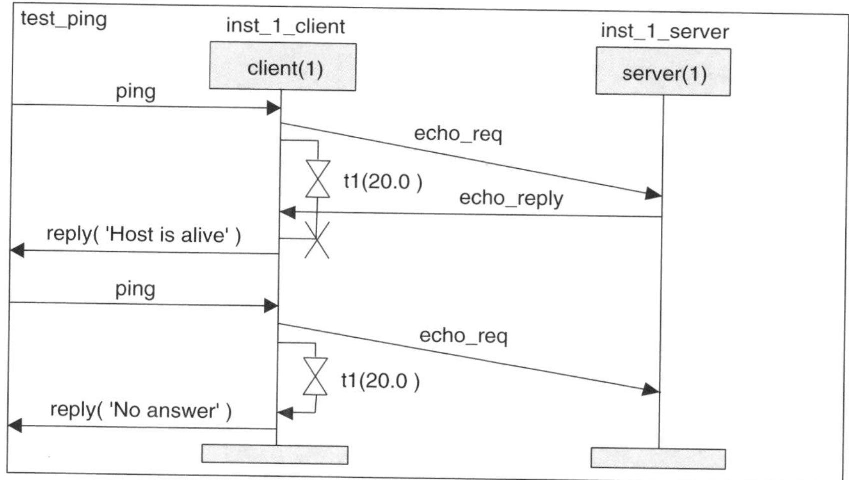

Figure 7.8 MSC generated by simulation of model *ping*

obtained with the Tau SDL Suite Validator (except that in Tau, the exploration is performed in depth-first mode). During exhaustive simulation, the Simulator executes all the transitions from every global state of the SDL model. After each transition execution, the Simulator compares the reached global state with the states previously reached, which have been stored in memory: if the global state is new, it is stored. When all possible transitions have been executed from every state, the simulation is terminated.

Two Simulator variables (*define states_dump 'sta.wri'* and *define edges_dump 'edg.wri'*) enable the generation of two textual files, one containing the transitions of the reachable states graph (here *edg.wri*) and the other containing the contents of each model's global state (here *sta.wri*). We have used those files to manually draw the reachable states graph of our SDL model, shown in Figure 7.9.

The exhaustive simulation is started from the global state S1:

- process *client* is in state *ready*, its input queue is empty
- process *server* is in state *idle*, its input queue is empty.

A. From global state S1, process *client* inputs the external signal *PING* (which was not present in the queue due to the *feed* ObjectGeode feature), outputs signal *echo_req*, starts timer *T1* and enters state *wait*. The next global state is S2.

B. From global state S2, process *server* inputs signal *echo_req*, and stops in the decision *'Alive ?'*. The next global state is S3.

C. From global state S3, the Simulator selects the answer *'Yes'* after the decision *'Alive ?'*. Process *server* outputs signal *echo_reply*. The next global state is S4.

D. From global state S4, process *client* in state *wait* inputs *echo_reply*, stops timer *T1* and outputs signal *REPLY* with parameter *'Host is alive'*. The next global state is S1.

E. The Simulator recognizes that S1 has already been explored, thus will not loop forever. It backtracks to global state S3, from where a transition has not been executed.

F. From global state S3, the Simulator selects the answer *'No'* after the decision *'Alive ?'*. Process *server* does not output signal *echo_reply*. The next global state is S5.

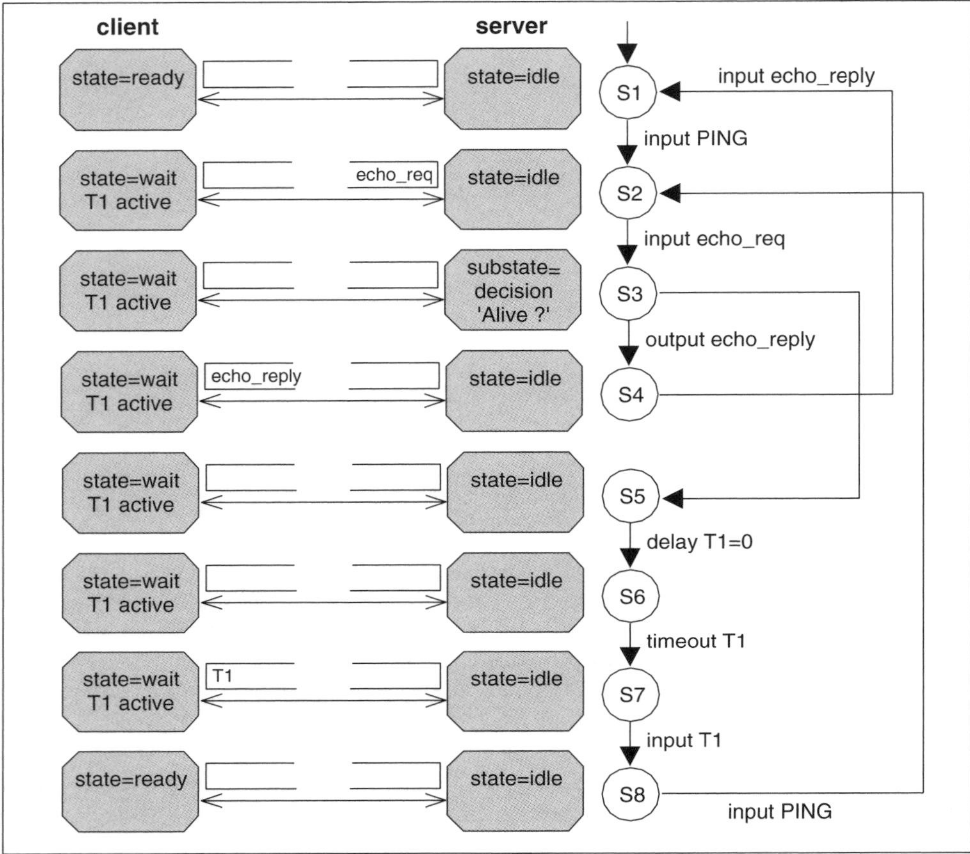

Figure 7.9 States graph of model *ping*

G. From global state S5, the Simulator makes the time progress of 20 units. The next global state is S6.

H. From global state S6, the Simulator times-out timer *T1* in process *client*. The timer signal *T1* is transmitted to *client*. The next global state is S7.

I. From global state S7, process *client* in state *wait* inputs timer signal *T1* and outputs signal *REPLY* with parameter *'No answer'*. The next global state is S8.

J. From global state S8 (similar to state S1), process *client* inputs the external signal *PING*, outputs signal *echo_req*, starts timer *T1* and enters state *wait*. The next global state is S2.

K. The Simulator finds that S2 has already been explored. As all the transitions from any global state have been executed, the exhaustive simulation stops, and the Simulator reports the number of global states found, eight, plus other statistics.

Figure 7.10 shows the MSC generated by interactive simulation of the *ping* model, where the global states numbers from Figure 7.9 have been manually added in the form of global conditions. You can follow the MSC to have another view of how exhaustive simulation works.

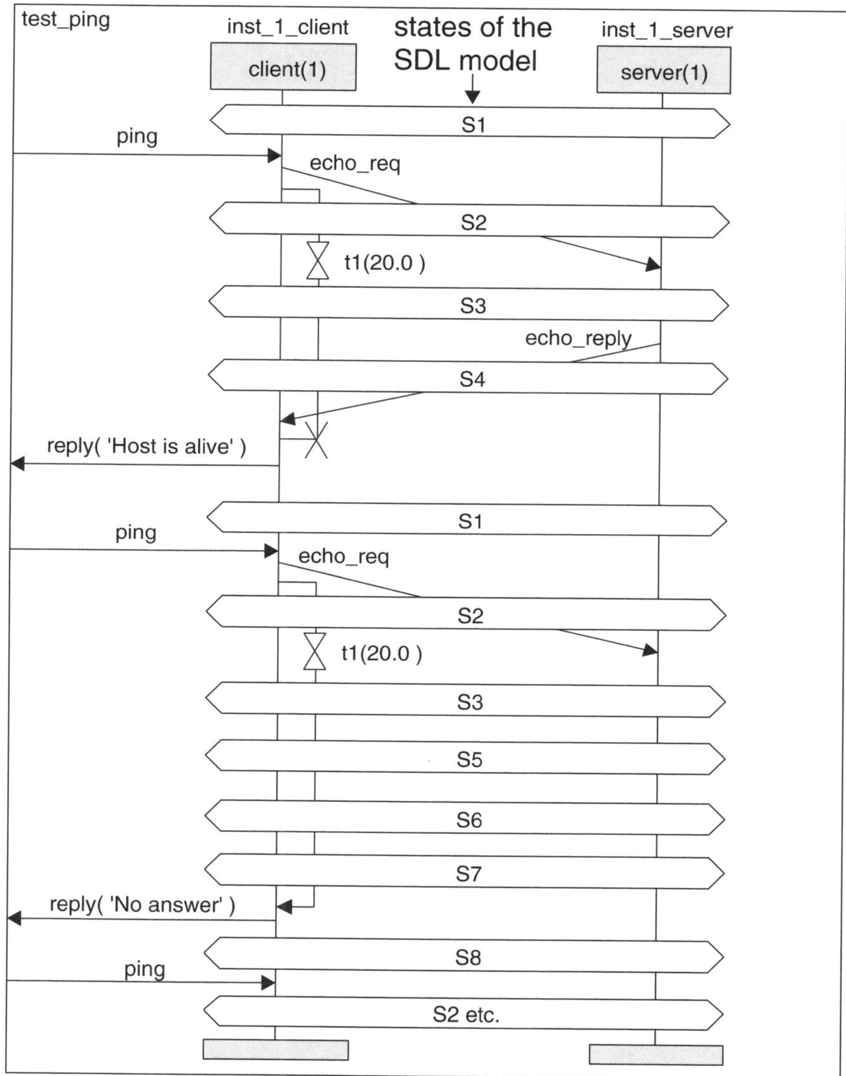

Figure 7.10 The *ping* model global states manually added to the MSC

7.2.1.3 Remarks concerning ping exhaustive simulation

The following Simulator options have been used, to generate the smallest number of global states, for better clarity:

- Reasonable environment on (priority to internal events, that is, a new *PING* can be transmitted only when the previous *PING* is finished).
- Loose time progression off (otherwise, timer *T1* can time-out more frequently).

Note that the global states S8 and S1 look identical, but the difference is that in S1, the expression *sender* in process *client* contains the Pid of process instance *server(1)* (because it has just consumed the signal *echo_reply*), while in S8, *sender* contains the Pid of process instance *client(1)* (because it has just consumed the timer signal *T1*).

Also, a *PING* sequence has been simulated in interactive mode before starting the exhaustive simulation; otherwise we get two more states where the expression *sender* contains *Null* as no signal has been consumed yet.

7.2.2 Exhaustive simulation of counters

To show a larger states graph, we simulate an SDL model containing incremented variables.

7.2.2.1 A model with counters

Figure 7.11 shows an SDL model containing two identical processes *count1* and *count2*, based on the block type *counter*.

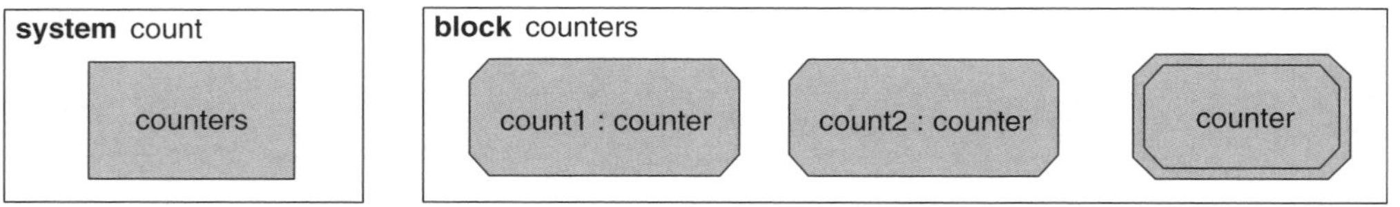

Figure 7.11 The SDL system *count*

Figure 7.12 shows the contents of process type *counter*: variable *n* is initialized to 1, then each time an input NONE is performed (spontaneous transition), *n* is incremented. When *n* is greater than 100, *n* is set back to 1, and the count starts again.

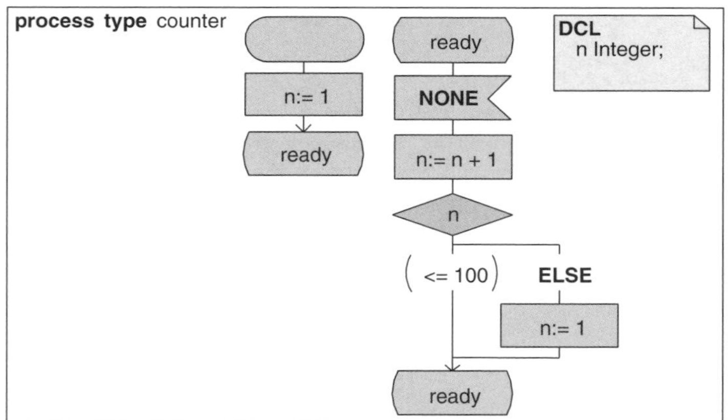

Figure 7.12 The process type *counter*

7.2.2.2 Exhaustive simulation of model count

As expected, running the exhaustive simulation on model *count* generates 10000 global states: each process *count1* and *count2* has 100 states; as they are not synchronized, the number of global states is the product 100 × 100. The tool used here is the ObjectGeode Simulator.

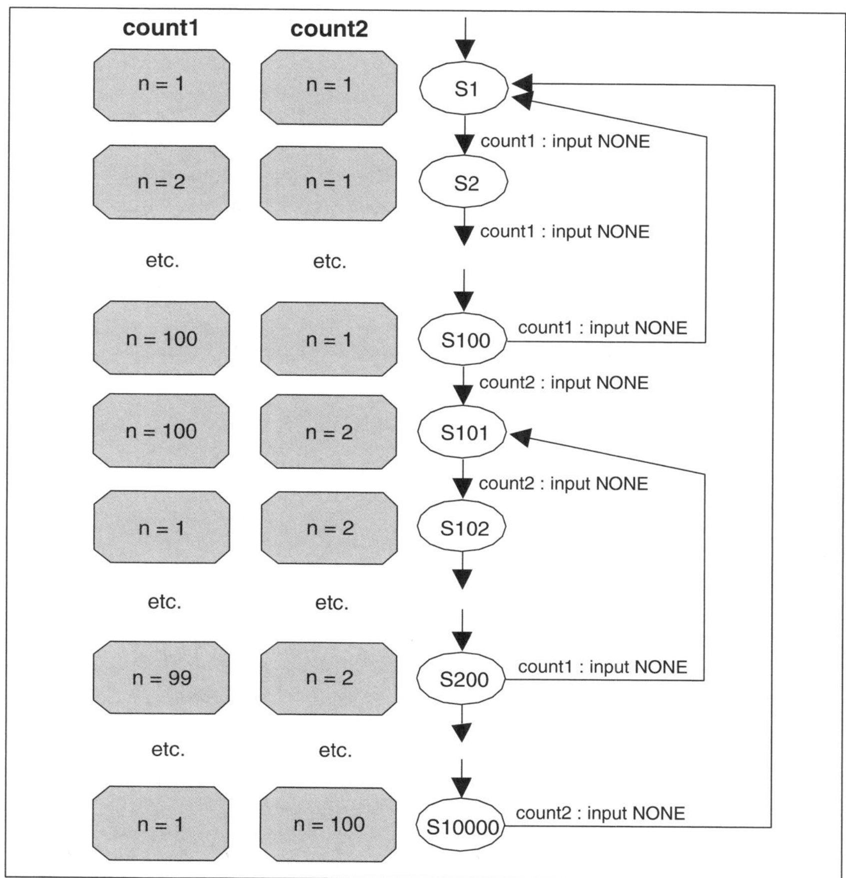

Figure 7.13 Partial states graph of model *count*

Using the two textual files optionally generated by the Simulator, we have manually drawn a part of the reachable states graph of the SDL model, as shown in Figure 7.13.

As we have used the *depth-first* simulation mode, the Simulator first executes process *count1* until $n = 100$, state S100, then n in *count2* takes the value 2, 100 transitions are executed in *count1* and so on.

From state S10000, n in *count2* is incremented and returns to 1, therefore the Simulator discovers that the obtained global state is S1.

Note that the exhaustive simulation stops after executing once all the transitions between the 10000 global states, the result being difficult to obtain by interactive or random simulation, where it is difficult to avoid repeating loops several times.

7.3 CASE STUDY WITH TAU SDL SUITE

You will run the exhaustive simulation on the V.76 SDL model to discover errors automatically, and much faster and with much better dynamic coverage than with interactive or random simulation.

7.3.1 One second to detect missing save of *v76frame*

7.3.1.1 Prepare the SDL model

Our V.76 SDL system is symmetric because two identical instances *DLCa* and *DLCb* of block type *V76_DLC* are used: the Validator allows to transmit signals to both instances, but it is not possible, for example, to define signal *L_DataReq* to be transmitted to *DLCa* and not to *DLCb*. This leads to a configuration with many global states, which is longer to simulate.

To reduce the number of global states, we split the channel *DLCbSU* in two, one carrying the signals not to be transmitted to the *DLCb*: *L_EstabReq*, *L_ReleaseReq*, L_*SetparmReq* and *L_DataReq*; a Validator command will be used to disable this channel.

A. In the Organizer, double-click the system *V76test* to open it in the SDL Editor. Use the version of *V76test* without observer process.

B. In the SDL Editor, select *File* > *Save As* and enter *v76test_dis.ssy*.

C. In the SDL Editor, modify *V76test* to obtain the configuration shown in Figure 7.14: rename one signallist *su2dlc1*, add the channel *dis* and declare the signallists *su2dlc1* and *disabled*.

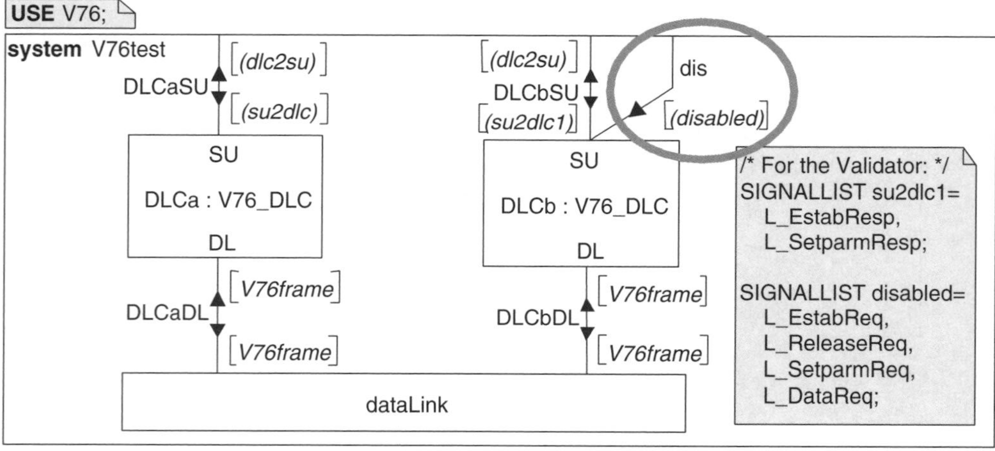

Figure 7.14 Configuration of V.76 for the Validator

D. In the Organizer, press the *save* button.

E. Be sure to use the last version of process *dispatch*, stored in *dispatch.spr*, including the two corrections mentioned in Chapter 4:
- input signal *DLCstopped* added under state *waitUA*,
- after transmitting *L_ReleaseInd*, go to state *ready* instead of state *waitUA*.

7.3.1.2 Start the Validator

A. In the Organizer, select the SDL system *V76test* and do *Generate* > *Make*.

B. In the SDL Make window, select *Microsoft Validation* or *Borland Validation*, and press *Full Make*.

C. In the Organizer, press the *Validate* button to start the Validator.

7.3.1.3 Define test values for external signals

We will define which external signals the Validator will transmit to the SDL model, and the values of their parameters.

A. In the Validator, check the group *TEST VALUES* and press on *List Value*. The Validator answers:

```
Command : List-Test-Values

Sort integer:
0
-55
55

Sort DLCident:
0
1
```

B. Press on *Clear Value*, select *integer*, press *OK*, enter *–55*, and press *OK*.

C. Repeat the operation for *integer*: *0* and *55*.

D. Press on *Def Value*, select *integer*, press *OK*, enter *86*, and press *OK*.

E. Repeat the operation for *integer*: *39*.

F. Save the signal definitions into a file, typing the command: *Save-Test-Values sig_defs.com*.

G. Open *sig_defs.com* with a text editor, and remove the following last lines from the file, which are duplicated (because the same signals are on two external channels in the SDL model):

```
clear-signal-definitions L_EstabResp
define-signal L_EstabResp
clear-signal-definitions L_SetparmResp
define-signal L_SetparmResp
clear-signal-definitions L_ReleaseReq
define-signal L_ReleaseReq 0
define-signal L_ReleaseReq 1
clear-signal-definitions L_EstabReq
define-signal L_EstabReq 0
define-signal L_EstabReq 1
clear-signal-definitions L_SetparmReq
define-signal L_SetparmReq
clear-signal-definitions L_DataReq
define-signal L_DataReq 0 86
define-signal L_DataReq 0 39
define-signal L_DataReq 1 86
define-signal L_DataReq 1 39
```

H. Remove the following lines from the file:

```
define-signal L_DataReq 0 39
define-signal L_DataReq 1 86
```

I. Check that *sig_defs.com* contains, mixed with *Clear-Test-Values* commands, the following test values:

```
define-test-value integer 86
define-test-value integer 39
define-test-value DLCident 0
define-test-value DLCident 1
```

J. Check that *sig_defs.com* contains the following signal definitions:

```
clear-signal-definitions L_EstabReq
define-signal L_EstabReq 0
define-signal L_EstabReq 1
clear-signal-definitions L_EstabResp
define-signal L_EstabResp
clear-signal-definitions L_SetparmReq
define-signal L_SetparmReq
clear-signal-definitions L_SetparmResp
define-signal L_SetparmResp
clear-signal-definitions L_ReleaseReq
define-signal L_ReleaseReq 0
define-signal L_ReleaseReq 1
clear-signal-definitions L_DataReq
define-signal L_DataReq 0 86
define-signal L_DataReq 1 39
```

K. Add the following line at the end of *sig_defs.com*, to disable the channel *dis* and save it:

```
channel-disable dis
```

L. Select *File > Restart*, and answer *No*. Select *Commands > Include Commands Script* and open *sig_defs.com*.

M. Press on *List Signal*, and check that you get the following result shown in Figure 7.15.

7.3.1.4 Run the exhaustive simulation

A. Press on *Exhaustive*, and after 150000 system states, press on *Break*; the Validator displays:

```
** Starting exhaustive exploration **
Search depth : 100
Passing 50000 system states
Passing 100000 system states
Passing 150000 system states
*** Break at user input ***

** Exhaustive exploration statistics **
No of reports: 9
Generated states: 276000
```

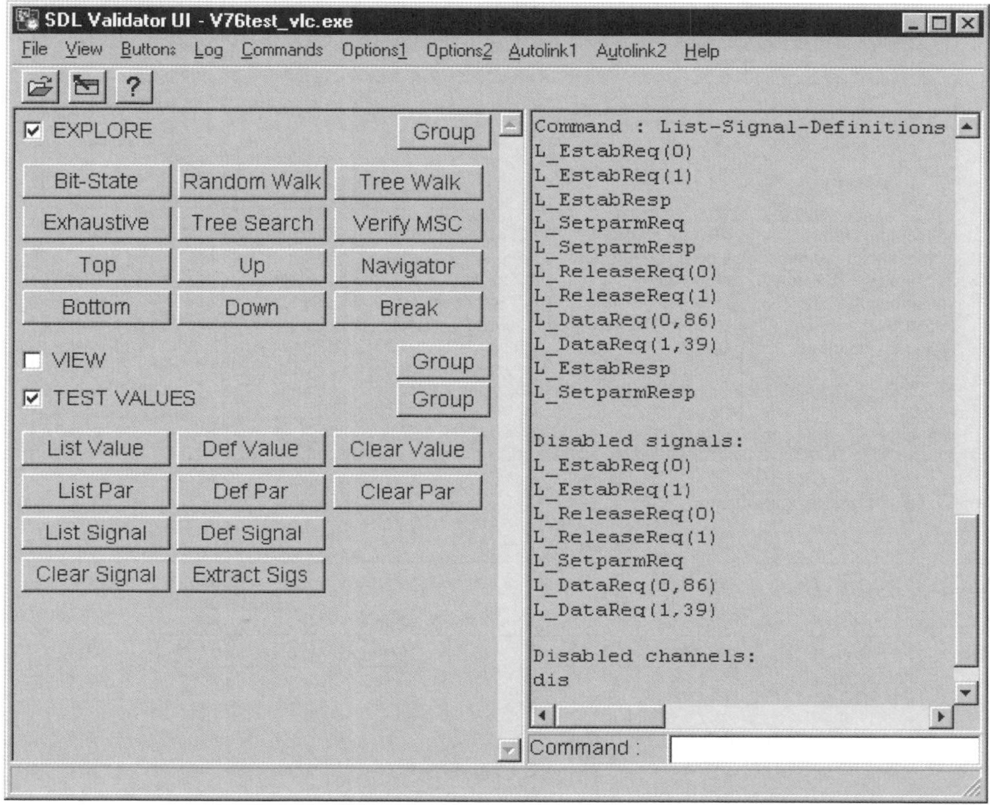

Figure 7.15 The Validator showing signal definitions

```
Truncated paths: 49564.
Unique system states: 213820.
Size of hash table: 100000 (400000 bytes)
Current depth: 73
Max depth: 100
Min state size: 212
Max state size: 692
Symbol coverage :  96.31
```

B. The *Report Viewer* appears. As indicated in the results above, nine reports have been generated[1]. Double-click on the *ImplSigCons* box to unfold it, as shown in Figure 7.16.

C. The left-most box shows that signal *V76frame*, transmitted by process *AtoB* to process *dispatch* in block *DLCb*, has been discarded (implicit signal consumption).

D. Double-click on this box: the MSC Editor displays the trace of the scenario leading to the error; the end of this trace is shown in Figure 7.17.

[1] The Validator automatically generates a file *reports.rep* containing the detected reports, including the path (the transition number at each step, similar to *.scn* files in ObjectGeode) leading from the initial state to the detected state.

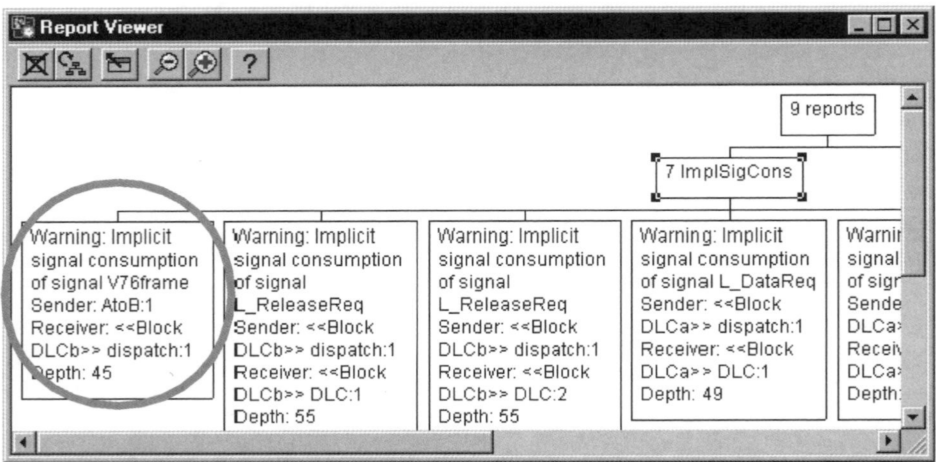

Figure 7.16 The Report Viewer

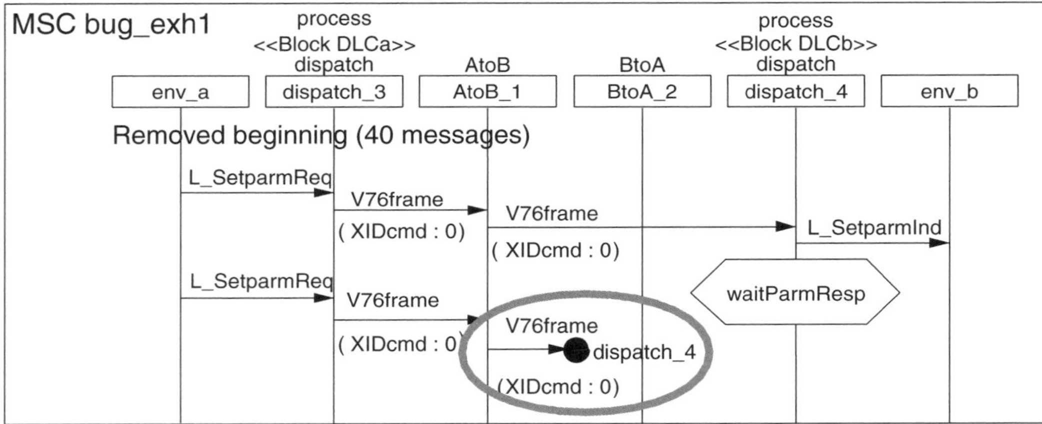

Figure 7.17 End of the first error MSC trace

We see that process *dispatch* in block *DLCb* is in state *waitParmResp*. If we look at the SDL model, under this state no input or save of signal *v76frame* are specified. Therefore, this signal has been discarded.

7.3.1.5 Correct the error

To prevent the signal from being lost, you will add a save of signal *v76frame* in state *waitParmResp* of process *dispatch*.

A. Exit from the Validator (answering *No* to the question).

B. In Windows (or Unix), make a copy of the file *dispatch.spr* (the file containing process *dispatch*) into *dispatch_v3.spr* (continue working on file *dispatch.spr*).

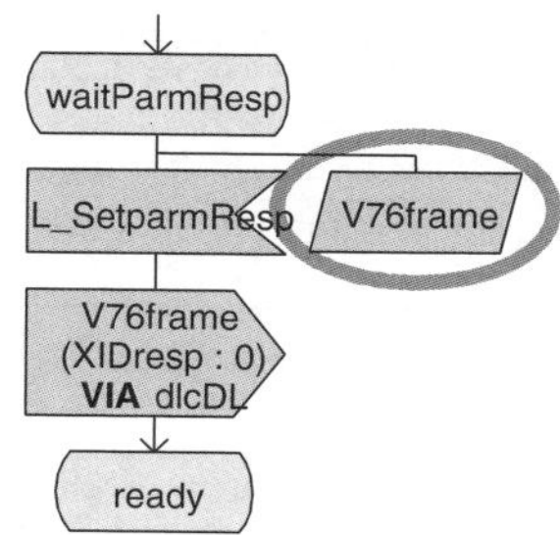

Figure 7.18 Missing save of *v76frame* added under *waitParmResp*

C. In process *dispatch*, page *part1*, add a save containing *v76frame* under the state *waitParm-Resp*, as illustrated in Figure 7.18.

D. Save the SDL model.

7.3.2 One second to detect missing input *L_ReleaseReq*

7.3.2.1 Run the exhaustive simulation

A. In the Organizer, select the SDL system *V76test* and press the *Validate* button.

B. In the Validator, select *Commands > Include Command Script*, and choose *sig_defs.com*.

C. Press on *List Signal*, and check that you get the same signals as previously.

D. Press on *Exhaustive*, and after 100000 system states, press on *Break*; the Validator displays:

```
*** Starting exhaustive exploration ***
Search depth : 100
Passing 50000 system states
Passing 100000 system states
*** Break at user input ***

** Exhaustive exploration statistics **
No of reports: 11
Generated states: 140000
Truncated paths: 21517.
Unique system states: 104940.
Size of hash table: 100000 (400000 bytes)
Current depth: 95
Max depth: 100
Min state size: 212
Max state size: 680
Symbol coverage :  88.00
```

E. The *Report Viewer* appears. Double-click on the *ImplSigCons* box to unfold it, as shown in Figure 7.19. The reports are stored in the file *reports.rep*.

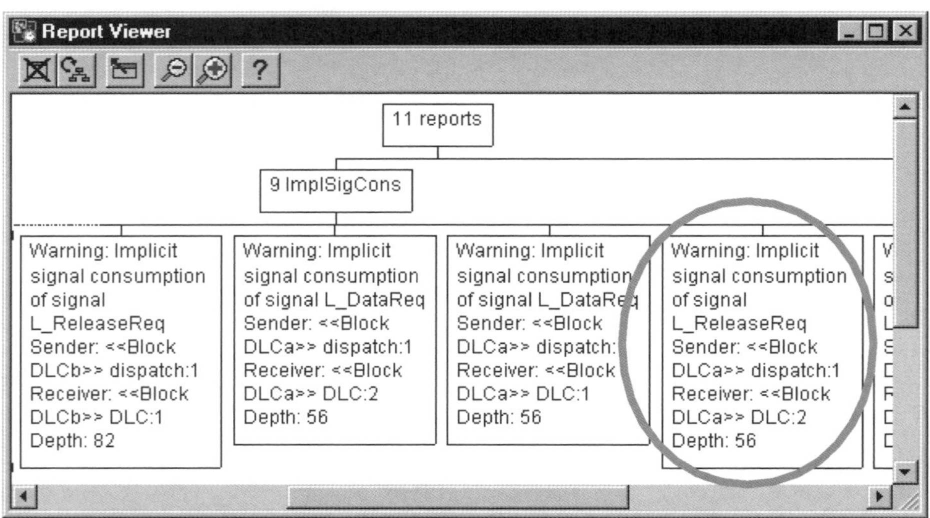

Figure 7.19 The Report Viewer (11 reports)

F. The seventh (or sixth) box from the left shows that signal *L_ReleaseReq*, transmitted by process *dispatch* to instance 2 of process *DLC*, has been discarded (implicit signal consumption). We have chosen this report because its depth is only 56, the left-most having a longer depth (such as 92). The MSC trace is shorter and easier to understand.

G. Double-click on this box: the MSC Editor displays the trace of the scenario leading to the error; the end of this trace is shown in Figure 7.20.

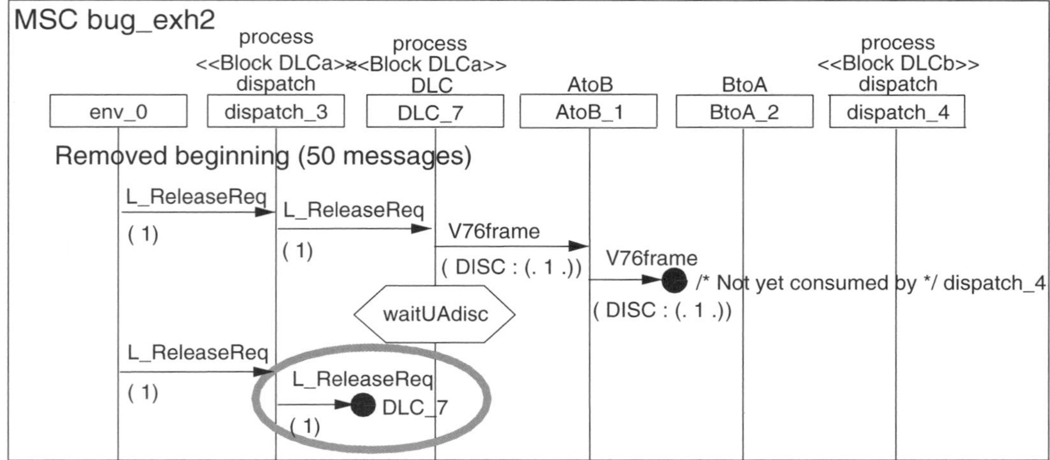

Figure 7.20 End of the error MSC trace

We see that process *DLC* in block *DLCa* is in state *waitUAdisc*. If we look at the SDL model, under this state no input or save of signal *L_ReleaseReq* are specified. Therefore, this signal has been discarded. The discarded *L_ReleaseReq* signal is shown in Figure 7.20.

7.3.2.2 Correct the error

After examining the error, we decide that it is better losing *L_ReleaseReq* than saving it. Therefore, you will add a dummy input of signal *L_ReleaseReq* in state *waitUAdisc* of process *DLC*.

A. Exit from the Validator (answering *No* to the question).

B. In Windows (or Unix), make a copy of the file *dlc.spr* into *dlc_v4.spr*.

C. In process *DLC*, page *part2*, add an input containing *L_ReleaseReq* followed by nextstate dash under the state *waitUAdisc*, as illustrated in Figure 7.21.

D. Save the SDL model.

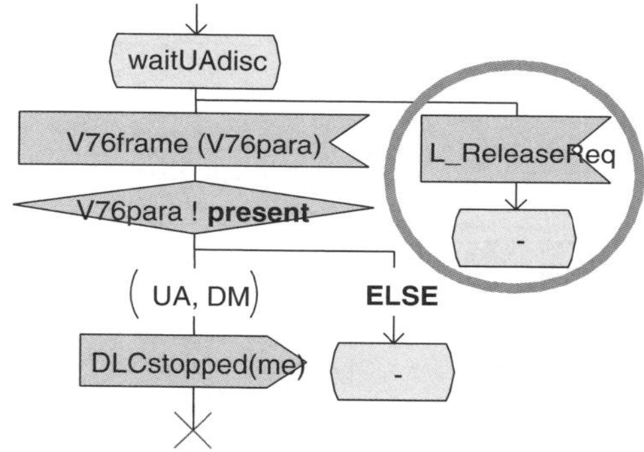

Figure 7.21 After adding input of signal *L_ReleaseReq*

7.3.3 One second to detect missing input *L_DataReq*

7.3.3.1 Run the exhaustive simulation

A. In the Organizer, select the SDL system *V76test* and press the *Validate* button.

B. In the Validator, select *Commands > Include Command Script*, and choose *sig_defs.com*.

C. Press on *List Signal*, and check that you get the same signals as previously.

D. Press on *Exhaustive*, and after 100000 system states, press on *Break*; the Validator displays:

```
*** Starting exhaustive exploration ***
Search depth : 100
Passing 50000 system states
Passing 100000 system states
*** Break at user input ***

** Exhaustive exploration statistics **
No of reports: 7
Generated states: 132000
Truncated paths: 21592.
```

```
Unique system states: 100262.
Size of hash table: 100000 (400000 bytes)
Current depth: 99
Max depth: 100
Min state size: 212
Max state size: 680
Symbol coverage :  87.20
```

E. The *Report Viewer* appears. Double-click on the *ImplSigCons* box to unfold it, as shown in Figure 7.22.

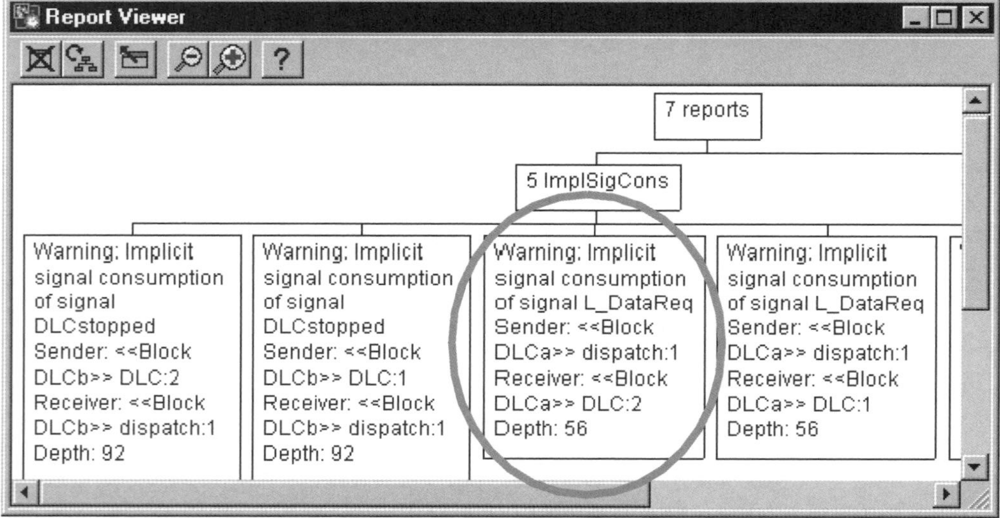

Figure 7.22 The Report Viewer (7 reports)

F. The third box from the left shows that signal *L_DataReq*, transmitted by process *dispatch* to instance 2 of process *DLC*, has been discarded (implicit signal consumption).

G. Double-click on this box: the MSC Editor displays the trace of the scenario leading to the error; the end of this trace is shown in Figure 7.23.

We see that process *DLC* in block *DLCa* is in state *waitUAdisc*. If we look at the SDL model, under this state no input or save of signal *L_DataReq* are specified. Therefore, this signal has been discarded. The discarded *L_DataReq* signal is shown in Figure 7.23.

7.3.3.2 Correct the error

After examining the error, we decide that as the connection is being released, it is better losing *L_DataReq* than saving it. Therefore, you will add a dummy input of signal *L_DataReq* in state *waitUAdisc* of process *DLC*.

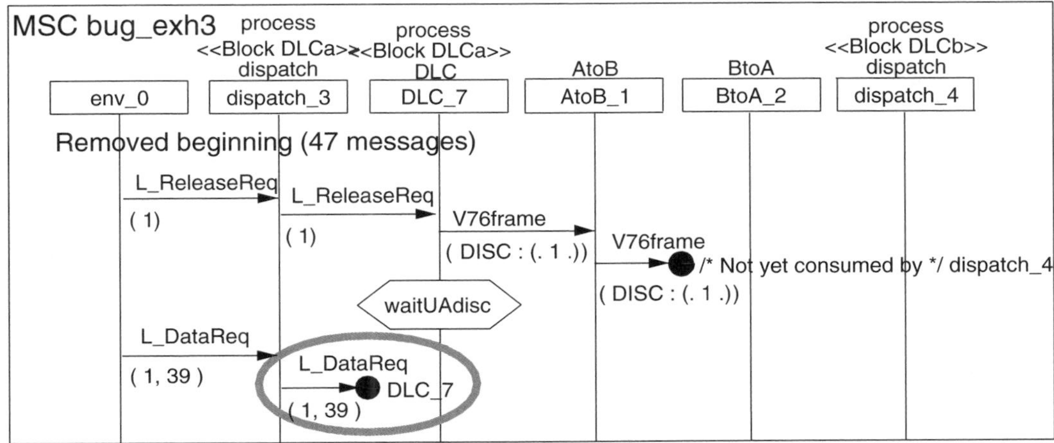

Figure 7.23 Last steps of the error MSC trace

A. Exit from the Validator (answering *No* to the question).

B. In Windows (or Unix), make a copy of the file *dlc.spr* into *dlc_v5.spr.*

C. In process *DLC*, page *part2*, insert a coma followed by *L_DataReq* in the input containing *L_ReleaseReq* previously added, as illustrated in Figure 7.24.

D. Save the SDL model.

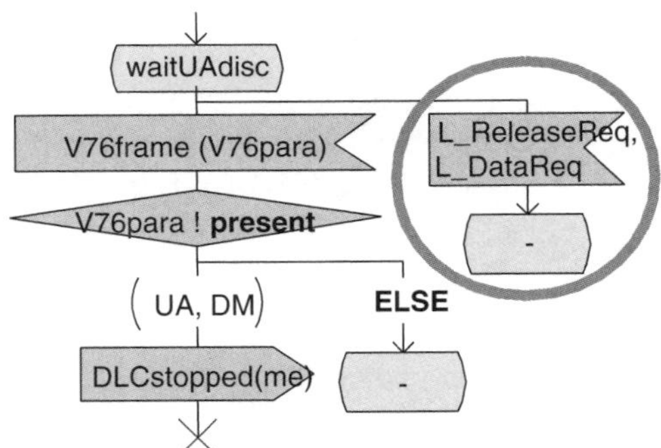

Figure 7.24 After adding input of signal *L_DataReq*

7.3.3.3 Run the exhaustive simulation

A. In the Organizer, select the SDL system *V76test* and press the *Validate* button.
B. In the Validator, select *Commands > Include Command Script*, and choose *sig_defs.com.*
C. Press on *List Signal*, and check that you get the same signals as previously.
D. In the Validator, select *Options2 > Exhaustive: Depth* and enter 30.

E. Press on *Exhaustive*; the Validator displays:

```
** Starting exhaustive exploration **
Search depth : 30

** Exhaustive exploration statistics **
No of reports: 3
Generated states: 8425
Truncated paths: 1708.
Unique system states: 6856.
Size of hash table: 100000 (400000 bytes)
Current depth: -1
Max depth: 30
Min state size: 212
Max state size: 572
Symbol coverage :  90.55
```

The exhaustive simulation has stopped and found 6856 unique system states (note that more states would have been found if the search depth was not limited to 30). The *Report Viewer* appears, showing that the only reports are three MaxQueueLength: the default limit of three signals in some process input queues has been exceeded. This is normal; more details are provided later.

In the 6856 explored global states of the SDL model, we are sure that we have no errors and no deadlocks. However, the global states not yet explored by the Simulator may contain errors.

7.3.4 Millions of states: detect output to Null

Now to test more features in the SDL model, we use a larger model configuration: again, one signal maximum in each queue, but the maximum exploration depth is no longer limited. To limit the number of states, we restrict the number of retransmissions in process *DLC* to 1, instead of 3.

7.3.4.1 Limit number of signals in input queue

To avoid an infinite number of global states, we need to limit the number of signals present in the input queue of each SDL process.

For example, in the V.76 SDL model, if you simulate the scenario shown in Figure 7.51, the queue of the instance 1 of process *DLC* in block *DLCa* contains 4 signals. If this process does not input the signals in its queue while other bursts of *L_DataReq* are transmitted to process *dispatch*, the number of *L_DataReq* stacked in the queue will grow rapidly. In addition, each new signal stacked in the queue generates a new global SDL model state during exhaustive simulation.

The Validator by default limits to three signals in each process instance input queue. To reduce the number of states, we will limit to one signal in each queue; note that some models might not work with such a limit, for example, if two signals are transmitted at the same time to a process queue.

7.3.4.2 Modify the SDL model

A. Exit from the Validator (answering *No* to the question).

B. Open process *DLC part1* and replace 3 by 1 in the declaration of *N320*, to obtain:

```
SYNONYM N320 Integer = 1;
```

C. Save the SDL model.

7.3.4.3 Run the bit-state simulation

After trying exhaustive simulation, we have found that it required 416 MB of RAM for 406049 unique global states of the SDL model. In ObjectGeode, we use exhaustive simulation because it compresses the global states (for example, storing once several identical input queues): in only 196 MB of RAM, ObjectGeode stores 2620001 states of the same model.

This is why instead of using exhaustive simulation we will use bit-state. Bit-state mode is similar to exhaustive mode, but it requires less memory, because instead of storing each new global model state, bit-state stores only one bit in an array. The index in the array is a hash-coding (a kind of checksum) of the global state contents. However, two different global states may have the same hash-code: they are considered as identical, therefore parts of the states graph may remain unexplored.

A. In the Organizer, select the SDL system *V76test* and press *Validate* .

B. In the Validator, select *Options1 > Input Port Length*, and enter 1.

C. Select *Options2 > Bit State: Hash Size* and enter 250000000 (250 millions of bytes). This is the size of the array of bits used to store the states hash-codes. If your machine is equipped, for example, with 128 MB of RAM, enter 80 millions.

D. Select *Options2 > Bit State: Depth* and enter 15000.

E. Select *Commands > Include Command Script*, and choose *sig_defs.com*.

F. Press on *List Signal*, and check that you get the same signals as previously.

G. Press on *Bit State*, the Validator displays:

```
** Starting bit state exploration **
Search depth    : 15000
Hash table size : 250000000 bytes
Transitions: 20000 States: 12408 Reports: 5 Depth: 376 Symbol
  coverage:  93.60 Time: 10:07:07
Transitions: 40000 States: 24847 Reports: 5 Depth: 300 Symbol
  coverage:  93.60 Time: 10:07:07
Transitions: 60000 States: 37274 Reports: 5 Depth: 138 Symbol
  coverage:  93.60 Time: 10:07:07
...
```

```
Transitions: 6940000 States: 4329979 Reports: 5 Depth: 215
  Symbol coverage: 93.60 Time: 10:09:13
Transitions: 6960000 States: 4342489 Reports: 5 Depth: 92
  Symbol coverage: 93.60 Time: 10:09:13
Transitions: 6980000 States: 4354917 Reports: 5 Depth: 172
  Symbol coverage: 93.60 Time: 10:09:13

** Bit state exploration statistics **
No of reports: 5.
Generated states: 6985039.
Truncated paths: 0.
Unique system states: 4358006.
Size of hash table: 2000000000 (250000000 bytes)
No of bits set in hash table: 8675533
Collision risk: 0 %
Max depth: 6530
Current depth: -1
Min state size: 212
Max state size: 584
Symbol coverage :  93.60
```

After only 2 min and 6 s, the bit-state simulation is terminated. 4358006 unique global states have been explored (you may get a different number), and the memory usage has been almost constant and equal to 255 MB only: the bits array plus a few megabytes. As the maximum depth indicated is equal to 6530, the search depth limit used, 15000, was enough.

Because the hash table used could store up to 250 millions $\times$ 8 = 2 billions of bits, the collision risk is evaluated at 0%.

H. The *Report Viewer* appears. Double-click on the *Output* box to unfold it, as shown in Figure 7.25.

I. The first box from the left shows that signal *V76frame* has been transmitted to a Null Pid by process *dispatch* in block *DLCa*.

J. Double-click on this box: the MSC Editor displays the trace of the scenario leading to the error; this trace is shown in Figure 7.26.

A attempts to establish DLC number 0; as the response *L_EstabResp* from B is too late, A has received an *L_ReleaseInd*, meaning failure of DLC establishment; the *L_EstabResp* from B finally arrives (E1 in the MSC), *dispatch* in B creates an instance of *DLC*, which transmits a *v76frame* containing a UA; reaching *dispatch* in A, the *v76frame* should have been transmitted to the instance of *DLC* by executing transition TR1 in Figure 7.27; unfortunately, the instance is dead; therefore, an output to a Null Pid is executed, detected by the Validator.

Remark: the error discovered by ObjectGeode in the same configuration is a bit different. The error scenario discovered by ObjectGeode cannot be replayed by the Validator, because in ObjectGeode the feed command transmits signals to the model without storing them in the input queues. When replaying the error discovered by ObjectGeode, the Tau Validator signals

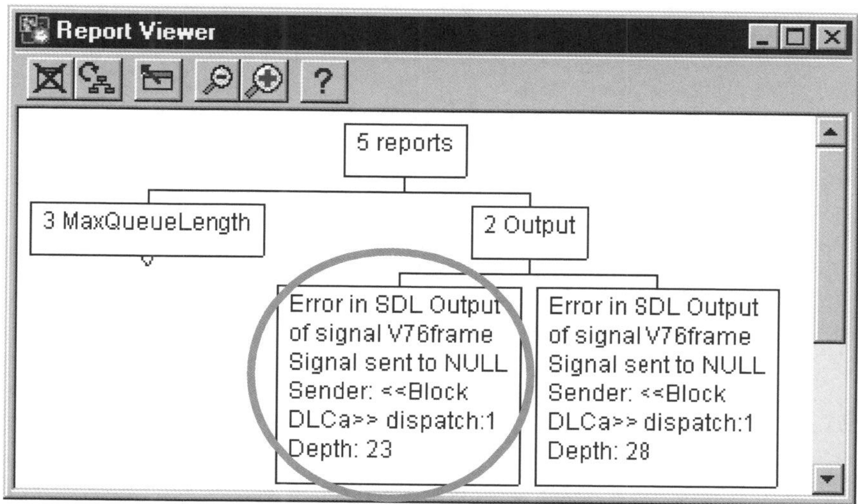

Figure 7.25 The Report Viewer (5 reports)

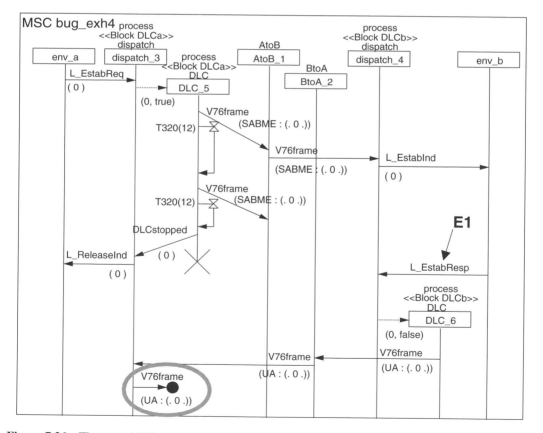

Figure 7.26 The error MSC trace

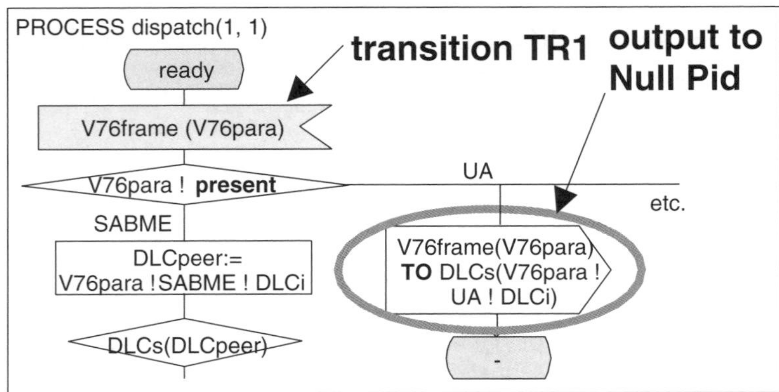

Figure 7.27 The output to Null in process *dispatch* part1 (extract)

that the input queue limit (of 1 signal here) is reached when transmitting the *L_EstabResp*: the input queue of *dispatch* already contains the saved *v76frame*.

7.3.4.4 Correct the error

The simulation has revealed that we must protect the expressions after TO in the output statements to avoid having a Null Pid. For that, you will add a decision to test the value of the expression: if Null, the output is not performed.

A. Exit from the Validator (answering *No* to the question).

B. In Windows (or Unix), make a copy of the file *dispatch.spr* into *dispatch_v6.spr*.

C. Open process *dispatch* in the SDL Editor, and create a new page *part1_2* and rename *part1 part1_1*.

D. Split the state machine in *part1_1* into two parts, one in *part1_1* and the other in *part1_2*, as illustrated in Figures 7.28 and 7.29.

E. Insert four decisions in *part1_1* as illustrated in Figure 7.28.

F. Insert one decision in *part2* after the answer *UA*, as shown in Figure 7.30.

G. Save the SDL model.

7.3.5 Forty seconds to detect missing save of *L_DataReq*

7.3.5.1 Run again the bit-state simulation

To save time, we will set the Validator to stop after discovering two exceptions, rather than finishing the whole reachable states exploration.

A. In the Organizer, select the SDL system *V76test* and press *Validate* .

B. Select *Options2 > Bit State: Depth* and enter 15000.

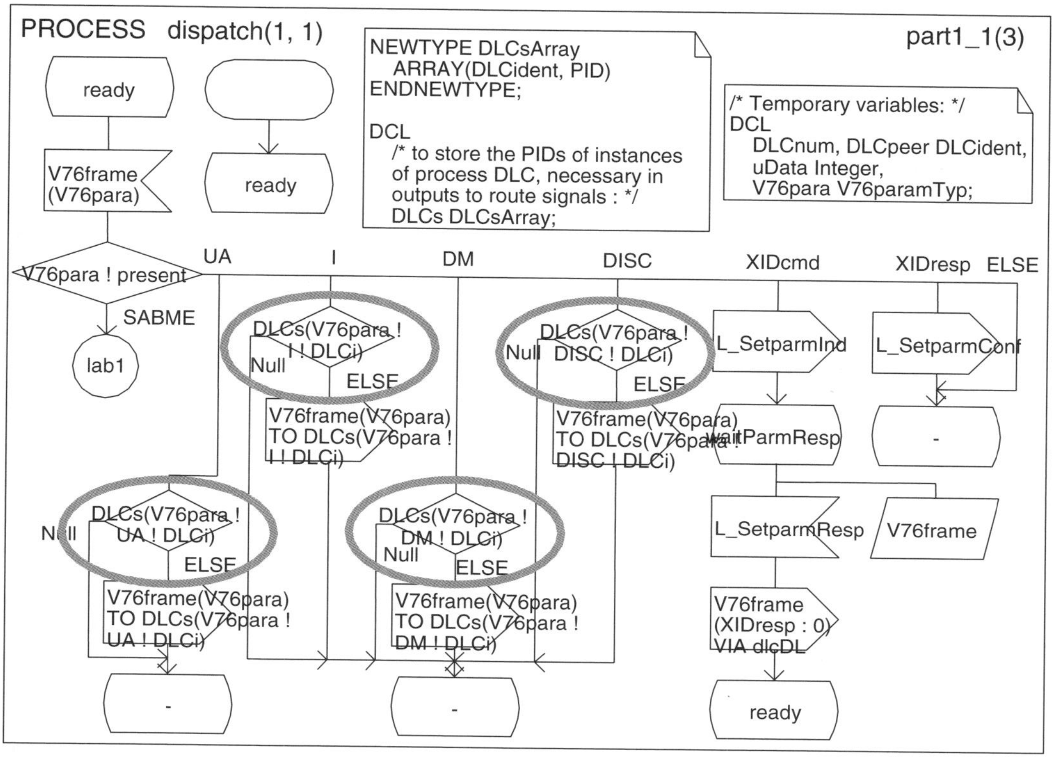

PROCESS dispatch(1, 1)
part1_1(3)
NEWTYPE DLCsArray
ARRAY(DLCident, PID)
ENDNEWTYPE;
DCL
/* to store the PIDs of instances of process DLC, necessary in outputs to route signals : */
DLCs DLCsArray;
/* Temporary variables: */
DCL
DLCnum, DLCpeer DLCident,
uData Integer,
V76para V76paramTyp;
ready
ready
V76frame
(V76para)
V76para ! present
UA
I
DM
DISC
XIDcmd
XIDresp
ELSE
SABME
lab1
DLCs(V76para ! I ! DLCi)
Null
ELSE
V76frame(V76para)
TO DLCs(V76para ! I ! DLCi)
DLCs(V76para ! DISC ! DLCi)
Null
ELSE
V76frame(V76para)
TO DLCs(V76para ! DISC ! DLCi)
L_SetparmInd
L_SetparmConf
waitParmResp
-
L_SetparmResp
V76frame
DLCs(V76para ! UA ! DLCi)
Null
ELSE
V76frame(V76para)
TO DLCs(V76para ! UA ! DLCi)
DLCs(V76para ! DM ! DLCi)
Null
ELSE
V76frame(V76para)
TO DLCs(V76para ! DM ! DLCi)
V76frame
(XIDresp : 0)
VIA dlcDL
-
-
ready

Figure 7.28 Process *dispatch* page *part1_1*

PROCESS dispatch(1, 1)
part1_2(3)
lab1
DLCpeer:=
V76para ! SABME ! DLCi
DLCs
(DLCpeer)
ELSE
Null
L_ReleaseInd
(DLCpeer)
L_EstabInd
(DLCpeer)
V76frame
(DM :(. DLCpeer .))
VIA dlcDL
waitEstabResp
-
waitEstabResp
L_EstabResp
V76frame
DLC
(DLCpeer, False)
Creates
instance of
process DLC
DLCs(DLCpeer)
:= OFFSPRING
Stores into the
table the PID of the
instance just created.
ready

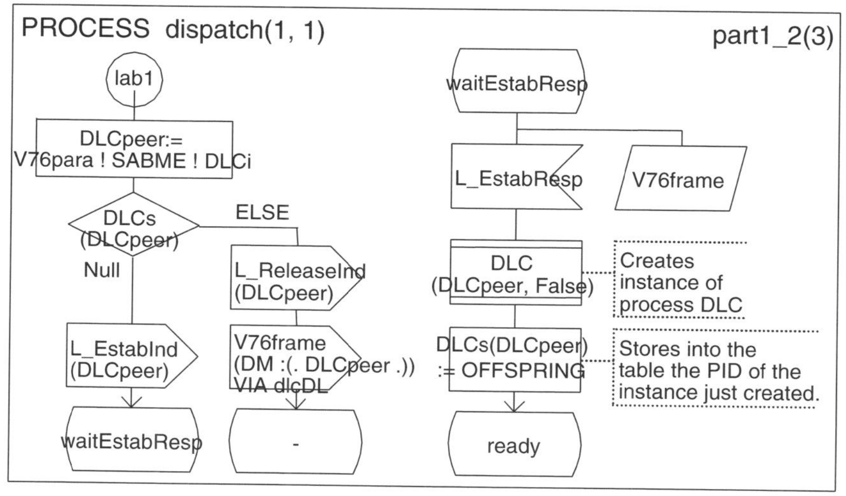

Figure 7.29 Process *dispatch* page *part1_2*

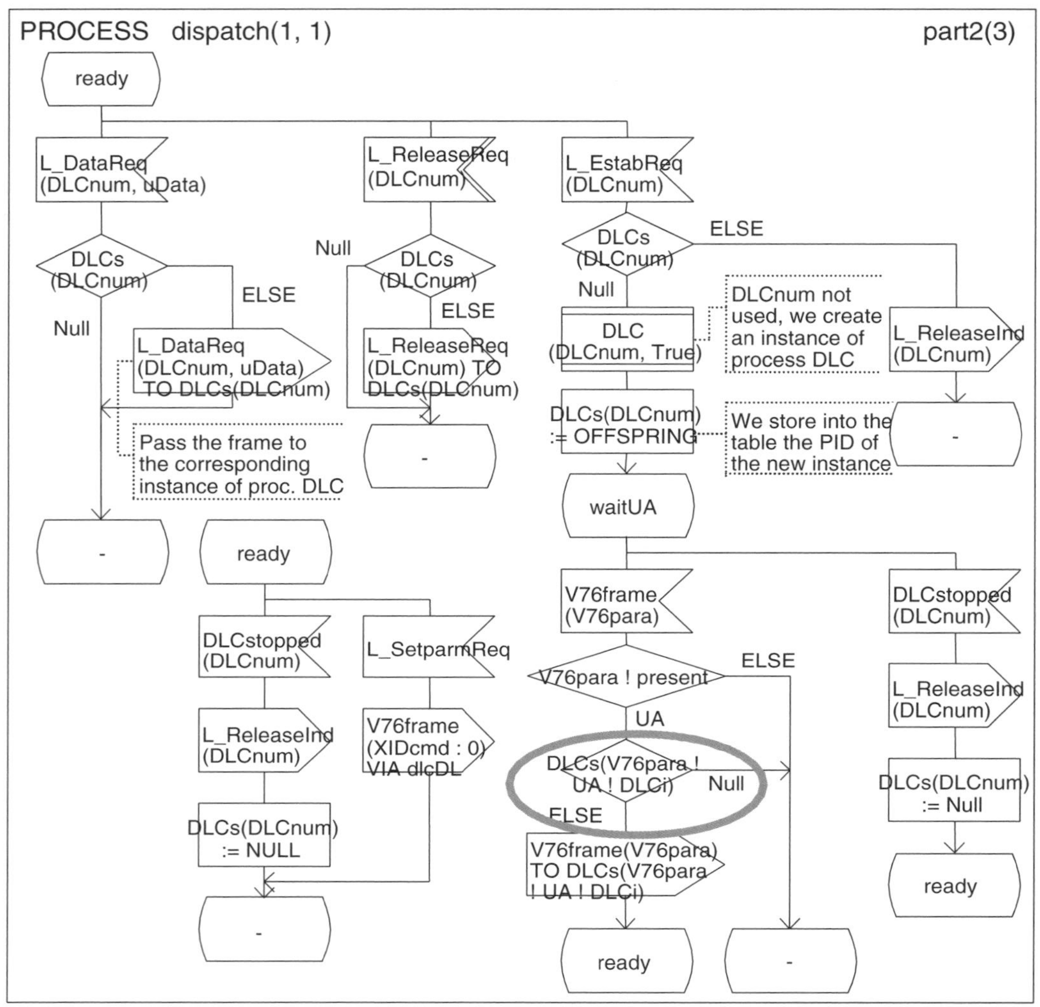

Figure 7.30 Process *dispatch* page *part2*

C. In the Validator, select *Options1 > Input Port Length*, and enter 2. We no longer limit to 1 because in each process queue, we need enough space for a saved signal plus an external signal.

D. Select *Options1 > Report: Report Log*, choose *MaxQueueLength* and select *Off*. The Validator will no longer generate any report when reaching the input port length limit.

E. Select *Commands > Include Command Script*, and choose *sig_defs.com*.

F. Press on *List Signal*, and check that you get the same signals as previously.

G. Press on *Bit State*, the Validator displays:

```
** Starting bit state exploration **
Search depth     : 15000
Hash table size : 1000000 bytes
```

```
Transitions: 20000 States: 12484 Reports: 0 Depth: 708
  Symbol coverage: 89.02 Time: 15:53:12
Transitions: 40000 States: 24892 Reports: 0 Depth: 604
  Symbol coverage: 96.44 Time: 15:53:12
...
Transitions: 1840000 States: 1136439 Reports: 2 Depth: 1783
  Symbol coverage: 98.22 Time: 15:53:51
Transitions: 1860000 States: 1148820 Reports: 2 Depth: 2262
  Symbol coverage: 98.22 Time: 15:53:51
Transitions: 1880000 States: 1160825 Reports: 2 Depth: 3279
  Symbol coverage: 98.22 Time: 15:53:51
```

H. When you see in the trace that the number of reports is no longer null, press on *Break*:

```
*** Break at user input ***

** Bit state exploration statistics **
No of reports: 2.
Generated states: 1888000.
Truncated paths: 0.
Unique system states: 1165580.
Size of hash table: 8000000 (1000000 bytes)
No of bits set in hash table: 2062758
Collision risk: 25 %
Max depth: 3639
Current depth: 3623
Min state size: 212
Max state size: 628
Symbol coverage :  98.22
```

I. In the *Report Viewer*, double-click on the *ImplSigCons* box to unfold it, as shown in Figure 7.31.

J. The first box from the left shows that signal *L_DataReq* has been discarded by process *DLC* in block *DLCa*.

K. Double-click on this box: the MSC Editor displays the trace of the scenario leading to the error; this trace is shown in Figure 7.32.

We see that the target instance of process *DLC* in block *DLCa* (named *DLC_25*) is in state *waitUA*. If we look at the SDL model, under this state no input or save of signal *L_DataReq* are specified. Thus, this signal has been discarded.

7.3.5.2 Correct the error

We decide to save signal *L_DataReq* in state *waitUA*, because once the connection is set up, the signal can be processed.

A. Exit from the Validator (answering *No* to the question).

B. In Windows (or Unix), make a copy of the file *dlc.spr* into *dlc_v7.spr*.

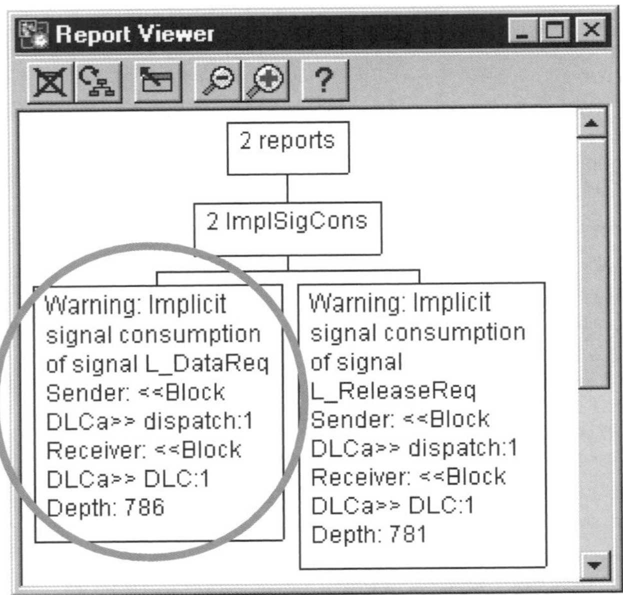

Figure 7.31 The Report Viewer (2 reports)

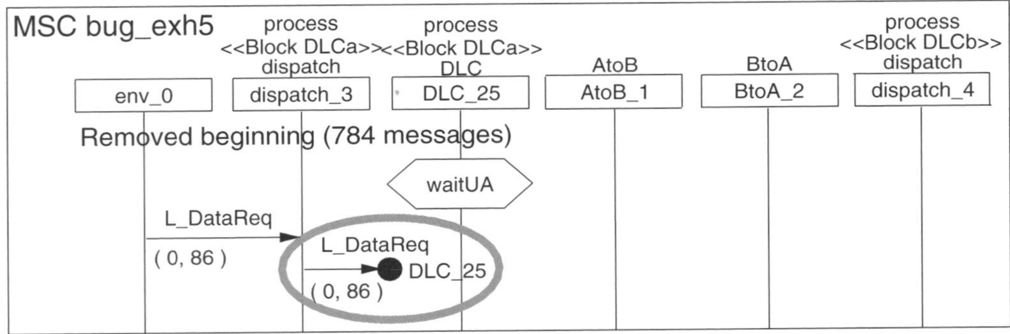

Figure 7.32 The end of the error MSC trace

C. In process *DLC*, page *part1*, add below state *waitUA* a save symbol containing signal *L_DataReq*, as shown in Figure 7.33.

D. Save the SDL model.

7.3.6 Two minutes to detect missing input *L_ReleaseReq* and answer *DM*

This time we will limit the input port length to 1 instead of 2, to finish more rapidly the bit-state simulation, to show how to detect never-executed SDL symbols.

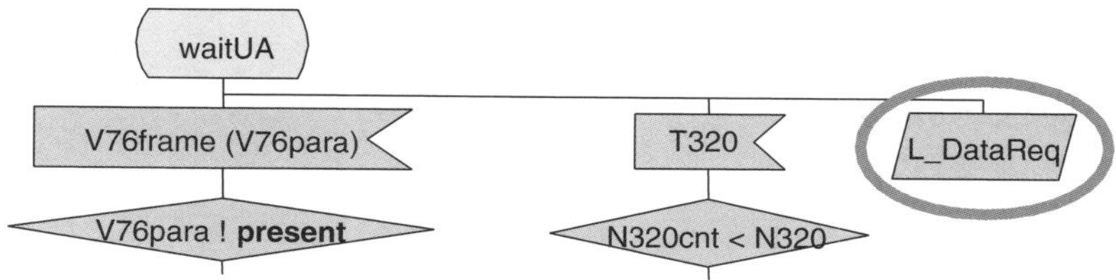

Figure 7.33 Process *DLC* after adding save *L_DataReq* under *waitUA*

7.3.6.1 Run again the bit-state simulation

A. In the Organizer, select the SDL system *V76test* and press *Validate* .

B. In the Validator, select *Options1 > Input Port Length*, and enter 1.

C. Select *Options2 > Bit State: Hash Size* and enter 250000000 (250 millions of bytes). If your machine is equipped, for example, with 128 MB of RAM, enter 80 millions.

D. Select *Options2 > Bit State: Depth* and enter 15000.

E. Select *Options1 > Report: Report Log*, choose *MaxQueueLength* and select *Off*.

F. Select *Commands > Include Command Script*, and choose *sig_defs.com*.

G. Press on *List Signal*, and check that you get the same signals as previously.

H. Press on *Bit State*, the Validator displays:

```
** Starting bit state exploration **
Search depth    : 15000
Hash table size : 250000000 bytes
Transitions: 20000 States: 12408 Reports: 0 Depth: 376
  Symbol coverage: 93.77 Time: 20:06:38
Transitions: 40000 States: 24847 Reports: 0 Depth: 300
  Symbol coverage: 93.77 Time: 20:06:38
...
Transitions: 7180000 States: 4479778 Reports: 0 Depth: 65
  Symbol coverage: 93.77 Time: 20:08:43
Transitions: 7200000 States: 4492191 Reports: 0 Depth: 150
  Symbol coverage: 93.77 Time: 20:08:43

** Bit state exploration statistics **
No of reports: 0.
Generated states: 7204384.
Truncated paths: 0.
Unique system states: 4494891.
Size of hash table: 2000000000 (250000000 bytes)
```

```
No of bits set in hash table: 8948021
Collision risk: 0 %
Max depth: 6530
Current depth: -1
Min state size: 212
Max state size: 584
Symbol coverage :   93.77
```

This time, no exception has been found, and the bit-state simulation has explored all the states of the SDL model reachable in the current test configuration (input ports limited to 1 etc.).

7.3.6.2 Analyze the nonexecuted SDL statements

After performing bit-state simulation, we must inspect the parts of the SDL model never executed. We see in the results displayed:

```
Symbol coverage :   93.77
```

Lets see exactly where the 6.23% never-executed symbols are.

A. In the Validator, select *Commands > Show Coverage Viewer*. The coverage viewer window appears as in Figure 7.34. If you double-click on the symbols marked with a zero, the SDL Editor opens the corresponding diagram and selects the symbol.

   The two uncovered symbols under process *dispatch* correspond to the reception of a *v76frame* containing a *DM*.

   The four symbols under process *DLC* correspond to two ELSE answers, supposed to never occur, and to the reception of a *v76frame* containing a *DM* under state *waitUA* shown in Figure 7.35.

   These two uncovered receptions of *v76frame* containing a *DM* cannot happen in our simulation, because signal *L_ReleaseReq* is never transmitted to side B (because the channel *dis* has been disabled in file *sig_defs.com*), but only to side A. Therefore, a connection established by A cannot be refused by B: the scenario shown in Figure 7.36 cannot happen.

   The MSC in Figure 7.36 shows the parts missing in the SDL model to refuse a connection: first, in process *dispatch* under state *waitEstabResp* the input of *L_ReleaseReq* is missing: Figure 7.37 shows this input added, followed by the transmission of *DM*. Second, when *DM* is received in *dispatch*, the answer *DM* is missing: Figure 7.38 shows this answer added, passing the *DM* to process *DLC*.

   Now, as process *DLC* can receive *DM*, the symbols shown in Figure 7.34 should be covered by the simulation.

B. Exit from the Validator (answering *No* to the question).

C. In Windows (or Unix), make a copy of the file *dispatch.spr* into *dispatch_v8.spr*.

D. Add the missing parts in process *dispatch*, as depicted in Figures 7.37 and 7.38.

E. Save the SDL model.

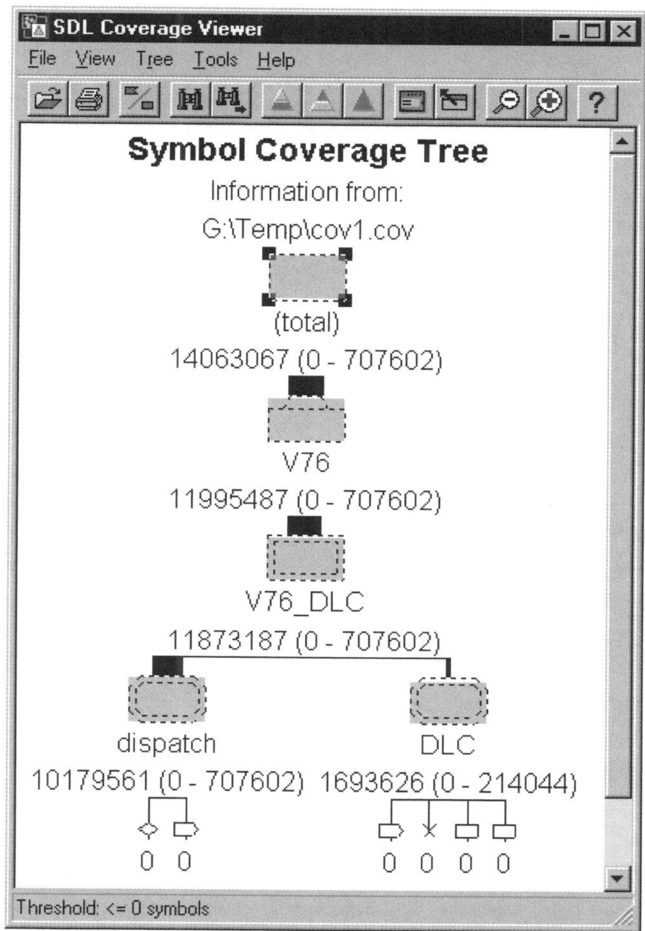

Figure 7.34 The six uncovered symbols in the coverage viewer

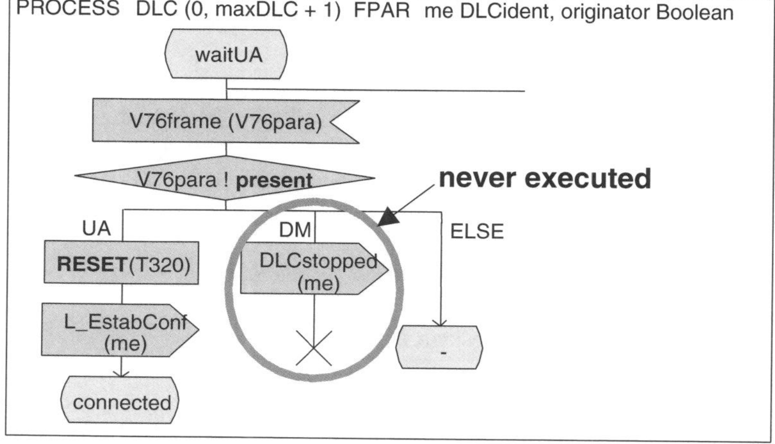

Figure 7.35 The branch never executed in process *DLC*

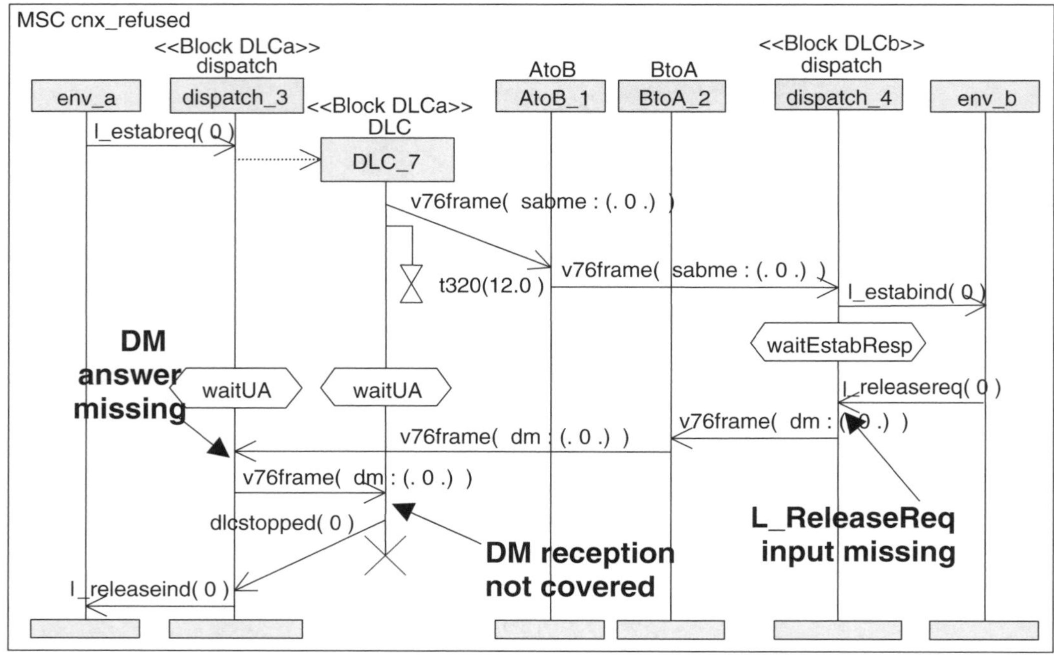

Figure 7.36 MSC showing connection establishment from A refused by B

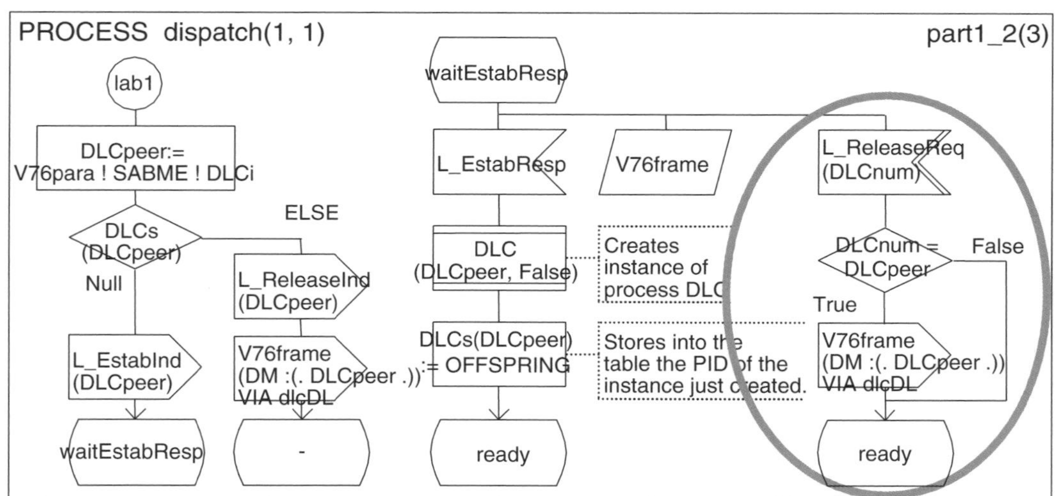

Figure 7.37 The input *L_ReleaseReq* added to process *dispatch*

7.3.7 Three minutes, 6.7 million states, no error

7.3.7.1 Run again the bit-state simulation

We simply rerun the bit-state simulation to check that no error has been introduced, and see if all the symbols are covered.

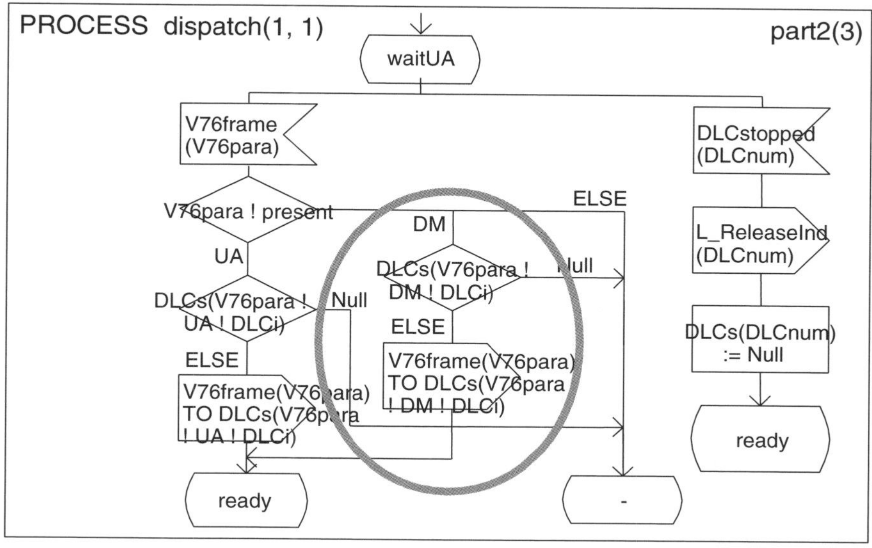

Figure 7.38 The answer *DM* added to process *dispatch*

A. In the Organizer, select the SDL system *V76test* and press *Validate* .

B. In the Validator, select *Options1 > Input Port Length*, and enter 2.

C. Select *Options2 > Bit State: Hash Size* and enter 250000000 (250 millions of bytes). If your machine is equipped, for example, with 128 MB of RAM, enter 80 millions.

D. Select *Options2 > Bit State: Depth* and enter 400.

E. Select *Options1 > Report: Report Log*, choose *MaxQueueLength* and select *Off*.

F. Select *Commands > Include Command Script*, and choose *sig_defs.com*.

G. Press on *List Signal*, and check that you get the same signals as previously.

H. Enter the command *Channel-Enable dis* to enable the Validator to transmit signal *L_ReleaseReq* to side B, to cover the SDL transitions previously added:

```
Command : Channel-Enable dis
Channel enabled.
```

I. Press on *Bit State*, the Validator displays:

```
** Starting bit state exploration **
Search depth     : 400
Hash table size : 250000000 bytes
Transitions: 20000 States: 15362 Reports: 0 Depth: 393
  Symbol coverage: 63.38 Time: 16:36:15
Transitions: 40000 States: 31214 Reports: 0 Depth: 398
  Symbol coverage: 63.38 Time: 16:36:15
...
```

```
Transitions: 10140000 States: 6736587 Reports: 0 Depth: 397
  Symbol coverage: 98.31 Time: 16:39:19
Transitions: 10160000 States: 6750862 Reports: 0 Depth: 380
  Symbol coverage: 98.31 Time: 16:39:19
```

J. After around six millions of states, press on *Break*; the Validator displays:

```
*** Break at user input ***

** Bit state exploration statistics **
No of reports: 0.
Generated states: 10168000.
Truncated paths: 794235.
Unique system states: 6756790.
Size of hash table: 2000000000 (250000000 bytes)
No of bits set in hash table: 13377019
Collision risk: 0 %
Max depth: 400
Current depth: 397
Min state size: 212
Max state size: 616
Symbol coverage :  98.31
```

No exception has been found. In 3 min and 4 s, the Validator has explored 6756790 of the reachable states of the SDL model. As we have enabled the SDL channel *dis*, more external signals are transmitted by the Validator to the SDL model than in the previous sessions: the simulation has been stopped before its end, the depth of which has been limited to 400 transitions. There are many truncated paths; therefore, this depth could be increased in order to explore all the states.

Remark: Tau generally finds more global states than ObjectGeode when the simulated SDL model expects signals coming from the environment, because in Tau Validator such signals are stored into the input queues of the process instances (like the other signals), whereas in ObjectGeode Simulator, such signals are input directly (like in a rendezvous).

7.3.7.2 Analyze the nonexecuted SDL statements

We must inspect again the parts of the SDL model that were never executed. We see in the results displayed:

```
Symbol coverage : 98.31
```

Lets see exactly where the 1.69% never-executed symbols are.

A. In the Validator, select *Commands > Show Coverage Viewer*. The coverage viewer window appears, as in Figure 7.39. Only one SDL symbol remains uncovered: it is an *ELSE* answer

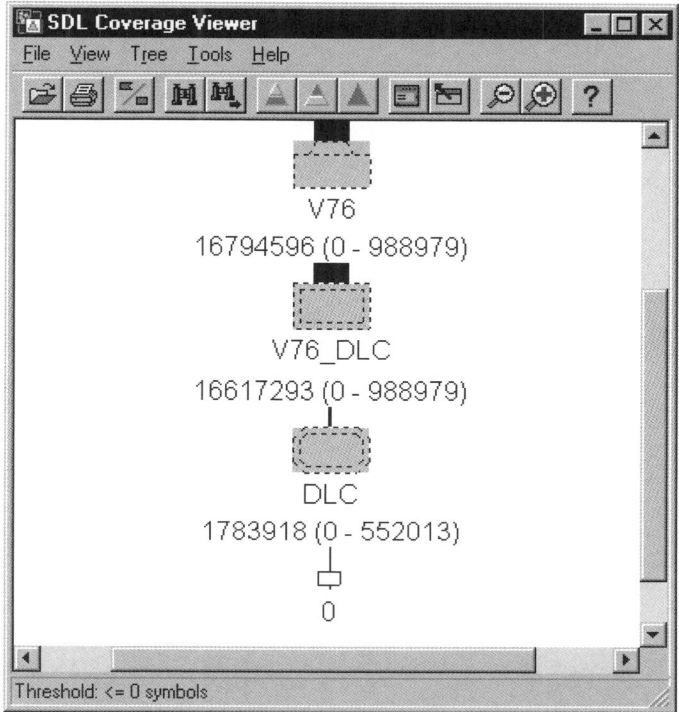

Figure 7.39 The symbol not covered in the coverage viewer

supposed to never occur. Therefore, this simulation session has covered all the symbols of the SDL model.

In a few minutes of simulation, you have:

- corrected all the discovered exceptions,
- covered all the SDL symbols, and
- proved that the SDL model contains no deadlock.

Note that this concerns a reduced model configuration. Bear in mind that millions of different scenarios have been executed here.

The next steps could be to simulate with other external signals configurations and other Validator settings such as different scheduling (default is *First*) or different priorities (default is higher priority to internal events than to signals from ENV).

7.3.8 Bit-state simulation with a user-defined rule

We want to detect that in our V.76 SDL model:

- instance 1 of process *AtoB* is in state *ready*,
- and instance 1 of process *BtoA* is in state *ready*.

More details on user-defined rules are provided in Chapter 5.

A. In the Organizer, select the SDL system *V76test* and press the *Validate* button.

B. In the Validator command line, enter:

```
Define-Rule state(AtoB:1)=ready and state(BtoA:1)=ready
```

C. Select *Options1 > Report : Abort*, and choose *UserSpecified*. After one report, the simulation will stop.

D. Press on *Bit State*, the Validator displays:

```
Search depth      : 100
Hash table size : 1000000 bytes

** Bit state exploration statistics **
No of reports: 1.
Generated states: 3.
Truncated paths: 0.
Unique system states: 2.
Size of hash table: 8000000 (1000000 bytes)
etc.
```

The simulation is terminated because the Validator has encountered a global state where the rule is true. The Report Viewer appears, as in Figure 7.40, showing that the rule is satisfied. To view the corresponding MSC trace, double-click on the lower box in the Report Viewer.

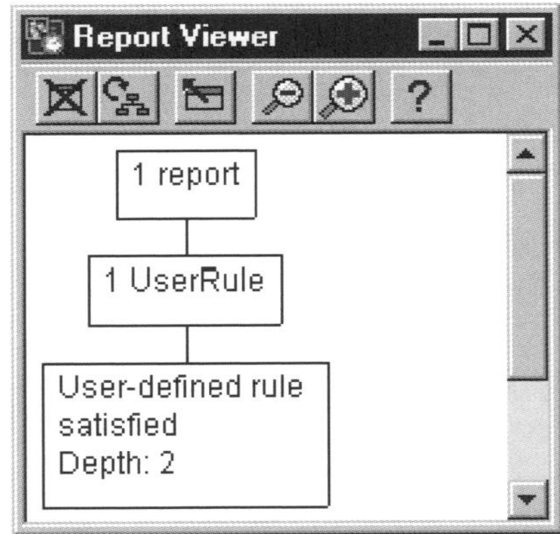

Figure 7.40 The Report Viewer showing that the rule is satisfied

7.3.9 Verifying an MSC with bit-state simulation

You will simulate the V.76 SDL model, observed by the basic MSC *test1.msc*. The Validator also accepts MSCs containing in-line operators and High-Level MSCs (HMSC).

Check that in *test1.msc* there is either a single environment instance named *env_0* or two environment instances named *DLCaSU* and *DLCbSU* (the names of the two external channels in the SDL model), otherwise the simulation would not match the loaded MSC.

More details on verifying an MSC are provided in Chapter 5.

A. In the Organizer, select the SDL system *V76test* and press the *Validate* button.

B. In the Validator, press on *Verify MSC*, and select the MSC *test1.msc*. The Validator starts a bit-state exploration:

```
MSC Test1 loaded.
Root of behavior tree set to current system state
Reports cleared
Bit state exploration started.
```

After less than one second, the Validator has finished:

```
** Bit state exploration statistics **
No of reports: 4.
Generated states: 69.
Truncated paths: 0.
Unique system states: 62.
Size of hash table: 8000000 (1000000 bytes)
No of bits set in hash table: 124
...
Symbol coverage :  78.03

** MSC Test1 verified **
```

The Report Viewer appears, as in Figure 7.41, showing four reports. There is one scenario satisfying the MSC and two scenarios violating the MSC. There are two violations because the MSC *test1* does not contain any possible behavior expected from the system. To view the corresponding MSC trace, double-click on one box in the Report Viewer.

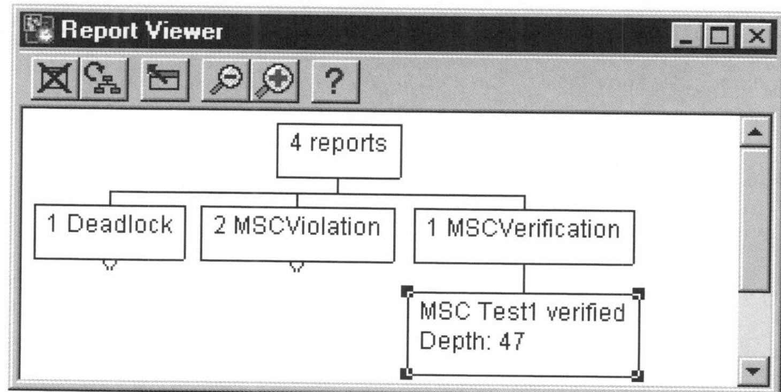

Figure 7.41 The Report Viewer showing that the MSC is satisfied

The textual command *Define-MSC-Verification-Algorithm* can be used to change the algorithm used by the Verify MSC command. The default is *BitState*, the other alternative is *TreeSearch*.

7.3.10 Bit-state simulation with observer processes

You will use the observer process created in Chapter 5, shown in Figure 5.36. This observer detects if the variable *uData* in process *dispatch* in block *DLCa* contains *55*.

A. In the Organizer, select *V76test*, choose *Edit > Connect*, choose *To an existing file*, press the folder-shaped icon and connect to the file *v76test_obs.ssy*.

B. In the Organizer, select *obs*, choose *Edit > Connect*, choose *To an existing file*, press the folder-shaped icon, select the file *obs.sbk*, check *Expand substructure* and press *Connect*.

C. In the Organizer, press the *save* button.

   The Organizer should now look like Figure 7.42.

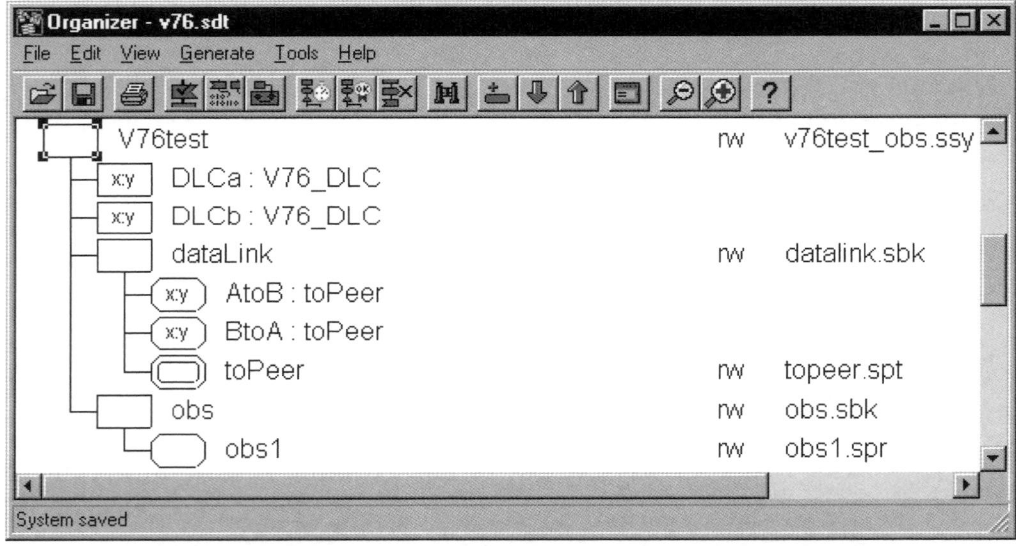

Figure 7.42 The Organizer showing the observer *obs*

D. In the Organizer, select the SDL system *V76test* and press the *Validate* button.

E. In the Validator, enter the command:

```
define-observer obs1
```

   This command tells the Validator to execute *obs1* as an observer instead of a regular process.

F. Select *Options1 > Report : Abort*, and choose *Assertion*. After one observer report, the simulation will stop.

G. Press on *Bit State*, the Validator displays:

```
Search depth    : 100
Hash table size : 1000000 bytes
```

```
** Bit state exploration statistics **
No of reports: 2.
Generated states: 195.
Truncated paths: 0.
Unique system states: 147.
Size of hash table: 8000000 (1000000 bytes)
etc.
```

The simulation is terminated because the Validator has encountered a global state where the observer process *obs1* has called the procedure *Report*. The Report Viewer appears, as illustrated in Figure 7.43, showing two reports. To view the corresponding MSC trace, double-click on the lower box in the Report Viewer.

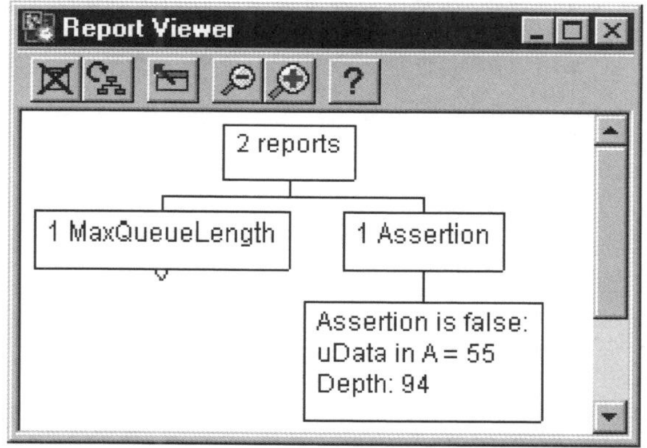

Figure 7.43 The Report Viewer showing that the assertion is reached

7.4 CASE STUDY WITH OBJECTGEODE

You will run the exhaustive simulation on the V.76 SDL model to discover errors automatically, and much faster and with much better dynamic coverage than with interactive or random simulation.

7.4.1 One second to detect missing save of *v76frame*

7.4.1.1 Start the Simulator

A. Open the model contained in *v76.pr* with the SDL Editor. Be sure to use the last version of *v76.pr*, including the two corrections mentioned in the previous chapters:

- Input signal *DLCstopped* added under state *waitUA* in process *dispatch.*
- Process *dispatch* goes to state *ready* instead of state *waitUA* after transmitting *L_Release-Ind.*

B. With a text editor, open the file *v76.startup* and remove the comment delimiter -- before the line *source v76_feed.wri*, preventing the feed commands execution, mentioned in Chapter 6.

C. Select *Tools > SDL & MSC Simulator.*

D. In the ObjectGeode Launcher, remove any file other than *v76.pr*, press the *Build* button, then, if there are no errors, press the *Execute* button. The Simulator starts.

E. If the Simulator has not executed automatically the four start transitions (step should be equal to 4), the file *v76.startup* is missing or incorrect (see Chapter 4). The exhaustive simulation starts from the current SDL model state, here Step 4.

F. Check that the feed commands (loaded by the file *v76.startup*) have been executed. See Chapter 4 if typing the Simulator command *list feed* does not give the following result:

```
> list feed
feed dlcbsu l_releasereq(0)
feed dlcbsu l_setparmresp()
feed dlcbsu l_estabresp()
feed dlcasu l_datareq(1 , 39)
feed dlcasu l_datareq(0 , 86)
feed dlcasu l_releasereq(1)
feed dlcasu l_setparmreq()
feed dlcasu l_estabreq(1)
feed dlcasu l_estabreq(0)
```

7.4.1.2 Run the exhaustive simulation

A. In the Simulator, select *Execute > Verify*: the *Verify Options* window appears, as shown in Figure 7.44. Do NOT use the Simulator button *Verify...*

B. In the *Exploration* part, enter 20000 for *States Limit* (you could also type the equivalent Simulator textual command *define states_limit 20000*).

C. Press the button *Verify* and confirm the verification startup: the Simulator displays the current options and starts the exhaustive simulation:

```
mode breadth
define edges_dump ''
define states_dump ''
deadlock limit 2
exception limit 2
stop limit 2
define stop_cut true
define states_limit 20000
define depth_limit 0
...
define verify_stats true
```

As expected, the exhaustive simulation stops after the exploration of 20000 global SDL model states. Only one second has been necessary to discover 568 exceptions, as indicated in the results:

```
(8192 states 19364 trans. 0 sec, depth=13, breadth=1897)
(16384 states 39243 trans. 1 sec, depth=15, breadth=3873)
verify stopped by states limit
```

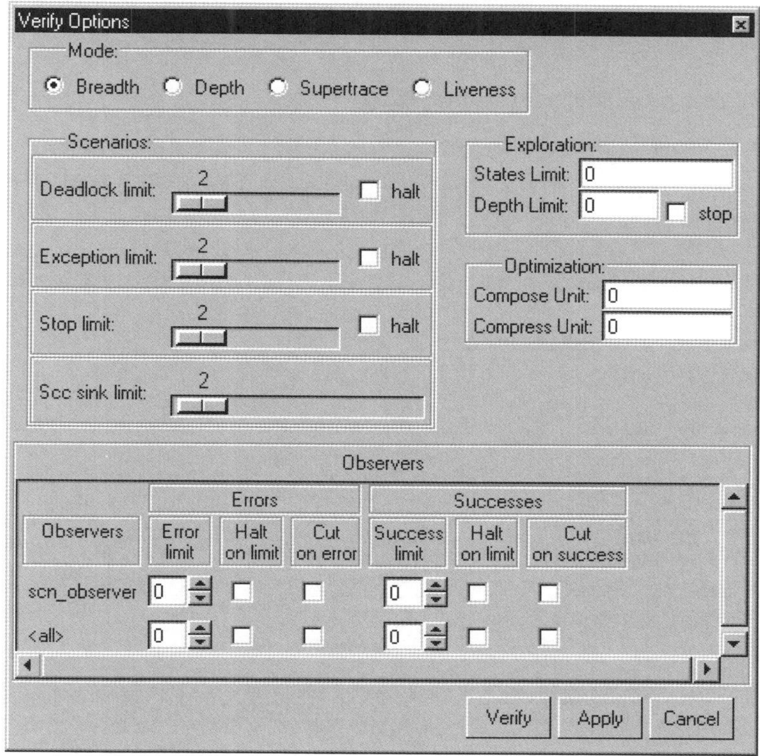

Figure 7.44 The *Verify Options* window

```
Number of states : 20000
Number of transitions : 47926
Maximum depth reached : 16
Maximum breadth reached : 6009
duration : 0 mn 1 s

Number of exceptions : 568
Number of deadlocks : 0
Number of stop conditions : 0
etc.
```

We see that the Simulator has executed 47 926 SDL transitions.

7.4.1.3 *Replay the exception scenario*

During the exhaustive simulation, as soon as the Simulator discovers a problem, it stores a scenario into a file. This scenario is the sequence of transitions that are to be executed to go from the initial state of the SDL model to the state where the problem has been discovered.

Here, the Simulator has generated two scenario files: *v76.x1.scn* and *v76.x2.scn*. The letter *x* stands for exception. The number after *x* is the order in which the exception was found. To get more files, you could enter a higher value for *Exception Limit* in the *Verify Options* window.

However, it is generally simpler to fix the first error, to recompile the SDL model and start a new simulation.

A. In the Simulator, press the button *Start MSC*.

B. Select *File > Scenario > Load*, and open *v76.x1.scn*.

C. Press the button *Redo: All*. The Simulator replays the scenario, reaches the exception, and displays:

```
Unexpected signal v76frame in dlcb!dispatch, line 312
  of v76.pr
11 transitions executed
end of scenario execution
```

The Editor displays the MSC trace of the exception scenario, shown in Figure 7.45.

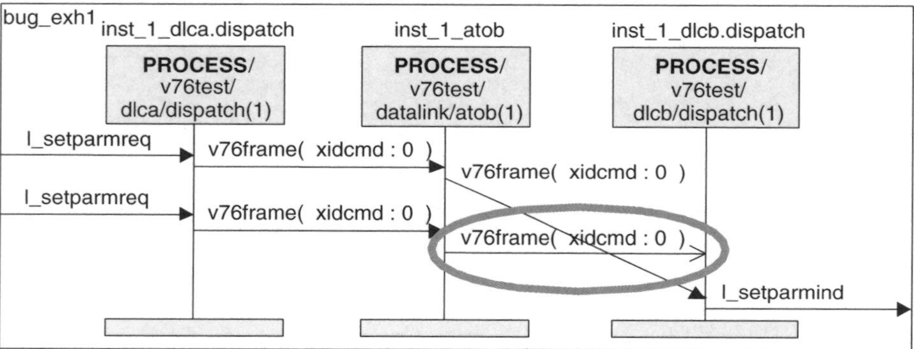

Figure 7.45 MSC trace of the first exception scenario

D. Enter the command *print state*; the Simulator answers:

```
> print state
btoa(1) ! state = ready
atob(1) ! state = ready
dlcb!dispatch ! state = waitparmresp
dlca!dispatch ! state = ready
```

We see that process *dispatch* in block *DLCb* is in state *waitParmResp*. If we look at the SDL model, under this state no input or save of signal *v76frame* are specified. Thus, this signal has been discarded.

7.4.1.4 Correct the exception

To prevent the signal from being lost, you will add a save of signal *v76frame* in state *waitParmResp* of process *dispatch*.

A. Exit from the Simulator (answering *No* to the question). Do not exit from the Editor.

B. In Windows (or Unix), make a copy of the file *v76.pr* into *v76_v3.pr*.

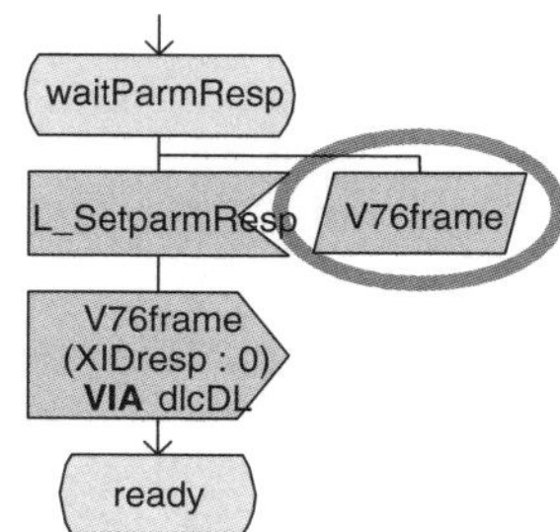

Figure 7.46 Missing save of signal *v76frame* added

C. In process *dispatch*, partition *part1*, select the input *L_SetParmResp* under the state *waitParmResp*, and add a save containing *v76frame*, as illustrated in Figure 7.46.

D. Save the SDL model.

7.4.1.5 Simulate to check the bug correction

To check that the bug has been corrected, you will load and automatically replay the scenario *v76.x1.scn*.

A. In the SDL Editor, unload all files except *v76.pr*.

B. If the ObjectGeode Launcher is not running, in the Editor select *Tools > SDL & MSC Simulator*.

C. In the ObjectGeode Launcher, remove any file other than *v76.pr*, press the *Build* button, then, if there are no errors, press the *Execute* button.

D. The Simulator starts: press on *Start MSC*.

E. In the Simulator, select *File > Scenario > Load*, open *v76.x1.scn*, and press the button *Redo: All*. The Simulator replays the scenario and the exception no longer exists.

   Do not exit from the Simulator.

7.4.2 One second to detect missing input *L_ReleaseReq*

7.4.2.1 Run the exhaustive simulation

A. In the Simulator, press on *init* ⏮, then press on *redo* ⏭ four times.

B. Select *Execute > Verify*: enter 20000 for *States Limit*, press *Verify* and confirm the verification startup.

This time, the exhaustive simulation has discovered 77 exceptions instead of 568, as indicated in the results:

```
(8192 states 19777 trans. 0 seconds, depth=13, breadth=1925)
(16384 states 40292 trans. 1 seconds, depth=15, breadth=3944)
verify stopped by states limit

Number of states : 20000
Number of transitions : 49140
```

```
Maximum depth reached : 16
Maximum breadth reached : 6115
duration : 0 mn 1 s
```

```
Number of exceptions : 77
Number of deadlocks : 0
Number of stop conditions : 0
etc.
```

It proves that the model has been improved by the previous correction.

7.4.2.2 Replay the exception scenario

Again, the Simulator has generated two scenario files: *v76.x1.scn* and *v76.x2.scn*.

A. In the Simulator, select *File > Scenario > Load*, and open *v76.x1.scn*.

B. Press the button *Redo: All*. The Simulator replays the scenario, reaches the exception, and displays:

```
Unexpected signal l_releasereq in dlcb!dlc(1), line 907
  of v76.pr
14 transitions executed
end of scenario execution
```

The Editor displays the MSC trace of the exception scenario, shown in Figure 7.47.

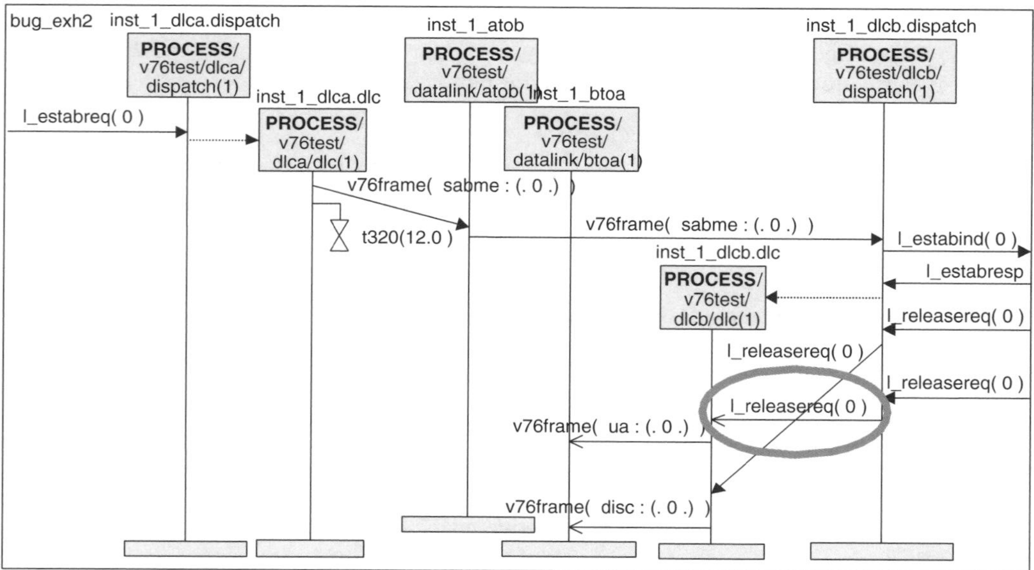

Figure 7.47 MSC trace of the exception scenario

C. Enter the command *print state*; the Simulator answers:

```
> print state
btoa(1) ! state = ready
```

```
atob(1) ! state = ready
dlcb!dlc(1) ! state = waituadisc
dlca!dlc(1) ! state = waitua
dlcb!dispatch ! state = ready
dlca!dispatch ! state = waitua
```

We see that process *DLC* in block *DLCb* is in state *waitUAdisc*. If we look at the SDL model, under this state no input or save of signal *L_ReleaseReq* are specified. Therefore, this signal has been discarded. The discarded *L_ReleaseReq* signal is shown in Figure 7.47.

7.4.2.3 Correct the exception

After examining the error, we decide that it is better losing *L_ReleaseReq* than saving it. Therefore, you will add a dummy input of signal *L_ReleaseReq* in state *waitUAdisc* of process *DLC*.

A. Exit from the Simulator (answering *No* to the question). Do not exit from the Editor.

B. In Windows (or Unix), make a copy of the file *v76.pr* into *v76_v4.pr*.

C. In process *DLC*, partition *part2*, select the input *v76frame* under the state *waitUAdisc*, and add an input containing *L_ReleaseReq* followed by nextstate dash, as illustrated in Figure 7.48.

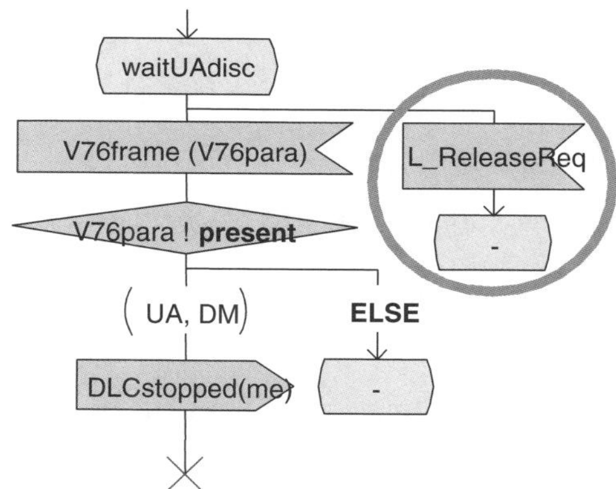

Figure 7.48 After adding input *L_ReleaseReq*

D. Save the SDL model.

7.4.3 One second to detect missing input *L_DataReq*

7.4.3.1 Run again the exhaustive simulation

A. In the SDL Editor, unload all files except *v76.pr*.

B. If the ObjectGeode Launcher is not running, in the Editor select *Tools > SDL & MSC Simulator*.

C. In the ObjectGeode Launcher, remove any file other than *v76.pr*, press the *Build* button, then, if there are no errors, press the *Execute* button.

D. In the Simulator, select *Execute* > *Verify*: enter 20 000 for *States Limit*, press *Verify* and confirm the verification startup.

   This time, the exhaustive simulation has discovered 1 exception instead of 77, as indicated in the results:

```
(8192 states 19791 trans. 0 seconds, depth=13, breadth=1929)
(16384 states 40370 trans. 1 seconds, depth=15, breadth=3964)
verify stopped by states limit

Number of states : 20000
Number of transitions : 49401
Maximum depth reached : 16
Maximum breadth reached : 6163
duration : 0 mn 1 s

Number of exceptions : 1
Number of deadlocks : 0
Number of stop conditions : 0
Etc.
```

It proves that the model has been improved again by the previous correction. Note that we do not know if the SDL model still provides the required functionalities: for that, we must use observers, as we will do in Sections 7.4.9 to 7.4.11.

7.4.3.2 Replay the exception scenario

The Simulator has generated one scenario file: *v76.x1.scn*.

A. In the Simulator, select *File* > *Scenario* > *Load*, and open *v76.x1.scn*.

B. In the Simulator, press on *Start MSC*.

C. Press the button *Redo: All*. The Simulator replays the scenario, reaches the exception, and displays:

```
Unexpected signal l_datareq in dlca!dlc(1), line 907
  of v76.pr
18 transitions executed
end of scenario execution
```

   The Editor displays the MSC trace of the exception scenario, depicted in Figure 7.49.

D. Enter the command *print state*; the Simulator answers:

```
> print state
btoa(1) ! state = ready
atob(1) ! state = ready
dlcb!dlc(1) ! state = connected
dlca!dlc(1) ! state = waituadisc
```

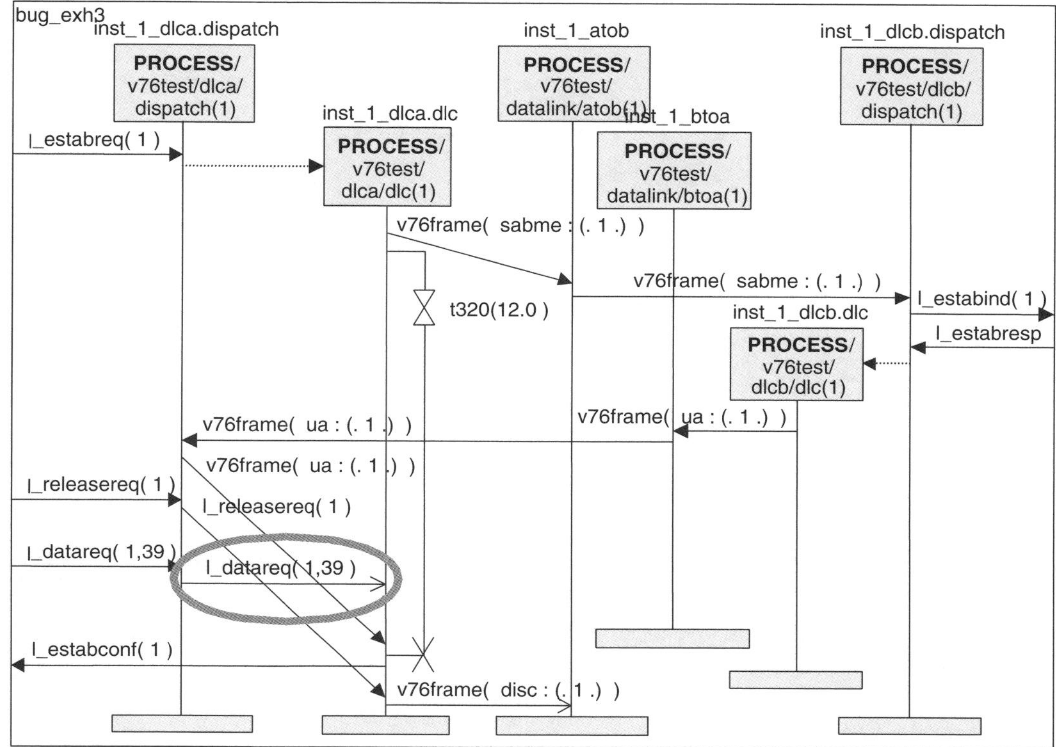

Figure 7.49 MSC trace of the exception scenario (*bug_exh3*)

```
dlcb!dispatch ! state = ready
dlca!dispatch ! state = ready
```

We see that process *DLC* in block *DLCa* is in state *waitUAdisc*. If we look at the SDL model, under this state no input or save of signal *L_DataReq* are specified. Thus, this signal has been discarded. The discarded *L_DataReq* signal is shown in Figure 7.49.

7.4.3.3 Correct the exception

After examining the error, we decide that as the connection is being released, it is better to lose *L_DataReq* than to save it. Therefore, you will add a dummy input of signal *L_DataReq* in state *waitUAdisc* of process *DLC*.

A. Exit from the Simulator (answering *No* to the question). Do not exit from the Editor.

B. In Windows (or Unix), make a copy of the file *v76.pr* into *v76_v5.pr*.

C. In process *DLC*, partition *part2*, insert a comma followed by *L_DataReq* in the input containing *L_ReleaseReq* previously added, as illustrated in Figure 7.50.

D. Save the SDL model.

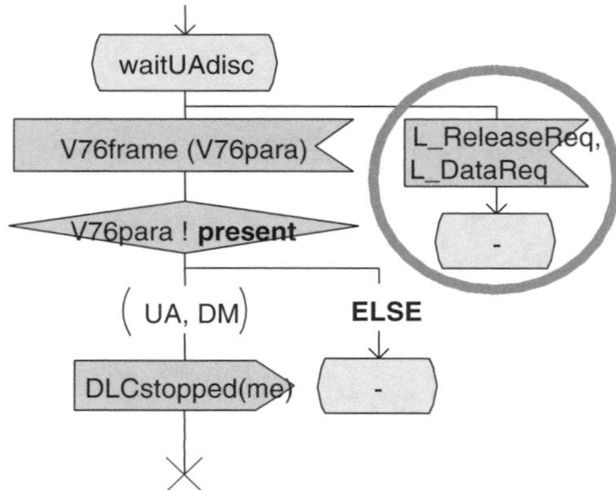

Figure 7.50 After adding input *L_DataReq*

7.4.3.4 Run again the exhaustive simulation

A. In the SDL Editor, unload all files except *v76.pr*.

B. If the ObjectGeode Launcher is not running, in the Editor select *Tools > SDL & MSC Simulator*.

C. In the ObjectGeode Launcher, remove any file other than *v76.pr*, press the *Build* button, then, if there are no errors, press the *Execute* button.

D. In the Simulator, select *Execute > Verify*: enter 20 000 for *States Limit*, press *Verify* and confirm the verification startup.

This time, the exhaustive simulation has not discovered any exceptions, as indicated in the results:

```
(8192 states 19791 trans. 0 seconds, depth=13, breadth=1929)
(16384 states 40370 trans. 1 seconds, depth=15, breadth=3964)
verify stopped by states limit

Number of states : 20000
Number of transitions : 49405
Maximum depth reached : 16
Maximum breadth reached : 6164
duration : 0 mn 1 s

Number of exceptions : 0
Number of deadlocks : 0
Number of stop conditions : 0
Etc.
```

In the 20 000 explored global states and 49 405 explored transitions of the SDL model, we are sure that we have no exceptions and no deadlocks. However, the global states not yet explored by the Simulator may contain errors.

7.4.4 Seventeen seconds to explore 87174 global states

To prove that the SDL model is correct, we will try to explore all its reachable states. We first use the following reduced configuration: one signal maximum in each queue, maximum two instances for each process *DLC*, and signals never lost in the block *dataLink*.

7.4.4.1 Start the Simulator

A. In the SDL Editor, unload all files except *v76.pr*, and select *Tools > SDL & MSC Simulator*.

B. In the ObjectGeode Launcher, remove any file other than *v76.pr*, press the *Build* button, then, if there are no errors, press the *Execute* button. The Simulator starts.

7.4.4.2 Add filters

To avoid an infinite number of global states, we need to limit the number of signals present in the input queue of each SDL process.

To detect which queues have too many signals, you must type (or put into your *geodesm.startup* file) *define verify_stats true*: after interrupting an exhaustive simulation, the Simulator will display the number of states taken by each process queue.

For example, in the V.76 SDL model, if you simulate the scenario shown in Figure 7.51, the queue of the instance 1 of process *DLC* in block *DLCa* contains four signals. If this process does not input the signals in its queue while other bursts of *L_DataReq* are transmitted to process *dispatch*, the number of *L_DataReq* stacked in the queue will grow rapidly. In addition, each new signal stacked in the queue generates a new global SDL model state during exhaustive simulation.

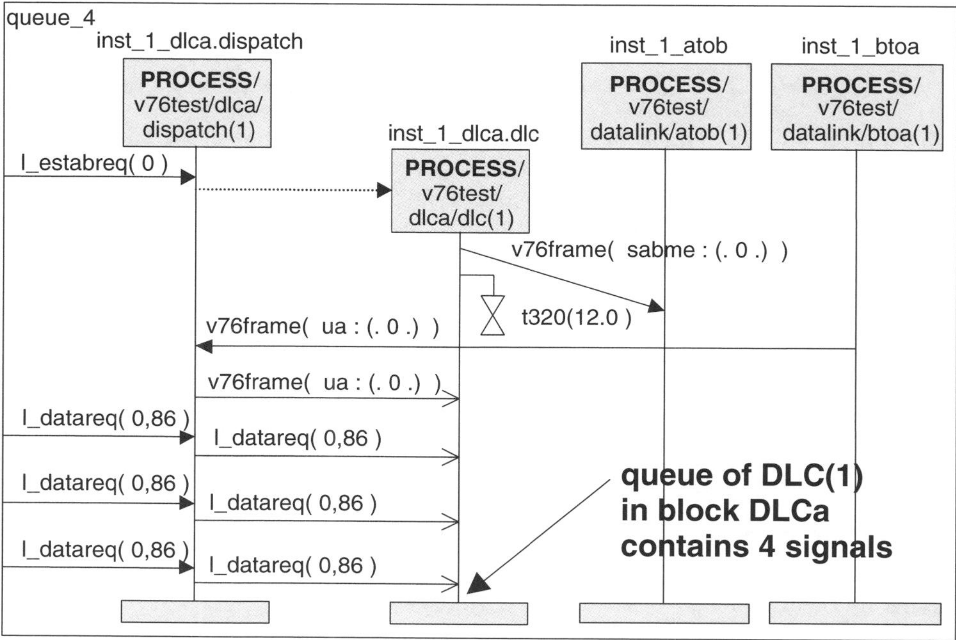

Figure 7.51 Scenario showing four signals in the queue of an instance of process *DLC*

To prevent such a situation, we use the *filter* command to limit to one signal in each queue. Note that some models might not work with such a limit, for example, if two signals are transmitted at the same time to a process queue.

A. In the Simulator, press on the button *Verify...*

B. In the *Verifying* window, press on *Limit a queue...*

C. In the *Limit a queue* window, press on *Instance*.

D. In the *Instance* window, select process *atob(1)* and press *OK*.

E. In the *Limit a queue* window, press on *Apply*.

F. Repeat the previous steps for instances *btoa(1), dlca!dispatch(1)* and *dlcb!dispatch(1)*.

G. Check that the *list filter* command gives the results below:

```
> list filter
filter is_active(dlcb!dispatch(1)) and length(dlcb!dispatch(1)
   ! queue) > 1
filter is_active(dlca!dispatch(1)) and length(dlca!dispatch(1)
   ! queue) > 1
filter is_active(btoa(1)) and length(btoa(1) ! queue) > 1
filter is_active(atob(1)) and length(atob(1) ! queue) > 1
```

H. We need filters for process instances *DLC(1)* and *DLC(2)*, in blocks *DLCa* and *DLCb*. Select *Edit* > *Filter Conditions* and add the following filters:

```
is_active(dlca!dlc(1)) and length(dlca!dlc(1)!queue) > 1
is_active(dlcb!dlc(1)) and length(dlcb!dlc(1)!queue) > 1
is_active(dlca!dlc(2)) and length(dlca!dlc(2)!queue) > 1
is_active(dlcb!dlc(2)) and length(dlcb!dlc(2)!queue) > 1
```

We must also limit the number of instances that can be created, because each new process instance gets its own new Pid; therefore, each new Pid generates a new global state. For example, if you simulate 50 times the sequence 'establish a DLC, release a DLC', you get 50 different Pids for process *DLC*. Remember that the limit indicated in the SDL model such as *DLC(0, 2)* only prevents having more than two instances of process *DLC* at the same time.

I. In the Simulator, select *Edit* > *Filter Conditions* and add the following filters:

```
create dlca!dlc(3)
create dlcb!dlc(3)
```

It means that the sequence 'establish a DLC, release a DLC' can be simulated two times only, because the transition leading to the creation of the third instance of process *DLC* is filtered. You can try the sequence in interactive mode, and see that after the sequence mentioned, it is not possible to establish a new DLC (signal *L_EstabReq*).

J. Finally, to simulate first a configuration where signals are not lost in the block *dataLink*, add the following filters:

```
trans btoa(1) : decision_lose_the_frame('Yes')
trans atob(1) : decision_lose_the_frame('Yes')
```

Now the only answer to the decision *'Lose the frame'* is *'No'*.

7.4.4.3 Save and tune the filters

To avoid entering again the filters at the next simulation session, we will save them into a file, automatically executed by the model startup file.

A. In the Simulator, type the command:

```
list filter >> v76_filter.wri
```

This creates the file *v76_filter.wri* and inserts the filter commands into it. We will simplify the filters: as the instances of process *dispatch, AtoB* and *BtoA* are static, that is, always exist, we can remove the expression *is_active* before them.

B. Open the file *v76_filter.wri* and remove *is_active* before *dispatch, AtoB* and *BtoA*. The file should now contain:

```
filter length(atob(1) ! queue) > 1
filter length(btoa(1) ! queue) > 1
filter length(dlca!dispatch(1) ! queue) > 1
filter length(dlcb!dispatch(1) ! queue) > 1
filter is_active(dlca!dlc(1)) and length(dlca!dlc(1)!queue)
  > 1
filter is_active(dlcb!dlc(1)) and length(dlcb!dlc(1)!queue)
  > 1
filter is_active(dlca!dlc(2)) and length(dlca!dlc(2)!queue)
  > 1
filter is_active(dlcb!dlc(2)) and length(dlcb!dlc(2)!queue)
  > 1
filter create dlca!dlc(3)
filter create dlcb!dlc(3)
trans btoa(1) : decision_lose_the_frame('Yes')
trans atob(1) : decision_lose_the_frame('Yes')
```

C. Open the file *v76.startup* and add *source v76_filter.wri*. The file *v76.startup* should now contain:

```
source v76_feed.wri
source start.scn
source v76_filter.wri
```

7.4.4.4 Set the configuration options

To get less global states, we will change the default settings of the Simulator. See Chapter 4 for details on *Edit > Configuration.*

A. Select *Edit > Configuration* and set *Reasonable environment* to on (box checked) and *Loose time progression* to off (box not checked).

7.4.4.5 Run the exhaustive simulation

In case the simulation never terminates, you can stop it by pressing the *halt* button.

A. Type the command *verify* to start the exhaustive simulation: the Simulator displays the current options and starts the exhaustive simulation:

```
mode breadth
...
deadlock limit 2
exception limit 2
stop limit 2
define stop_cut true
define states_limit 20000
define depth_limit 0
...
define verify_stats true
```

Then, after every 8192 global model states, the Simulator displays a line showing the simulation progression: number of (unique) global states, number of transitions executed, time elapsed since the beginning of simulation, maximum depth reached in the states graph and maximum breadth reached in the states graph.

```
(8192 states 11676 trans. 1 seconds, depth=34, breadth=934)
(16384 states 23502 trans. 2 seconds, depth=40, breadth=1615)
(24576 states 35716 trans. 4 seconds, depth=44, breadth=2086)
...
(65536 states 99823 trans. 12 seconds, depth=58, breadth=3269)
(73728 states 113177 trans. 14 seconds, depth=61, breadth=3269)
(81920 states 126678 trans. 15 seconds, depth=65, breadth=3269)
```

After 17 seconds, the exhaustive simulation stops and the Simulator displays the results:

```
Number of states : 87174
Number of transitions : 135912
Maximum depth reached : 79
Maximum breadth reached : 3269
duration : 0 mn 17 s

Number of exceptions : 0
Number of deadlocks : 0
Number of stop conditions : 0
Transitions coverage rate : 100.00 (0 transitions not covered)
States coverage rate : 100.00 (0 states not covered)
Basic blocks coverage rate : 92.98 (4 basic blocks not covered)
etc.
```

The simulation has covered all the 87174 reachable states of the reduced configuration of our SDL model. Obtaining such a coverage of the behavior would take weeks of interactive simulation.

No exceptions or deadlocks have been found.

7.4.5 Add faults in block *dataLink* : detect output to Null

Now to test more features in the SDL model, we use a larger model configuration: again one signal maximum in each queue and maximum two instances for each process *DLC*, but now signals can be lost in the block *dataLink*. To limit the number of states, we restrict the number of retransmissions in process *DLC* to 1, instead of 3.

7.4.5.1 Modify the SDL model

A. Exit from the Simulator. In the SDL Editor, unload all files except *v76.pr*; use the last corrected version of *v76.pr* obtained previously.

B. Open process *DLC part1* and replace 3 by 1 in the declaration of *N320*, to obtain:

```
SYNONYM N320 Integer = 1;
```

C. Save the SDL model and select *Tools > SDL & MSC Simulator.*

7.4.5.2 Run the exhaustive simulation

A. To enable block *dataLink* to lose signals, open the file *v76_filter.wri* with a text editor, and insert a comment symbol -- as shown:

```
-- trans atob(1)  : decision_lose_the_frame('Yes')
-- trans btoa(1)  : decision_lose_the_frame('Yes')
```

B. In the ObjectGeode Launcher, remove any file other than *v76.pr*, press the *Build* button, then, if there are no errors, press the *Execute* button.

C. Check that the Simulator has executed automatically the four start transitions.

D. In the Simulator, select *Edit > Configuration* and set *Reasonable environment* to on (box checked) and *Loose time progression* to off (box not checked)[2].

E. Select *Edit > Filter Conditions* and check that the following filters remain:

```
filter length(atob(1) ! queue) > 1
filter length(btoa(1) ! queue) > 1
filter length(dlca!dispatch(1) ! queue) > 1
filter length(dlcb!dispatch(1) ! queue) > 1
filter is_active(dlca!dlc(1)) and length(dlca!dlc(1)!queue) > 1
filter is_active(dlcb!dlc(1)) and length(dlcb!dlc(1)!queue) > 1
filter is_active(dlca!dlc(2)) and length(dlca!dlc(2)!queue) > 1
filter is_active(dlcb!dlc(2)) and length(dlcb!dlc(2)!queue) > 1
filter create dlca!dlc(3)
filter create dlcb!dlc(3)
```

[2] To avoid repeating this manual operation, you could add *define reasonable_feed 'true'* and *define loose_time 'false'* into the file *v76.startup*.

F. Type the command *verify* to start the exhaustive simulation: the Simulator displays the current options and starts the exhaustive simulation. After every 8192 global model states, the Simulator displays a line showing the simulation progression:

```
(8192 states 12156 trans. 1 s., depth=23, breadth=1432)
(16384 states 25067 trans. 2 s., depth=26, breadth=2612)
(24576 states 37704 trans. 3 s., depth=28, breadth=3829)
...
(1277952 states 2369806 trans. 278 s., depth=55, breadth=101778)
(1286144 states 2385398 trans. 279 s., depth=55, breadth=101778)
(1294336 states 2400944 trans. 281 s., depth=55, breadth=101778)
...
(2596864 states 5238512 trans. 635 s., depth=78, breadth=103218)
(2605056 states 5259095 trans. 637 s., depth=79, breadth=103218)
(2613248 states 5280374 trans. 640 s., depth=81, breadth=103218)
```

After 10 minutes and 42 seconds (on a cheap PC with a 950-MHz processor and 512 MB of RAM), the exhaustive simulation is completed and the Simulator displays the results:

```
Number of states : 2620001
Number of transitions : 5298932
Maximum depth reached : 92
Maximum breadth reached : 103218
duration : 10 mn 42 s

Number of exceptions : 23139
Number of deadlocks : 0
Number of stop conditions : 0
Transitions coverage rate : 100.00 (0 transitions not covered)
States coverage rate : 100.00 (0 states not covered)
Basic blocks coverage rate : 94.74 (3 basic blocks not covered)
etc.
```

The simulation has covered all the 2620001 reachable states of the current configuration of our SDL model. Obtaining such a coverage of the behaviors would take months of interactive simulation.

23139 exception states have been reached.

To estimate the size of a global state of the SDL model, type *tree*:

```
> tree
system v76test
   block datalink
      process atob ( 40 bytes )
      process btoa ( 40 bytes )
   block dlca
      process dispatch ( 60 bytes )
      process dlc ( 96 bytes )
   block dlcb
      process dispatch ( 60 bytes )
      process dlc ( 96 bytes )
```

Thus, the size of one global state is at minimum (no instance of *DLC* exists):

```
40 + 40 + 60 + 60 = 200 bytes
```

And the maximum is (when two instances of *DLC* exist on each side):

```
200 + (96 x 4) = 584 bytes
```

Therefore, the average state size is (not counting the input queues):

```
(200 + 584) / 2 = 392 bytes
```

The memory occupied by the states graph would have been (not counting its edges):

```
392 x 2 620 001 = 1027 megabytes
```

As the executable simulation file *v76.sim* has consumed a maximum of 196 MB of RAM instead of 1027, we see that the Simulator has compressed the states in a factor of approximately:

```
1027 / 196 = 5.24
```

Note that this compression does not lose any state, as opposed to algorithms such as bit-state or supertrace (which consume less memory).

7.4.5.3 Replay an exception scenario

The Simulator has generated two scenario files: *v76.x1.scn* and *v76.x2.scn*.

A. In the Simulator, select *File > Scenario > Load*, and open *v76.x1.scn*.

B. In the Simulator, press on *Start MSC*.

C. Press the button *Redo: All*. The Simulator replays the scenario, reaches the exception, and displays:

```
exception in transition
   dlca!dispatch : from_ready_input_v76frame :
No receiver for output v76frame from dlca!dispatch, line 468
  of v76.pr
23 transitions executed
time progressed from 0 to 24
end of scenario execution
```

The Editor displays the MSC trace corresponding to the exception scenario, depicted in Figure 7.52: A attempts to establish DLC number 0; as the response *L_EstabResp* from B is too late, A has received an *L_ReleaseInd*, meaning failure of DLC establishment; the *L_EstabResp* from B finally arrives (E1 in the MSC), *dispatch* in B creates an instance of *DLC*; signal *v76frame* containing a SABME was saved, thus as *dispatch* in B is back to state *ready*, it inputs the signal; *dispatch* in B transmits an *L_ReleaseInd* and a *v76frame* containing a DM; reaching *dispatch* in A, the *v76frame* should have been transmitted to the instance of *DLC* by executing the transition TR1 shown in Figure 7.53; unfortunately, the instance is dead; therefore an output to a Null Pid is executed, detected by the simulator.

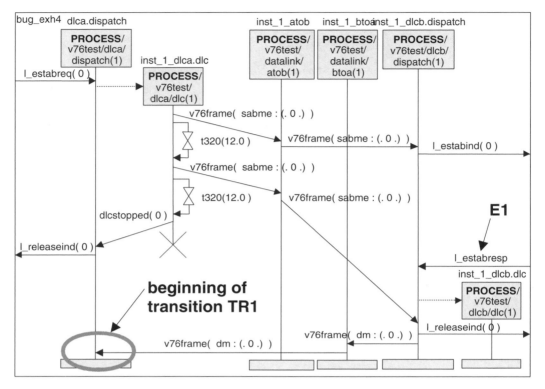

Figure 7.52 MSC trace of the exception scenario (*bug_exh4*)

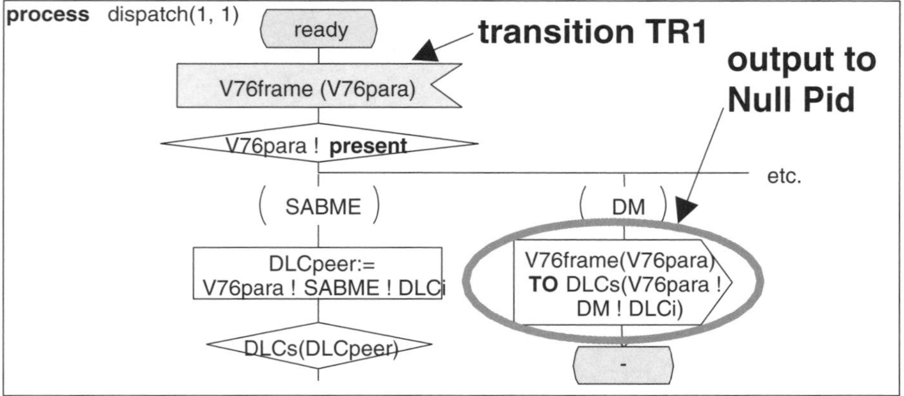

Figure 7.53 The process *dispatch* part1 (extract)

7.4.5.4 Correct the exception

The simulation has revealed that we must protect the expressions after TO in the output statements to avoid having a Null Pid. For that, you will add a decision to test the value of the expression: if Null, the output is not performed.

A. Exit from the Simulator (answering *No* to the question). Do not exit from the Editor.

B. In Windows (or Unix), make a copy of the file *v76.pr* into *v76_v6.pr*.

C. In process *dispatch*, from the Framework window, create a new partition *part1_2* and rename *part1 part1_1*.

D. Split the state machine in *part1_1* into two parts, one in *part1_1* and the other in *part1_2*, as illustrated in Figures 7.54 and 7.55.

E. Insert four decisions in *part1_1* as illustrated in Figure 7.54.

F. Insert a decision in *part2* after answer *UA*, as shown in Figure 7.56. Take care of staying in state *waitUA* when the answer is *Null* (*nextstate* -). Save the SDL model.

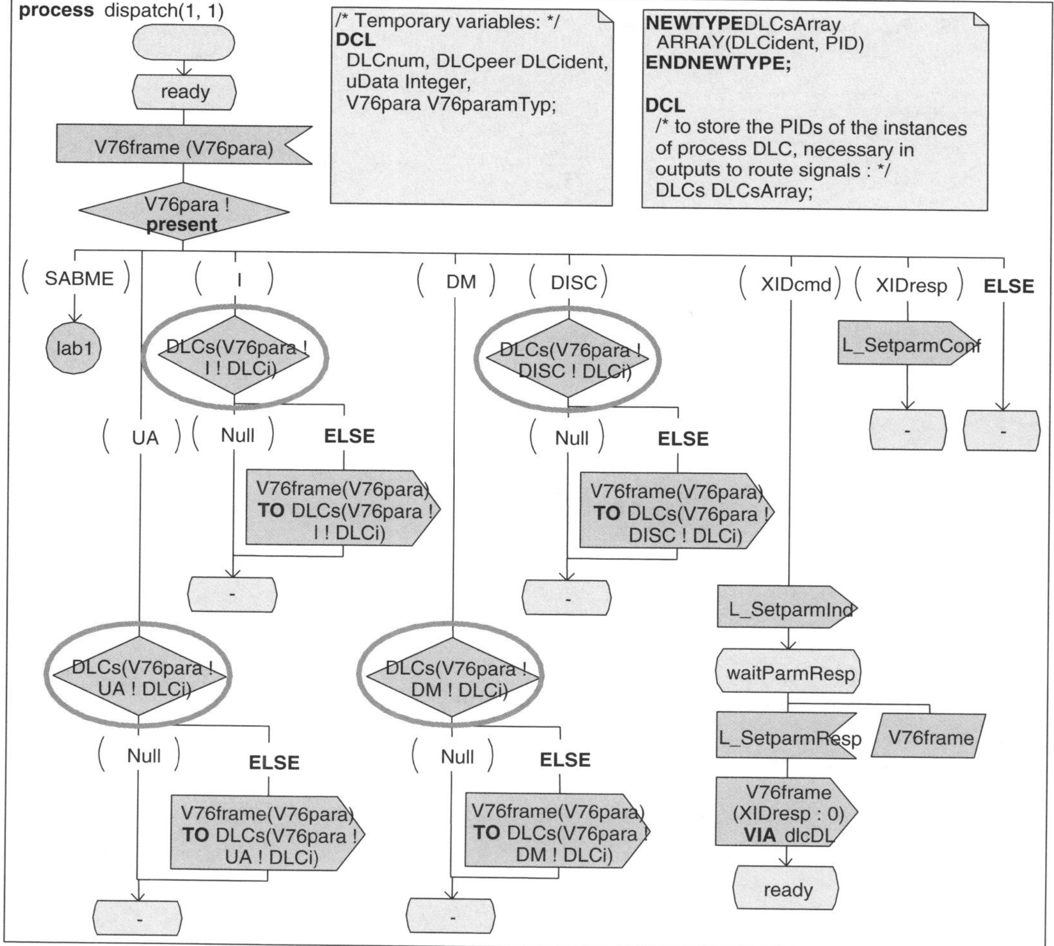

Figure 7.54 Process *dispatch* partition *part1_1*

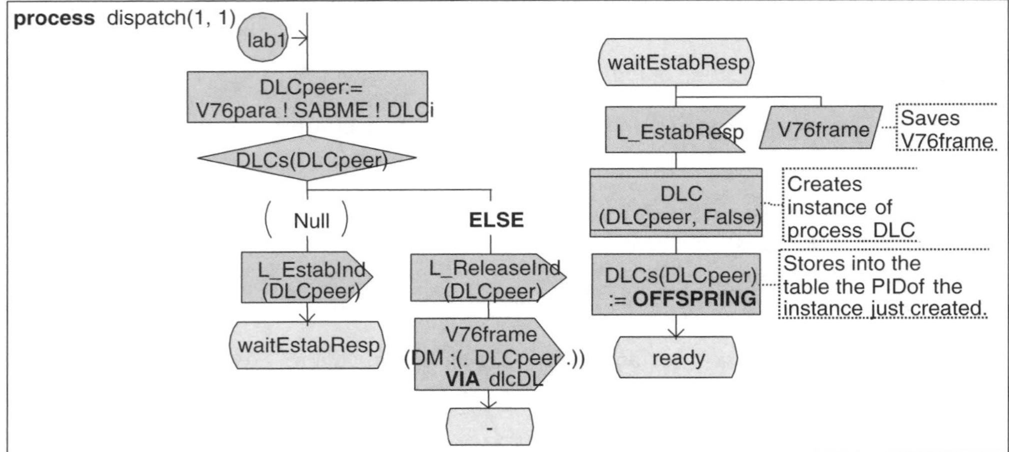

Figure 7.55 Process *dispatch* partition *part1_2*

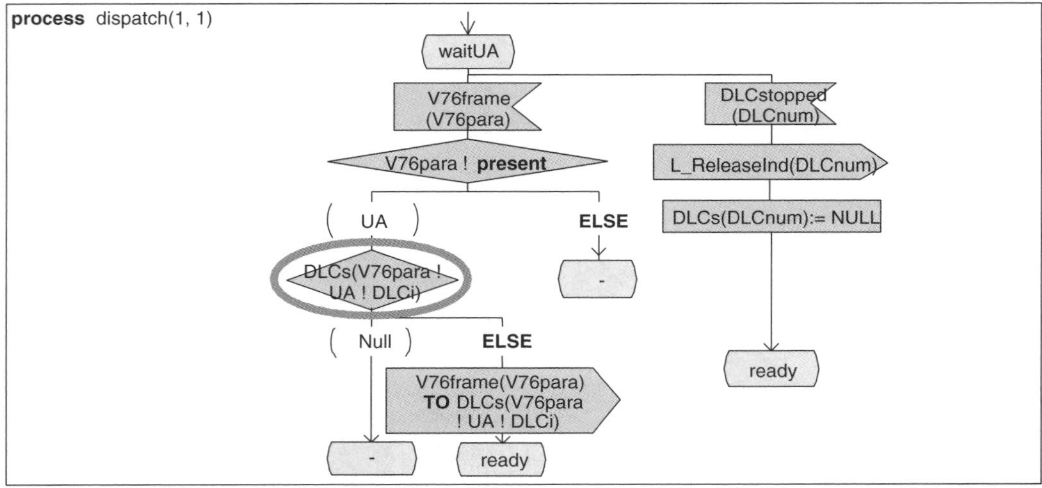

Figure 7.56 Process *dispatch* partition *part2*

7.4.6 Twenty-two seconds to detect missing save of *L_DataReq*

7.4.6.1 Run again the exhaustive simulation

To save time, we will set the simulator to stop after discovering two exceptions, rather than finishing the whole reachable states exploration.

A. In the SDL Editor, unload all files except *v76.pr*.

B. If the ObjectGeode Launcher is not running, in the Editor select *Tools* > *SDL & MSC Simulator*.

C. In the ObjectGeode Launcher, remove any file other than *v76.pr*, press the *Build* button, then, if there are no errors, press the *Execute* button.

D. In the Simulator, select *Edit > Configuration* and set *Reasonable environment* to on (box checked) and *Loose time progression* to off (box not checked).

E. Check that the Simulator has executed automatically the four start transitions.

F. Select *Edit > Filter Conditions* and check that the filters are the same as in Section 7.4.5.2 (especially the signal loss is no longer filtered).

G. Select *Execute > Verify*: in *Exception Limit*, enter 2 and check the *halt* box; the simulation will stop after discovering two exceptions.

H. Press *Verify* and confirm the verification startup. As expected, the exhaustive simulation stops after finding two exceptions in the SDL model, as indicated in the results:

```
(8192 states 12209 trans. 1 s., depth=23, breadth=1502)
(16384 states 25489 trans. 3 s., depth=26, breadth=2863)
...
(122880 states 194497 trans. 21 s., depth=37, breadth=18248)
(131072 states 207854 trans. 22 s., depth=37, breadth=18248)
verify stopped by an exception state

Number of states : 131367
Number of transitions : 208355
Maximum depth reached : 37
Maximum breadth reached : 18248
duration : 0 mn 22 s

Number of exceptions : 2
Number of deadlocks : 0
etc.
```

7.4.6.2 Replay the exception scenario

Again, the Simulator has generated two files containing the exception scenarios: *v76.x1.scn* and *v76.x2.scn*.

A. In the Simulator, select *File > Scenario > Load*, and open *v76.x1.scn*.

B. Press on *Start MSC*.

C. Press the button *Redo: All*. The Simulator replays the scenario, reaches the exception, and displays:

```
Unexpected signal l_datareq in dlca!dlc(2), line 1037
  of v76.pr
39 transitions executed
end of scenario execution
```

The Editor displays the MSC trace of the exception scenario, shown in Figure 7.57.

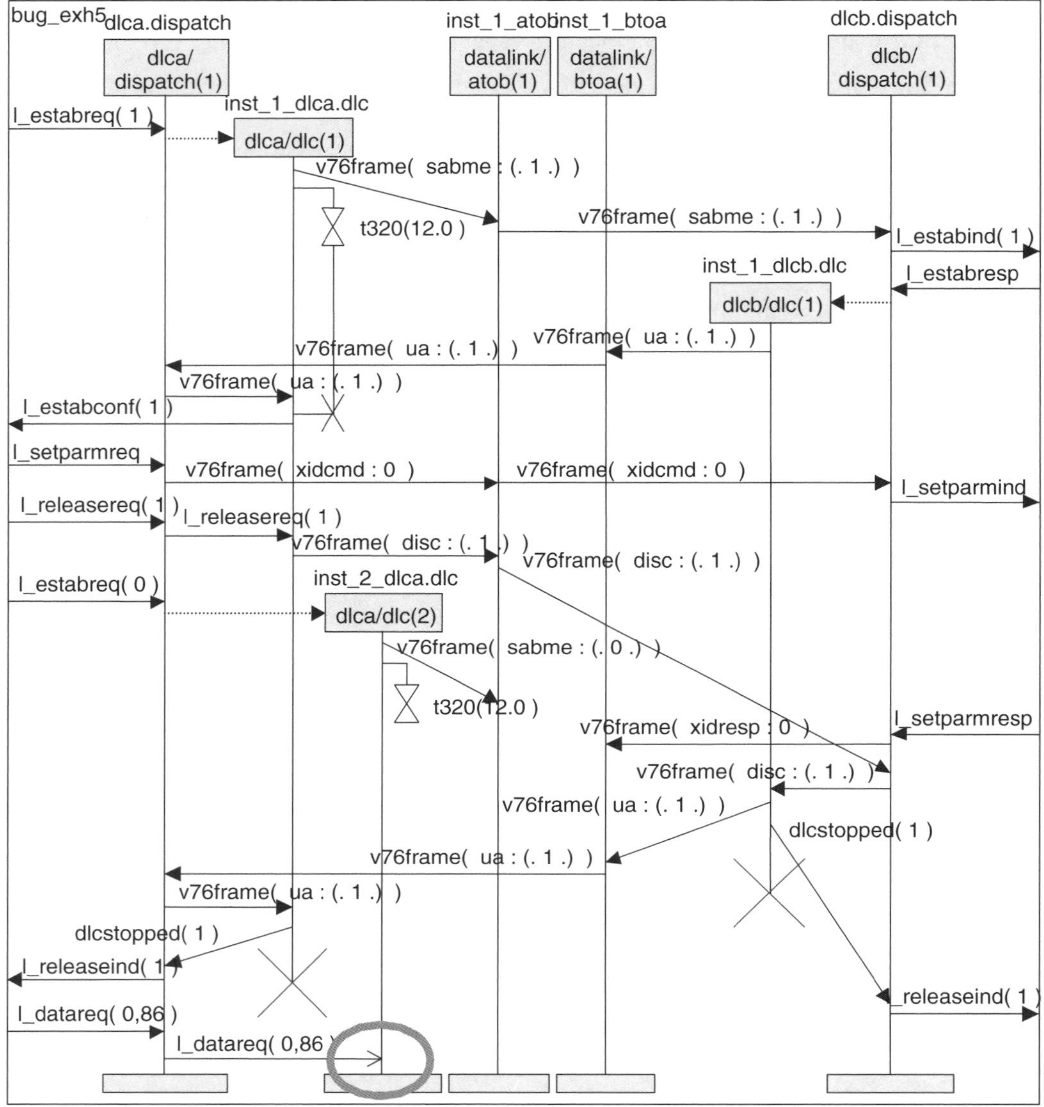

Figure 7.57 MSC trace of the exception scenario (*bug_exh5*)

D. Enter the command *print state*; the Simulator answers:

```
> print state
btoa(1) ! state = ready
atob(1) ! state = ready
dlca!dlc(2) ! state = waitua
dlcb!dispatch ! state = ready
dlca!dispatch ! state = ready
```

We see that instance 2 of process *DLC* in block *DLCa* is in state *waitUA*. If we look at the SDL model, under this state no input or save of signal *L_DataReq* are specified. Thus, this signal has been discarded.

7.4.6.3 Correct the exception

We decide to save signal *L_DataReq* in state *waitUA*, because once the connection is set up, the signal can be processed.

A. Exit from the Simulator (answering *No* to the question). Do not exit from the Editor.
B. In Windows (or Unix), make a copy of the file *v76.pr* into *v76_v7.pr*.
C. In process *DLC*, partition *part1*, add below state *waitUA* a save symbol containing signal *L_DataReq*, as shown in Figure 7.58.
D. Save the SDL model.

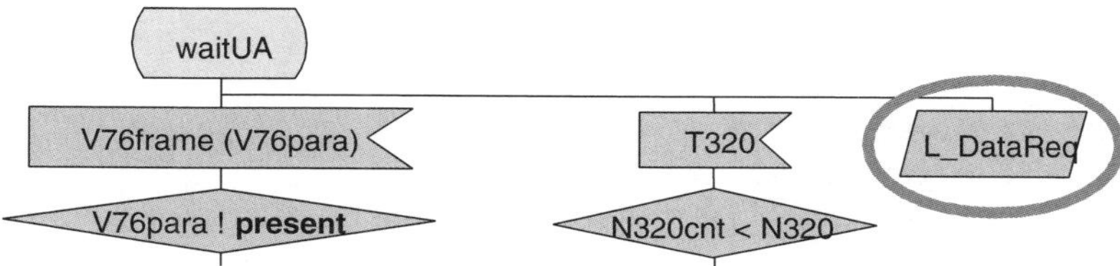

Figure 7.58 Process *DLC* after adding save *L_DataReq*

7.4.7 Eleven minutes to detect missing input *L_ReleaseReq* and answer *DM*

7.4.7.1 Run again the exhaustive simulation

A. In the SDL Editor, unload all files except *v76.pr*.
B. If the ObjectGeode Launcher is not running, in the Editor select *Tools > SDL & MSC Simulator*.
C. In the ObjectGeode Launcher, remove any file other than *v76.pr*, press the *Build* button, then, if there are no errors, press the *Execute* button.
D. In the Simulator, select *Edit > Configuration* and set *Reasonable environment* to on (box checked) and *Loose time progression* to off (box not checked).
E. Select *Edit > Filter Conditions* and check that the filters are the same as in Section 7.4.5.2 (especially the signal loss is no longer filtered).
F. Select *Execute > Verify*, press *Verify* and confirm the verification startup. The exhaustive simulation starts:

```
mode breadth
...
deadlock limit 2
exception limit 2 , halt
stop limit 2
define stop_cut true
define states_limit 0
define depth_limit 0
```

```
...
define verify_stats true

(8192 states 12209 trans. 1 s, depth=23, breadth=1502)
(16384 states 25489 trans. 2 s, depth=26, breadth=2863)
...
(974848 states 1779784 trans. 205 s, depth=52, breadth=94997)
(983040 states 1797464 trans. 207 s, depth=52, breadth=94997)
...
(2703360 states 5491611 trans. 679 s, depth=80, breadth=106711)
(2711552 states 5513338 trans. 682 s, depth=84, breadth=106711)

Number of states : 2713338
Number of transitions : 5518588
Maximum depth reached : 92
Maximum breadth reached : 106711
duration : 11 mn 22 s

Number of exceptions : 0
Number of deadlocks : 0
Number of stop conditions : 0
Transitions coverage rate : 100.00 (0 transitions not covered)
States coverage rate : 100.00 (0 states not covered)
Basic blocks coverage rate : 94.12 (4 basic blocks not covered)
```

G. Do not exit from the Simulator, you will need the coverage results for the next step.

This time, no exception has been found, and the exhaustive simulation has explored all the states of the SDL model reachable in the current test configuration (input queues limited to 1 etc.). The model is much better than the version where 23139 exceptions were found.

7.4.7.2 Analyze the nonexecuted SDL statements

After performing an exhaustive simulation, we must inspect the parts of the SDL model never executed. We see in the results displayed:

```
Transitions coverage rate : 100.00 (0 transitions not covered)
States coverage rate : 100.00 (0 states not covered)
Basic blocks coverage rate  : 94.12 (4 basic blocks not covered)
```

Lets see exactly where these four basic blocks are.

A. In the Simulator, type *cover bblocks all 0:0*; the result is (we have removed the 100% covered entities and the duplicate results for side *DLCb*):

```
>cover bblocks all 0:0

basic blocks coverage of dlca!dispatch : rate 95.00
  from_ready_input_v76frame
    29   D_presentextract(v76para).A_=i.D_extract(dlcs,
                dlciextract(iextract(v76para))).A_=null. : 0
    39   D_presentextract(v76para).A_else. : 0
```

```
basic blocks coverage of dlca!dlc : rate 91.30
  from_connected_input_v76frame
    11   D_presentextract(v76para).A_=i.D__vrp_1.A_=false. : 0
  from_waitua_input_v76frame
    15   D_presentextract(v76para).A_=dm. : 0
```

The basic block labeled 29 in the results above is located in process *dispatch*, in a transition from state *ready* beginning with the input of signal *v76frame. D_* means decision, *presentextract(v76para)* means *v76para ! present*, *.A_=i* means answer *=i*. Finally, the answer is followed by another decision: *D_extract(dlcs, dlciextract (iextract(v76para))). A_=null* means a decision containing *DLCs (V76para !I ! DLCi)=Null*. The corresponding basic block, plus the basic block number 39, are shown in Figure 7.59.

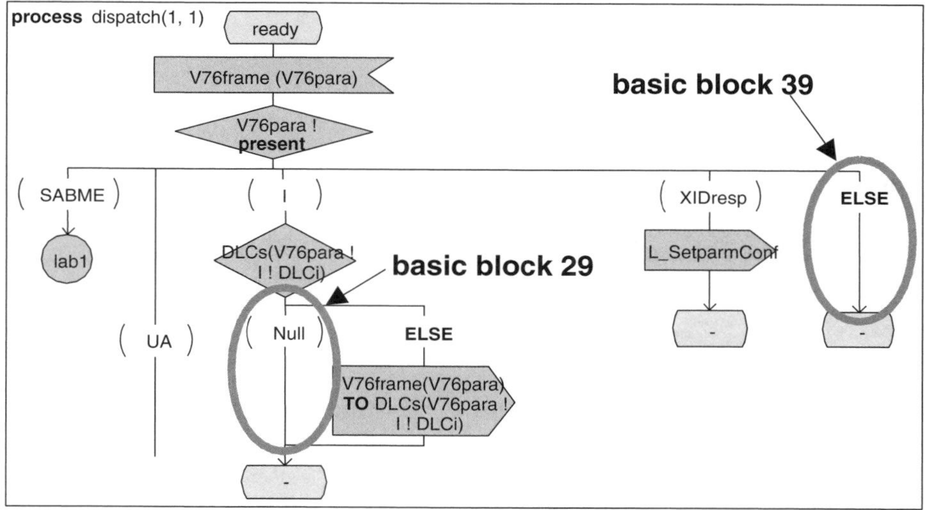

Figure 7.59 The two basic blocks never executed in process *dispatch*

The basic block number 29 corresponds to one of the five decisions added previously, to protect against output to Null. The fact that it has never been executed is not a problem. The basic block number 39 also corresponds to a case that should never happen.

The basic block number 11, shown in Figure 7.60, corresponds to the result *False* to the procedure *CRCok*: it is normal for this basic block to have never been executed, as the procedure always returns *True*.

The basic block number 15, shown in Figure 7.61, corresponds to the answer *DM* to the decision *V76para ! present*. It means that a connection establishment has never been refused by the peer Service User. After checking that our feed commands contain the transmission of *L_ReleaseReq* to block *DLCb*, we see that an input of *L_ReleaseReq* is missing in state *waitEstabResp,* corresponding to the case where the connection is refused.

Note that if we had used a process to model each Service User instead of using the Simulator feed command, the SU process would have transmitted *L_ReleaseReq* to block *DLCb*; then the Simulator would have detected an exception because the signal would have been discarded.

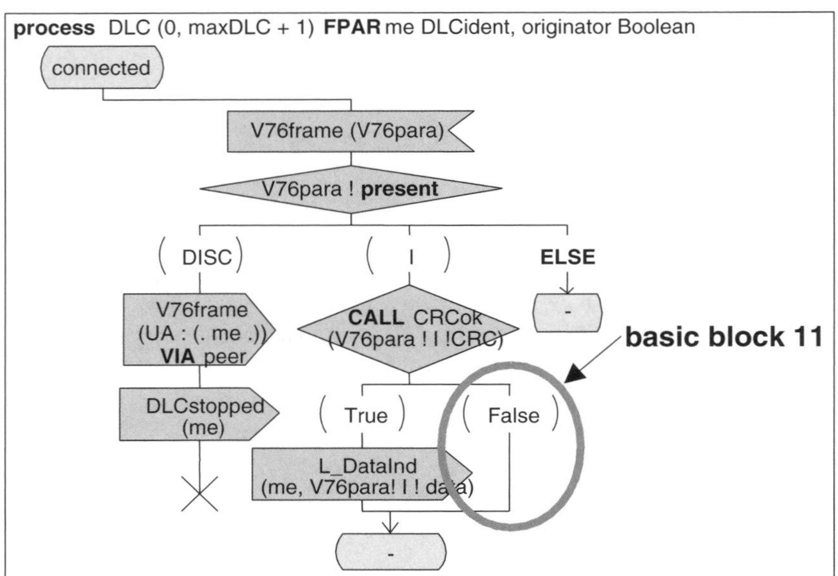

Figure 7.60 The basic block number 11 never executed in process *DLC*

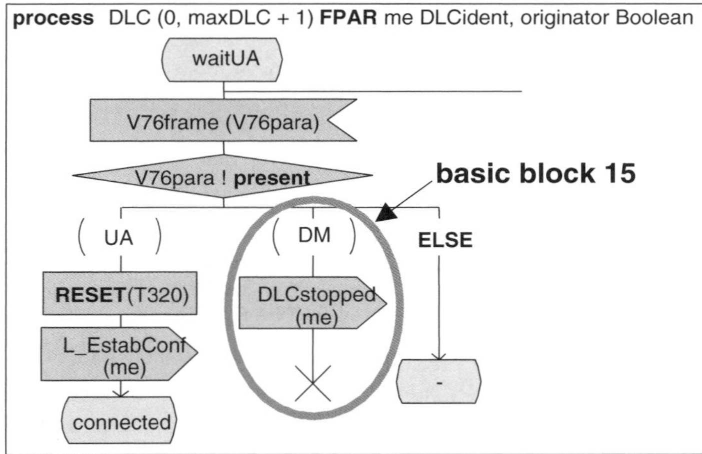

Figure 7.61 The basic block number 15 never executed in process *DLC*

7.4.7.3 Add missing transitions

The MSC in Figure 7.62 shows the parts missing in the SDL model to refuse a connection: first, in process *dispatch* under state *waitEstabResp* the input of *L_ReleaseReq* is missing: Figure 7.63 shows this input added, followed by the transmission of *DM*. Second, when *DM* is received in *dispatch*, the answer *DM* is missing: Figure 7.64 shows this answer added, passing the *DM* to process *DLC*.

Now, as process *DLC* can receive *DM*, the basic block 15 should be covered by the exhaustive simulation.

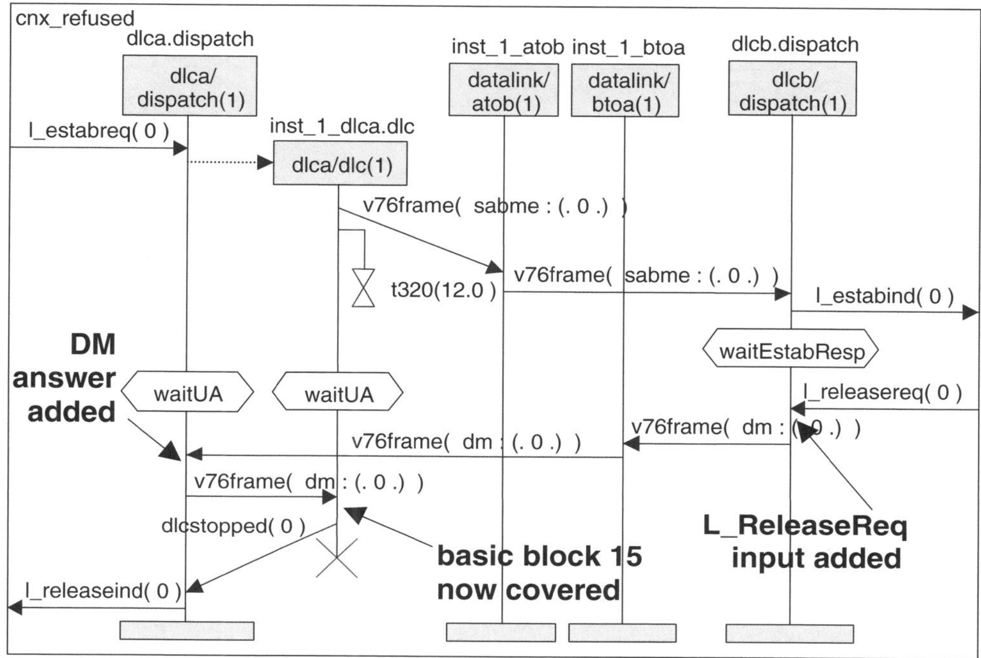

Figure 7.62 Connection establishment from A refused by B

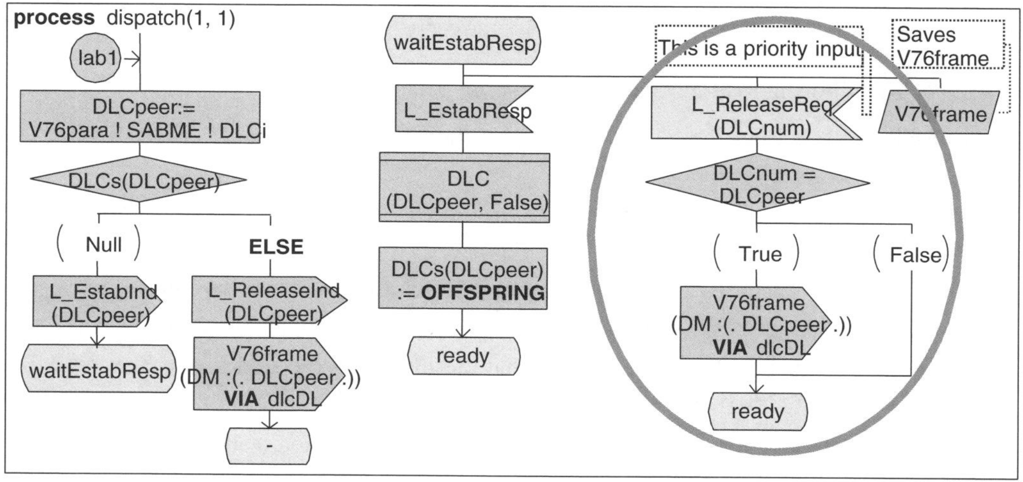

Figure 7.63 The input *L_ReleaseReq* added to *dispatch*

A. Exit from the Simulator (answering No to the question). Do not exit from the Editor.

B. In Windows (or Unix), make a copy of the file *v76.pr* into *v76_v8.pr*.

C. Add the missing parts in process *dispatch*, as depicted in Figures 7.63 and 7.64.

D. Save the SDL model.

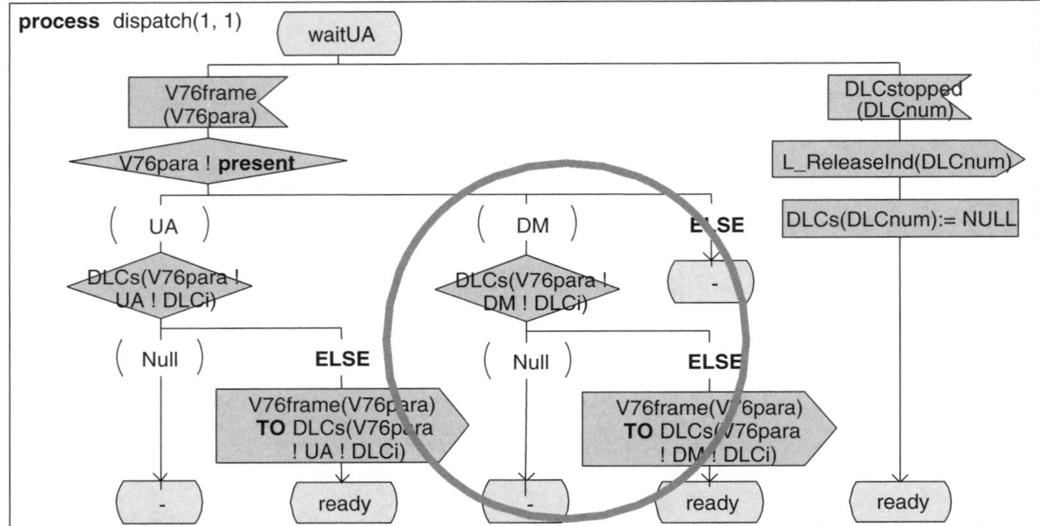

Figure 7.64 The answer *DM* added to *dispatch*

7.4.8 Eleven minutes, 2.8 million states, no error

7.4.8.1 Run again the exhaustive simulation

We simply rerun the exhaustive simulation to check that no error has been introduced and to see if the basic block previously numbered 15 is covered.

A. In the SDL Editor, unload all files except *v76.pr*.

B. If the ObjectGeode Launcher is not running, in the Editor select *Tools > SDL & MSC Simulator*.

C. In the ObjectGeode Launcher, remove any file other than *v76.pr*, press the *Build* button, then, if there are no errors, press the *Execute* button.

D. In the Simulator, select *Edit > Configuration* and set *Reasonable environment* to on (box checked) and *Loose time progression* to off (box not checked).

E. Select *Edit > Filter Conditions* and check that the filters are the same as in Section 7.4.5.2 (especially the signal loss is no longer filtered).

F. Select *Execute > Verify*: press *Verify* and confirm the verification startup.
The exhaustive simulation starts:

```
mode breadth
...
deadlock limit 2
exception limit 2 , halt
stop limit 2
define stop_cut true
define states_limit 0
```

```
define depth_limit 0
...
define verify_stats true

(8192 states 12178 trans. 1 s., depth=22, breadth=1607)
(16384 states 25318 trans. 2 s., depth=25, breadth=3084)
...
(1368064 states 2561762 trans. 296 s., depth=54, breadth=106925)
(1376256 states 2578827 trans. 299 s., depth=54, breadth=106925)
...
(2842624 states 5769663 trans. 699 s., depth=80, breadth=109861)
(2850816 states 5791131 trans. 701 s., depth=82, breadth=109861)

Number of states : 2855711
Number of transitions : 5804732
Maximum depth reached : 92
Maximum breadth reached : 109861
duration : 11 mn 43 s

Number of exceptions : 0
Number of deadlocks : 0
Number of stop conditions : 0
Transitions coverage rate  : 100.00 (0 transitions not covered)
States coverage rate : 100.00 (0 states not covered)
Basic blocks coverage rate  : 95.95 (3 basic blocks not covered)
```

Again, no exception has been found, and the exhaustive simulation has explored all the reachable states of the SDL model in its current configuration (nearly 3 millions).

7.4.8.2 Analyze the nonexecuted SDL statements

We must inspect again the parts of the SDL model never executed. We have three basic blocks not covered instead of four.

A. In the Simulator, type *cover bblocks all 0:0*; the result is (we have removed the 100% covered entities and the duplicate results for side *DLCb*):

```
>cover bblocks all 0:0

basic blocks coverage of dlca!dispatch : rate 95.65
  from_ready_input_v76frame
    35   D_presentextract(v76para).A_=i.D_extract(dlcs,
               dlciextract(iextract(v76para))).A_=null. : 0
    45   D_presentextract(v76para).A_else. : 0

basic blocks coverage of dlca!dlc : rate 95.65
  from_connected_input_v76frame
    11  D_presentextract(v76para).A_=i.D__vrp_1.A_=false. : 0
```

We see that the three basic blocks not covered are the normal ones identified previously and that the previous pathologically nonexecuted fourth basic block is now executed.

In a few minutes of simulation, we have:

- corrected all the discovered exceptions,
- covered all the SDL symbols,
- and proved that the SDL model contains no deadlock.

Note that this concerns a reduced model configuration. Bear in mind that millions of different scenarios have been executed here.

The next steps could be to simulate with other feed configurations, and other Simulator settings such as reasonable environment off, or two signals maximum in each input queue instead of one.

7.4.9 Exhaustive simulation with stop conditions

More details on stop conditions are provided in Chapter 5. We want to detect that a DLC is established: it means, in our V.76 SDL model, that:

- instance 1 of process *DLC* in block *DLCa* is in state *connected*,
- and instance 1 of process *DLC* in block *DLCb* is in state *connected*.

7.4.9.1 Run the exhaustive simulation

A. Start the Simulator as indicated previously (be sure not to compile any MSC with the SDL model). Check that the feeds are correct and that the step number is 4.

B. Select *Edit* > *Stop Conditions* and enter:

```
(DLCa!DLC(1)!state= connected) and (DLCb!DLC(1)!state=
   connected)
```

C. Select *Execute* > *Verify*: in *Stop Limit*, enter 2 and check the *halt* box; the simulation will stop after finding two scenarios leading to a global state where the stop condition is satisfied. To stop after one scenario instead of two, you should type the textual command *stop limit 1, halt*.

D. Press *Verify* and confirm the verification startup.

The exhaustive simulation stops after finding two stop condition scenarios:

```
mode breadth
...
deadlock limit 2
exception limit 2
stop limit 2 , halt
define stop_cut true
define states_limit 0
define depth_limit 0
...
```

```
define verify_stats true
verify stopped by a stop condition
Number of states : 1554
Number of transitions : 3550
Maximum depth reached : 11
Maximum breadth reached : 325
duration : 0 mn 0 s
Number of exceptions : 1
Number of deadlocks : 0
Number of stop conditions : 2
etc.
```

An exception has been found because we have not set *Reasonable environment* to on.

7.4.9.2 Replay the exception scenario

The Simulator has generated two files containing the stop condition scenarios: *v76.b1.scn* and *v76.b2.scn*. The letter *b* means break, being used for means success.

A. In the Simulator, select *File > Scenario > Load*, and open *v76.b1.scn*.

B. Press on *Start MSC*.

C. Press the button *Redo: All*. The Simulator replays the scenario, reaches the stop condition, and displays:

```
stop condition: (dlca!dlc(1)! state = connected)
  and (dlcb!dlc(1) ! state = connected)

15 transitions executed
end of scenario execution
```

The Editor displays the MSC trace corresponding to the stop condition scenario.

D. Enter the command *print state*; the Simulator answers:

```
> print state
btoa(1) ! state = ready
atob(1) ! state = ready
dlcb!dlc(1) ! state = connected
dlca!dlc(1) ! state = connected
etc.
```

We see that the two instances of process *DLC* are in state *connected*. Apart from proving properties, stop conditions are very handy for quickly finding a situation in an SDL model.

7.4.10 Exhaustive simulation with MSC observers

More details on MSC observers are provided in Chapter 5. You will simulate the V.76 SDL model observed by the MSC *test1.msc*. If necessary, several MSCs can observe the SDL model, together with stop conditions and GOAL observers.

7.4.10.1 Compile the SDL model plus the MSC

A. With a text editor, open the file *v76.startup* and insert a comment delimiter -- as below in front of the feed source line (because here we use the feeds generated by the MSC):

```
-- source v76_feed.wri
```

B. In the SDL (and MSC) Editor, load the V.76 SDL model plus the MSC *test1.msc*.

C. In the SDL Editor, select the file *test1.msc* and do *Edit > MSC Simulation Properties*: select *verify* and press *OK*. The Simulator will find more rapidly a scenario complying with the MSC. Using *search*, the simulation would take more time (especially with *Reasonable environment* off).

D. Save the MSC.

E. Unload any other files from the SDL Editor, and quit (do NOT minimize) the ObjectGeode Launcher if running.

F. In the SDL Editor, select *Tools > SDL & MSC Simulator*.

G. In the ObjectGeode Launcher, check that the left area only contains *v76.pr* and *test1.msc*.

H. Press the *Build* button.

7.4.10.2 Run the exhaustive simulation

A. Press the *Execute* button to start the Simulator. The Simulator main window appears.

B. Select *Execute > Verify*: in front of *test1*, set *Success Limit* to 1 and check the *halt* box, as illustrated in Figure 7.65; the simulation will stop after finding one scenario identical to the MSC *test1*.

C. Press *Verify* and confirm the verification startup.
One second later, the exhaustive simulation stops:

```
mode breadth
...
error limit test1 2
success limit test1 1 , halt
error cut test1 = true
...
define verify_stats true

verify stopped by a success state

Number of states : 65
Number of transitions : 80
Maximum depth reached : 37
Maximum breadth reached : 4
duration : 0 mn 0 s
```

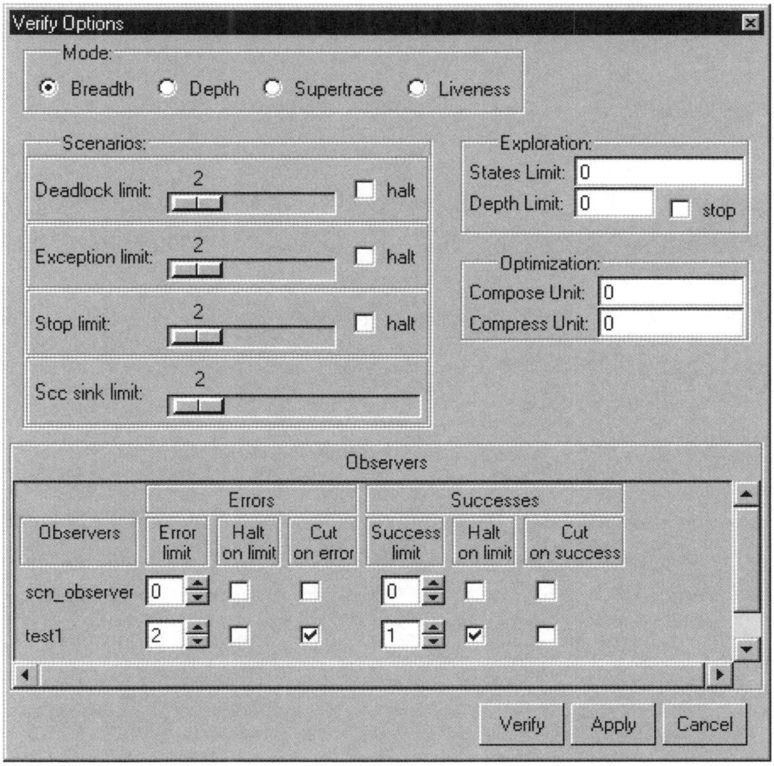

Figure 7.65 Setting success limit for MSC *test1*

```
Number of exceptions : 4
Number of deadlocks : 0
Number of stop conditions : 0
...
Number of errors : 132
Number of success : 1
etc.
```

The Simulator has found one success scenario and stopped, as required. In addition, it has discovered 132 errors: an error here is a scenario that is different from the expected MSC *test1*. As *test1* describes only one of the many possible execution scenarios, it is normal to get so many errors. To get no error here, you should create an MSC (using operators) describing all the possible behaviors of the system: this is extremely difficult on an actual system.

The success scenario discovered is contained in the file: *v76.s1.scn*. It could be loaded into the Simulator and replayed.

7.4.11 Exhaustive simulation with GOAL observers

More details on GOAL observers are provided in Chapter 5. You will simulate the V.76 SDL model observed by the GOAL observer *obs_ex2.obs*, built in Chapter 5. If necessary, this file could contain more than one observer.

7.4.11.1 Compile the SDL model plus the GOAL observer

A. With a text editor, open *v76.startup* and remove the comment delimiter – in front of the feed source line, added in the previous exercise, to get:

```
source v76_feed.wri
```

B. In the SDL Editor, load the V.76 SDL model plus the GOAL observer *obs_ex2.obs*.

C. Unload any other files from the SDL Editor, and quit (do NOT minimize) the ObjectGeode Launcher if running.

D. In the SDL Editor, select *Tools > SDL & MSC Simulator*.

E. In the ObjectGeode Launcher, check that the left area only contains *v76.pr* and *obs_ex2.obs*. Press the *Build* button.

7.4.11.2 Run the exhaustive simulation

A. Press the *Execute* button to start the Simulator. The Simulator main window appears.

B. Select *Execute > Verify*: in front of *obs1*, set *Success Limit* to 1 and *Error Limit* to 1 (to get only one scenario for each). Enter 100000 in *Exploration: States Limit*.

C. Press *Verify* and confirm the verification startup. The exhaustive simulation starts:

```
mode breadth
...
error limit obs1 1
success limit obs1 1
...
define states_limit 100000
...
(8192 states 20063 transitions 2 s., depth=16, breadth=1936)
(16384 states 42553 transitions 4 s., depth=18, breadth=3983)
...
```

Several seconds later, the exhaustive simulation stops:

```
verify stopped by states limit

Number of states : 100000
Number of transitions : 279118
Maximum depth reached : 21
Maximum breadth reached : 22119
duration : 0 mn 31 s

Number of exceptions : 1739
Number of deadlocks : 0
Number of stop conditions : 0
...
Number of errors : 3
Number of success : 4324
...
```

As expected, the Simulator has stopped after exploring 100000 global states. It has discovered 3 error scenarios and 4324 success scenarios.

The first success scenario discovered by observer *obs1* is contained in the file: *v76.obs1.s1.scn*. Figure 7.66 shows the MSC generated (at block levels) after loading the success scenario into the Simulator.

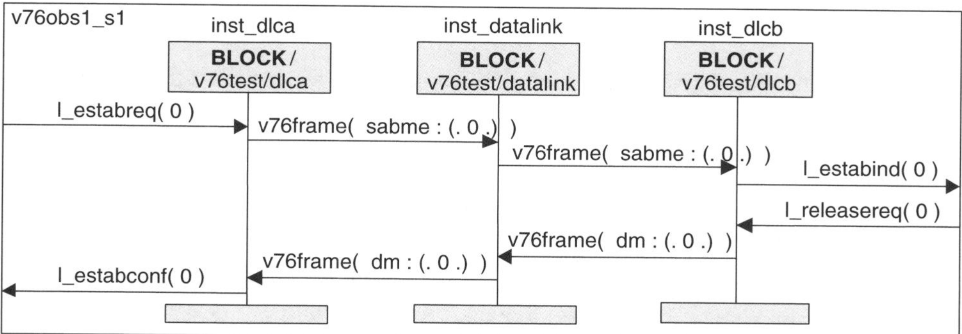

Figure 7.66 The first success scenario discovered by the Simulator

The first error scenario discovered by observer *obs1* is contained in the file: *v76.obs1.e1.scn*. Figure 7.67 shows the MSC generated (at block levels) after loading the error scenario into the Simulator.

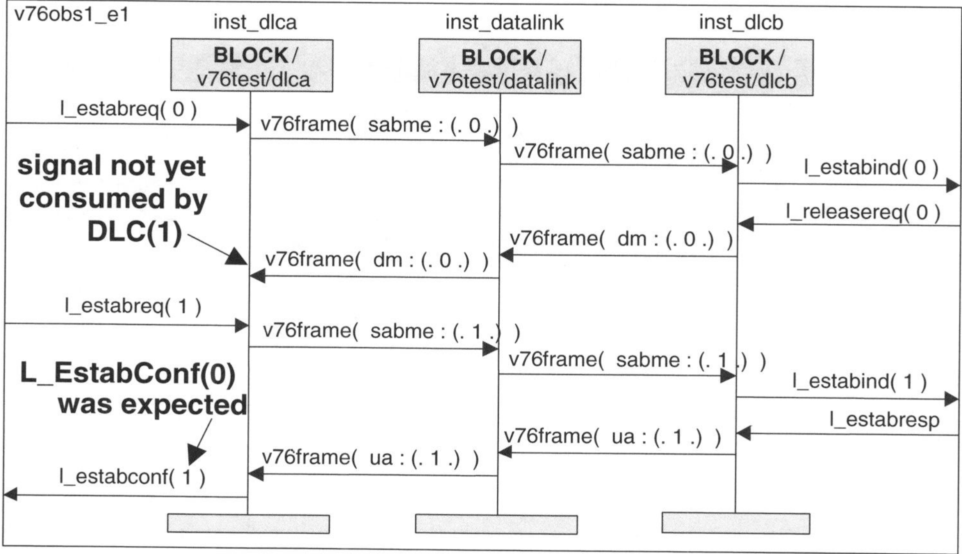

Figure 7.67 The error scenario discovered by the Simulator

We see that the observer has detected an error because the parameter of *L_EstabConf*, 1, is not equal to the parameter of *L_EstabReq*, 0: because *Reasonable environment* was off, the Simulator has transmitted a new *L_EstabReq* to the model before the end of the internal events.

The error scenario shows that our model could be improved by transmitting an *L_ReleaseInd* as soon as the *DM* is received. Then the observer could be improved to be reset by the observation of an *L_ReleaseInd* when an *L_EstabConf* is expected.

7.5 OTHER SIMULATION ALGORITHMS

7.5.1 Tau SDL Suite

Besides bit-state and exhaustive modes, Tau SDL Suite Validator proposes the following exploration modes:

- tree search: this is a free exploration of the states graph where the global states are not stored; thus the Validator cannot detect if a state has already been explored. In this mode, the exploration never stops, even on a model with few global states.
- power walk, tree walk: these algorithms are designed for TTCN test case automatic generation for Autolink, to maximize the SDL symbols coverage; we will not describe them.

7.5.2 ObjectGeode: supertrace

The supertrace exploration mode in ObjectGeode Simulator is equivalent to bit-state [Holz91] in Tau SDL Suite Validator. A presentation of bit-state is given in Section 7.1.2.

To use supertrace, you must do *Execute* > *Verify*, select *Supertrace*, and enter the size of the bits array (the hash table) in the field *States Limit*, as shown in Figure 7.68.

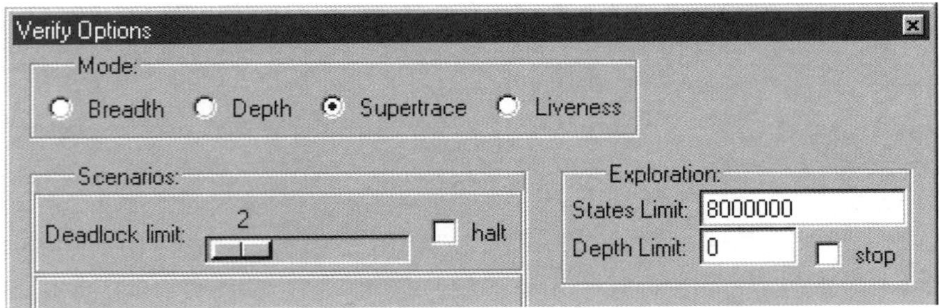

Figure 7.68 The top of the *Verify Options* window in supertrace mode

In this example, we have set the size of the bits array to 8 million bits, enough to contain 8 million states.

We have run the *count* SDL model, presented in Section 7.2.2, supposed to have 10000 unique states. As the size of the bits array is 800 times greater than the number of states, the collision risk is very low. ObjectGeode found only 830 unique states, whereas Tau SDL Suite Validator using the same model with the same tool configuration found 10000 unique states. This is because ObjectGeode uses only one hashing function, and Tau uses two hashing functions: if one function gives the same hash-code for two different states, the other function has a chance to give two different hash-codes, and therefore to distinguish that the states are not identical.

7.5.3 ObjectGeode: liveness

7.5.3.1 Safety and liveness properties

After the end of an exhaustive simulation (in modes *breadth* or *depth*) with observers, if you get the following results:

```
Number of errors : 0
Number of success : 350
```

it proves that the SDL model does not contain any error and that it can reach the success states defined by the observers. This is a safety property.

To prove that the SDL model always reaches a success state (a liveness property), you must use an exhaustive simulation mode called *liveness*: this mode, a variant of the *depth* mode, detects the loops in the states graph that do not lead to a success state of the observers.

7.5.3.2 The maze model

To illustrate the notions of safety and liveness, lets consider the simple maze represented in Figure 7.69(a): you enter by the top and you try to reach the exit; suppose you are not allowed to backtrack.

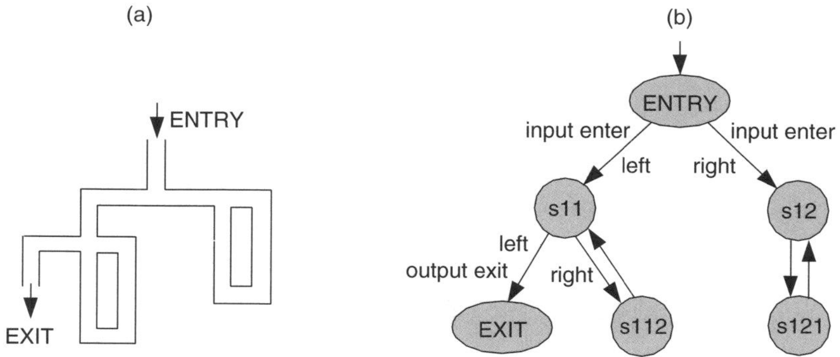

Figure 7.69 (a) A maze and (b) a state machine modeling it

Figure 7.69(b) shows a state machine modeling the maze: from state *ENTRY* the left path leads to state *s11*. From *s11* the left path leads to state *EXIT* and so on.

To simulate the state machine, we have translated it into the SDL model shown in Figures 7.70 and 7.71.

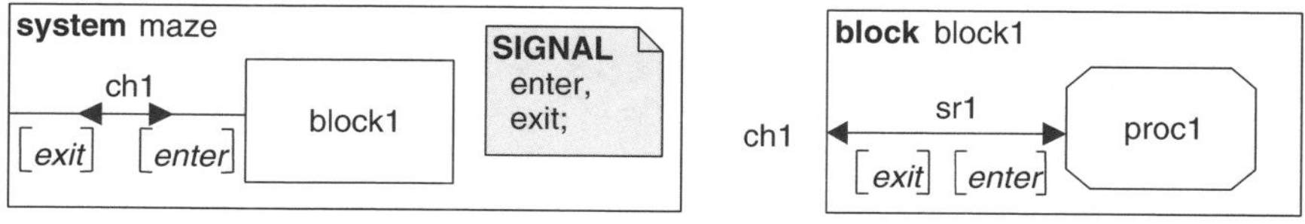

Figure 7.70 System *maze* and block *block1*

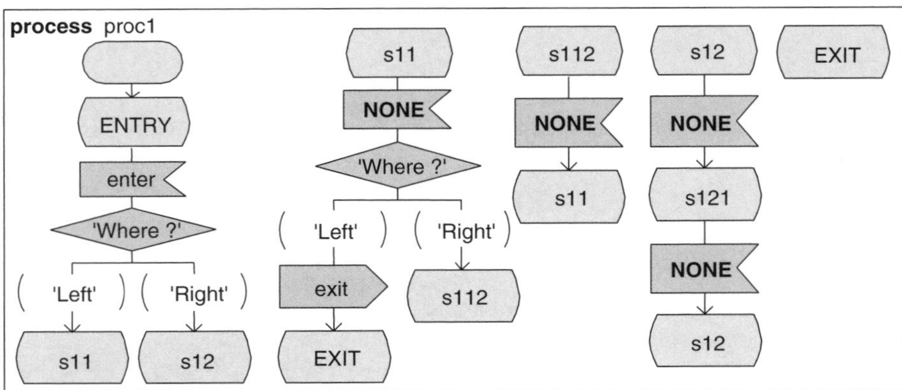

Figure 7.71 Process *proc1* modeling the maze

We have also created an MSC named *OK* represented in Figure 7.72, to observe the SDL model: after the input of signal *enter*, meaning that someone has entered the maze, the MSC waits for the output of signal *exit*, meaning that the person has found the exit. The *MSC Simulation Property* of this MSC is set to *verify* (in the MSC Editor) to detect errors and successes.

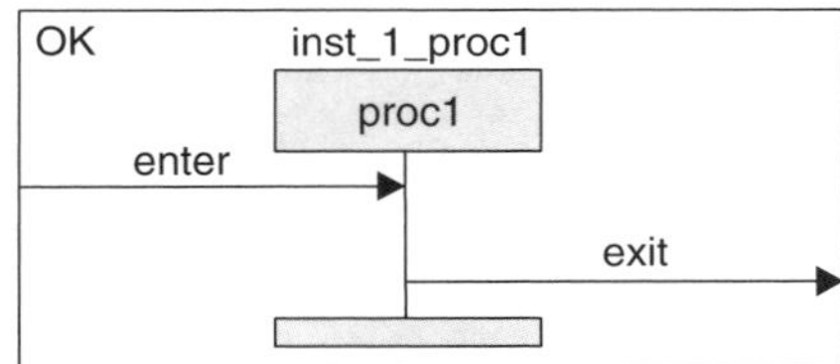

Figure 7.72 MSC to observe the maze model

If we run an exhaustive simulation in *depth* mode (*breadth* does not detect livelocks) of the model *maze* observed by the MSC *OK*, we get the following results:

```
Number of scc : 6
Number of transitory scc : 4
Number of sink scc : 1
5 intra-scc edges, 5 inter-scc edges

Number of states : 9
Number of transitions : 10
Maximum depth reached : 5
duration : 0 mn 0 s

Number of exceptions : 0
Number of deadlocks : 1
Number of stop conditions : 0
...
Number of errors : 0
Number of success : 1
```

The Simulator has discovered no errors, one success (generated in the file *maze.ok.s1.scn*), one deadlock (generated in the file *maze.d1.scn*) and one livelock (sink scc[3]) (generated in the file *maze.sc3.scn*).

By loading the file *maze.ok.s1.scn* into the Simulator and replaying it, the MSC trace shown in Figure 7.73 has been generated: it matches the expected signal sequence represented in Figure 7.72. The states reached are displayed: *entry*, *s11* and *exit*.

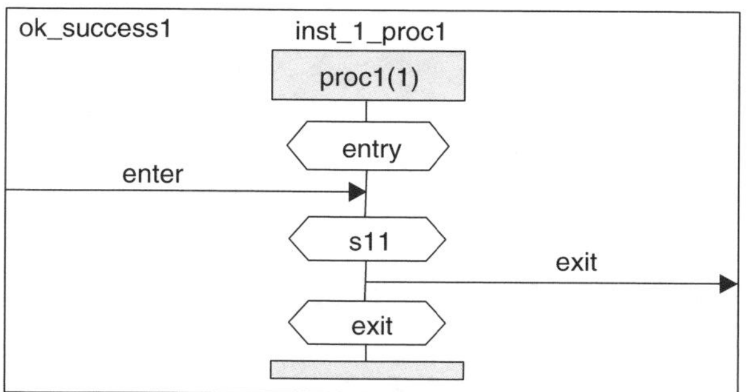

Figure 7.73 The success scenario detected

By loading the file *maze.sc3.scn* into the Simulator and replaying it, the MSC trace shown in Figure 7.74 has been generated: it shows the path leading to the livelock: states *entry* and *s12*. Once state *s12* is reached, it is impossible to exit from the loop between states *s12* and *s121* (as in the real maze).

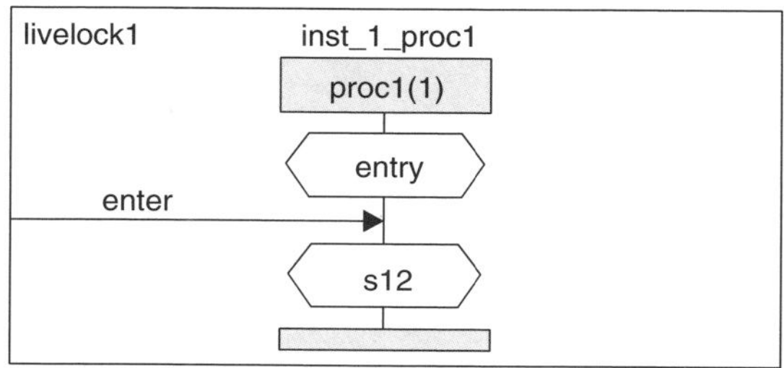

Figure 7.74 The scenario detected, leading to the livelock

The deadlock detected is normal because there is no transition to exit from state *exit* (Figure 7.75).

Now, lets try the liveness mode: choose *Execute* > *Verify* and select *Liveness* as shown in Figure 7.76.

We get the following results:

```
Number of states : 9
Number of transitions : 10
```

[3] Strongly connected component.

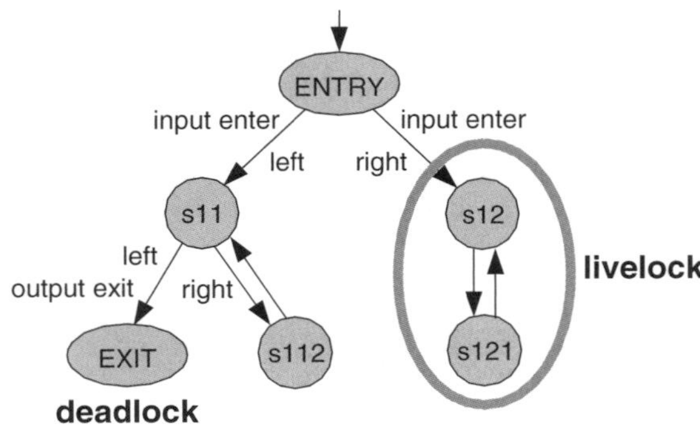

Figure 7.75 The deadlock and livelock detected

Verify Options
Mode:
Breadth Depth Supertrace Liveness
Scenarios:
Deadlock limit: 2 halt
Exploration:
States Limit: 0
Depth Limit: 0 stop

Figure 7.76 The top of the *Verify Options* window in liveness mode

```
Number of sub-exploration : 2
Maximum depth reached : 5
duration : 0 mn 0 s
Number of exceptions : 0
Number of deadlocks : 1
Number of stop conditions : 0
Transitions coverage rate : 100.00 (0 transitions not covered)
States coverage rate : 100.00 (0 states not covered)
Basic blocks coverage rate : 100.00 (0 basic blocks not covered)
Number of errors : 0
Number of non-success loops: 2
```

The Simulator has discovered two nonsuccess loops, generated in the files *maze.l1.scn* and *maze.l2.scn*.

By loading the file *maze.l2.scn* into the Simulator and replaying it, the MSC trace shown in Figure 7.77 has been generated: this nonsuccess loop corresponds to the livelock detected in *depth* mode, shown in Figure 7.75. The execution can continue forever, trapped between states *s12* and *s121*, never reaching any success state.

By loading the file *maze.l1.scn* into the Simulator and replaying it, the MSC trace shown in Figure 7.78 has been generated: this nonsuccess loop is shown in the left part of Figure 7.79. As for the previous nonsuccess loop, the execution can continue forever between states *s11* and *s112*, never reaching any success state. But this nonsuccess loop had not been detected in

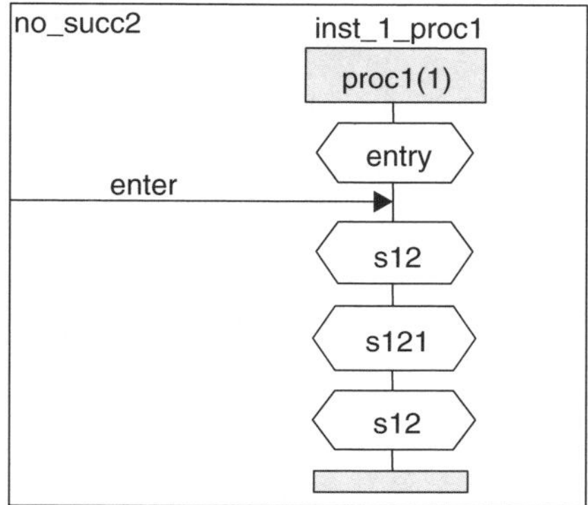

Figure 7.77 The second nonsuccess loop detected

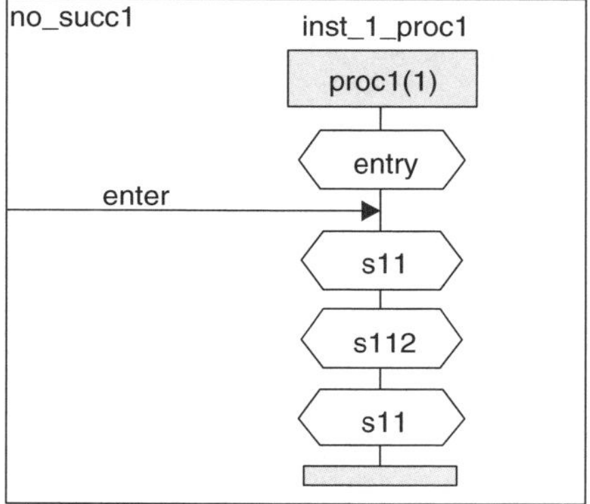

Figure 7.78 The first nonsuccess loop detected

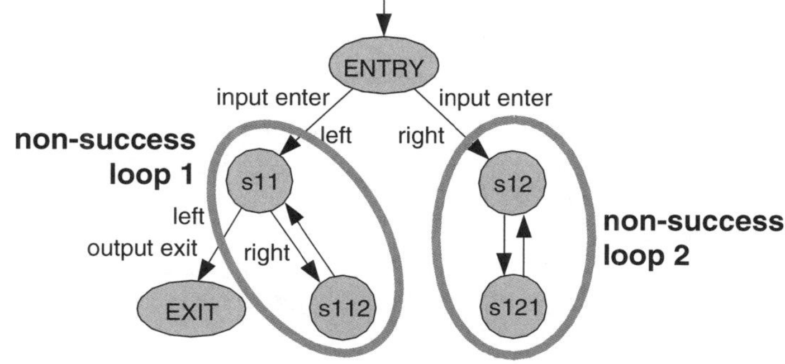

Figure 7.79 The two nonsuccess loops detected

depth mode because it does not correspond to any livelock: from *s11* the system can exit from the loop.

Examples of applications where liveness is helpful:

- for a communication protocol, to prove that any data transmitted are received, or if they were not transmitted the sender has been informed (this has been done with VEDA2, a tool based on the Estelle language, in [Doldi92]);
- for an aircraft or a car, to prove that several on-board systems always synchronize upon power-up or that the reconfiguration mechanism after a failure always works; this is very difficult to check manually because such systems are complicated by the requirement to remain operational after one (hard or soft) failure and safe after a second failure; as such systems especially in aircraft are safety-critical, a bug in the failure detection and recovery mechanisms could provoke a crash.

7.5.3.3 Application to V.76

If you simulate in *liveness* mode the V.76 SDL model observed by the MSC represented in Figure 7.80 (with the MSC Simulation property set to *search*), the simulator detects nonsuccess loops.

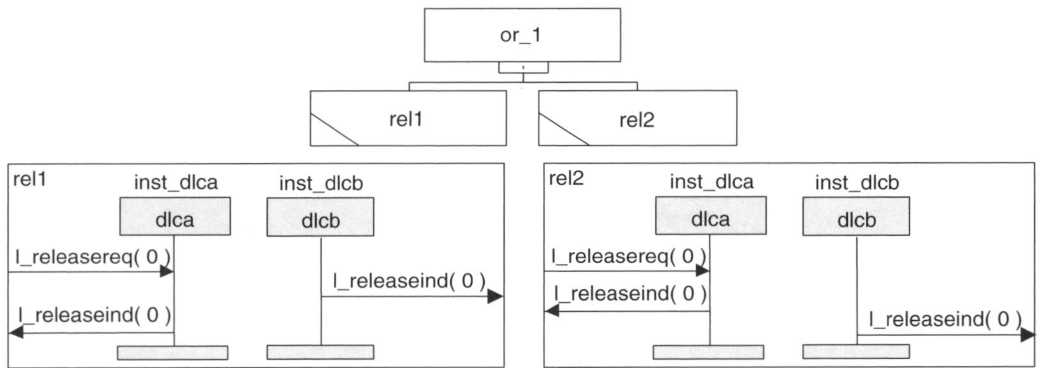

Figure 7.80 The MSC to observe the disconnection

Compile the SDL model plus the MSC, and launch the Simulator. Replay the scenario *cnx1.scn*. Set *Reasonable environment* to on and *Loose time progression* to off. Delete all the feeds except the *L_ReleaseReq(0)* on side A. Run the *liveness* simulation. After one second, it stops and you get three nonsuccess loops. Source *v76_feed.wri* to restore the feeds, and start the MSC trace. Load and replay the generated scenario *v76.l1.scn*: you see that after a UA response is lost by *dataLink*, process *DLC* side A is stuck in state *waitUAdisc*.

To correct the protocol, a timer could be added, or the disconnected state could be entered before receiving the response of the peer.

7.6 STRATEGY TO MASTER EXHAUSTIVE SIMULATION

7.6.1 Which simulation modes should be used

In Tau SDL Suite Validator, use the bit-state mode, working better than supertrace in ObjectGeode, rather than exhaustive.

In ObjectGeode, use the true exhaustive simulation modes (breadth, depth or liveness) rather than supertrace (named bit-state in Tau SDL Suite Validator).

7.6.2 If simulation never terminates

You will certainly be faced with SDL models whose simulation in exhaustive or bit-state modes never terminates, because the model has too many global states. In exhaustive mode, when you see that the RAM memory of your computer is full, you can stop the simulation, because it becomes extremely slow. In bit-state, the memory is allocated at the beginning (the size of the bits array plus a few extra megabytes) and no longer increases.

7.6.2.1 Tau SDL Suite Validator

Lets suppose we use the bit-state exploration mode.

1. Do not activate the *Advanced* mode, where all kinds of events have the same priority.
2. Use the command *Define-Max-Input-Port-Length* to limit the number of signals in each process input queue to one (the default is two).
3. For each dynamically created process, there is no command to limit the process instance creations to a certain number (the command *Define-Max-Instance* limits the number of instances at a certain moment, but not the successive number of instances created: for example, if you use *Define-Max-Instance proc1 2*, the loop 'process proc2 creates an instance of proc1, proc1 stops (dies)' can be simulated 100 times, triggering the creation of 100 instances). However, it seems that the Validator avoids creating 100 different global states in such a scenario.
4. Reduce the number of instances of entities (block types etc.), and the number of repetitions such as retries to the minimum required for simulation.
5. Use the command *Define-Variable-Mode* to remove temporary variables (such as a variable receiving the input parameter of a signal, provided the value is not used outside the transition) or variables not influencing the behavior from the global states. This reduces the number of global states.
6. Limit the number of external signals and the number of test values they carry in their parameters. Try to simulate separately independent features.
7. Replace the transmission of external signals by one or several test processes representing external entities (such as a layer above a protocol): the behavior of such test stubs will be more realistic than external signals, generally reducing the number of global states.
8. Reduce the depth limit of the exploration, for example, to 1000 (the default limit is 100, but this is too small in general). In this case, a part of the states graph is not explored.

7.6.2.2 ObjectGeode Simulator

Lets suppose we use the exhaustive modes (breadth, depth, liveness).

1. Select *Edit > Configuration* and set *Reasonable environment* to on and *Loose time progression* to off.

2. Enter the command *define verify_stats true* to see the number of states for each process and each queue. Run again the simulation, and after one minute interrupt it to see which input queue must be limited: use the *filter* command as in the case study to limit the number of signals in each process queue to one or two.
3. For each dynamically created process, use the command *filter create process_name(n)* to limit the successive process instance creations to n (do not confuse this limit with the declaration process_name(0, k), which only prevents having k process instances simultaneously, but does not limit the number of successive creation and stop).
4. Reduce the number of instances of each entity (block types etc.) and the number of repetitions such as retries to the minimum required for simulation.
5. Remove the declaration of temporary variables: an extension to SDL specific to ObjectGeode allows to omit the declaration of variables local to a transition, such as a variable receiving the input parameter of a signal. For example, writing *sig1(x)* in an input symbol, *x* does not need to be declared, if not used outside of the transition. Therefore, temporary variables are not stored in each global state of the system during simulation, reducing the number of global states.
6. Limit the number of external signals (feed) and the number of values you transmit as their parameters. Try to simulate separately independent features.
7. Replace the feed commands by one or several test processes representing external entities (such as a layer above a protocol): the behavior of such test stubs will be more realistic than feed, generally reducing the number of global states.
8. Limit the depth of the exploration, for example, to 1000. In this case, the simulation is no longer exhaustive.

7.7 ERRORS DETECTABLE BY EXHAUSTIVE SIMULATION

In addition to the errors enumerated in Chapter 4, exhaustive simulation detects the errors described in this chapter.

7.7.1 Errors detected by Tau SDL Suite

Tau SDL Suite Validator can detect the following errors:

- Deadlocks.
- Nonprogress loops: a subset of livelocks where if the loop contains inputs or outputs, it is not considered as nonprogress – in the *maze* example, the two loops are detected only if the option *Define-Spontaneous-Transition-Progress off* is used, otherwise input NONE is considered as progress.
- Success: conformance to behaviors described by a rule, an MSC or an observer process.
- Errors: violation of behaviors described by a rule, an MSC or an observer process.
- Never-simulated symbols.

- Process queues overflow.
- Infinite number of global states (for models small enough to finish the simulation).

7.7.2 Errors detected by ObjectGeode

ObjectGeode can detect the following errors:

- Deadlocks.
- Livelocks (in *depth* mode).
- Nonsuccess loops (in *liveness* mode).
- Success: conformance to behaviors described by observers (stop conditions, MSCs or GOAL modules).
- Errors: nonconformance to behaviors described by observers.
- Never-simulated symbols.
- Process queues overflow (if a stop condition is used, otherwise the number of states for each queue is displayed in the simulation results if *define verify_stats true* is set).
- Infinite number of global states (for models small enough to finish the simulation).

8

Other Simulator Features

8.1 TAU SDL SUITE

8.1.1 Writing in the Simulator trace

To write a message in the Simulator trace (only if you do not use its graphical interface, i.e. if you launch the executable directly in a DOS or Unix shell), you can call the C function printf. Executing the example shown in Figure 8.1 produces the trace:

```
*** n = 0 ***
```

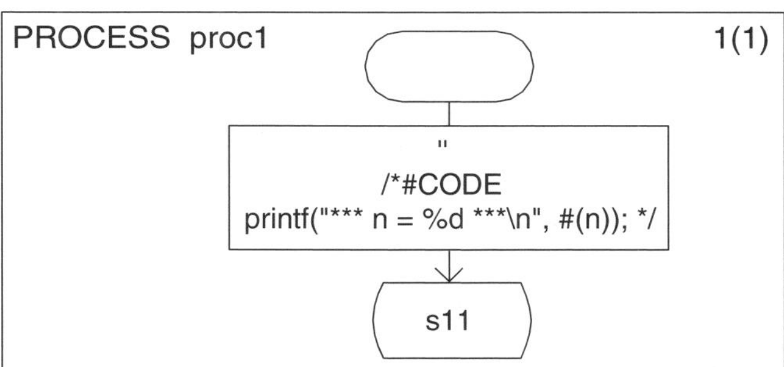

Figure 8.1 Calling printf

The task starts with two single quotes, to create an informal empty task. Then the C code is placed inside an SDL comment /* */. Note that the SDL variables in the generated code such as *n* are accessed by the expression #(*n*).

8.1.2 Calling external C code

8.1.2.1 Introduction

You may want to reuse existing C code: for example, in the V.76 SDL model, instead of writing a complex CRC (a kind of checksum) computation in the procedure *CRCok*, you could call an existing C function.

Validation of Communications Systems with SDL: The Art of SDL Simulation and Reachability Analysis.
Laurent Doldi © 2003 John Wiley & Sons, Ltd ISBN: 0-470-85286-0

The Tau SDL Suite Simulator provides several ways to call external C (or C++) code: SDL operators and SDL procedures can be implemented as C functions. The same interfacing mechanisms are provided in the C application generators.

By just inserting a *.h* file into the Tau SDL Suite Organizer, the H2SDL (or CPP2SDL) utility translates the C (or C++) definitions into an SDL package, which can be used in the SDL model.

8.1.2.2 Example of SDL procedure implemented as a C function

We will modify our SDL V.76 model to replace the SDL procedure *CRCok* by the C function *CRCok*.

A. Create a new directory, and copy all the files (except the MSCs) of the V.76 example into it.

B. Load *v76.sdt* in the Organizer.

C. If you added an observer process to the model as specified in Chapter 5, go back to the version without observer process: in the Organizer, select *Edit* > *Connect*, choose *To an existing file*, press the folder-shaped icon and connect to the file *v76test.ssy*.

D. In the Organizer, select the procedure *CRCok* and choose *Edit* > *Disconnect*, and press *Disconnect*.

E. Open the package *V76* and remove the procedure *CRCok*.

F. In the Organizer, press the *Save* button.

G. Open the process *DLC*, go to the page *part2*, and transform the call to *CRCok* as indicated in Figure 8.2: the return value becomes an Integer instead of a Boolean, to simplify the example.

H. With a text editor, create a file *my_c.h* containing the line shown in Figure 8.3.

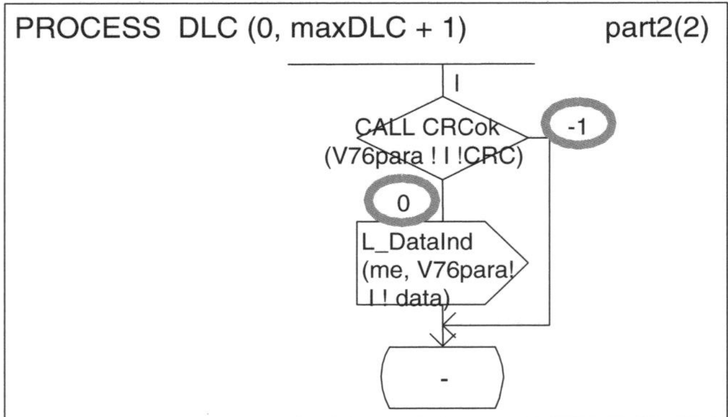

Figure 8.2 Modified call to procedure *CRCok*

```
extern int CRCok(int crc1);
```

Figure 8.3 The file *my_c.h*

I. In the Organizer, select the bar *Used Files* and press several times the *Move Up* button to move it above the bar *SDL System Structure*.

J. Select the bar *Used Files*, choose *Edit > Add Existing*, select *my_c.h* and press *Add*.

K. Open the package *V76* and add *use ctypes; use my_c;* as illustrated in Figure 8.4.

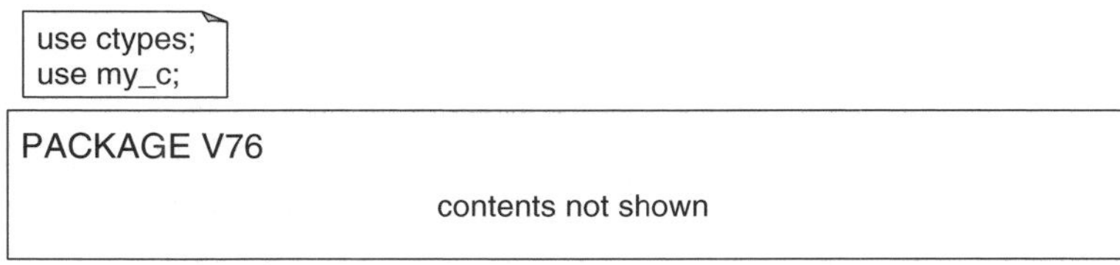

Figure 8.4 The package *V76* modified

The clause *use my_c* imports the SDL package *my_c* generated by H2SDL (or by CPP2SDL) because the file *my_c.h* has been added in the Organizer. The clause *use ctypes* imports the SDL package *ctypes* required to import C into the SDL model.

L. With a text editor, create a file *my_c.c* as shown in Figure 8.5. The function *CRCok* is defined, performing the same behavior as the SDL version. Here you could paste an actual CRC computation.

```
#include "my_c.h"

int CRCok(int crc1)
{
if (crc1<0) return -1;
else return 0;
}
```

Figure 8.5 The C function *CRCok* in the file *my_c.c*

M. With a text editor, create a file *my_c.tpm* containing the lines shown in Figure 8.6. This is a template makefile, to compile the file *my_c.c*. Remember that the three last lines begin with a tab character, not with spaces.

```
USERTARGET = my_c$(sctOEXTENSION)
my_c$(sctOEXTENSION): my_c.c
	$(sctCC) $(sctCPPFLAGS) $(sctCCFLAGS) \
	$(sctIFDEF) /Fomy_c$(sctOEXTENSION) \
	my_c.c
```

Figure 8.6 The file *my_c.tpm*

N. In the Organizer, select the system *V76test* and choose *Generate > Make*. Select *Generate makefile and use template* and enter *my_c.tpm* as indicated in Figure 8.7. Select *Microsoft* (or other) *Simulation*. Press *Full Make*: the executable *V76test_smc.exe* is generated.

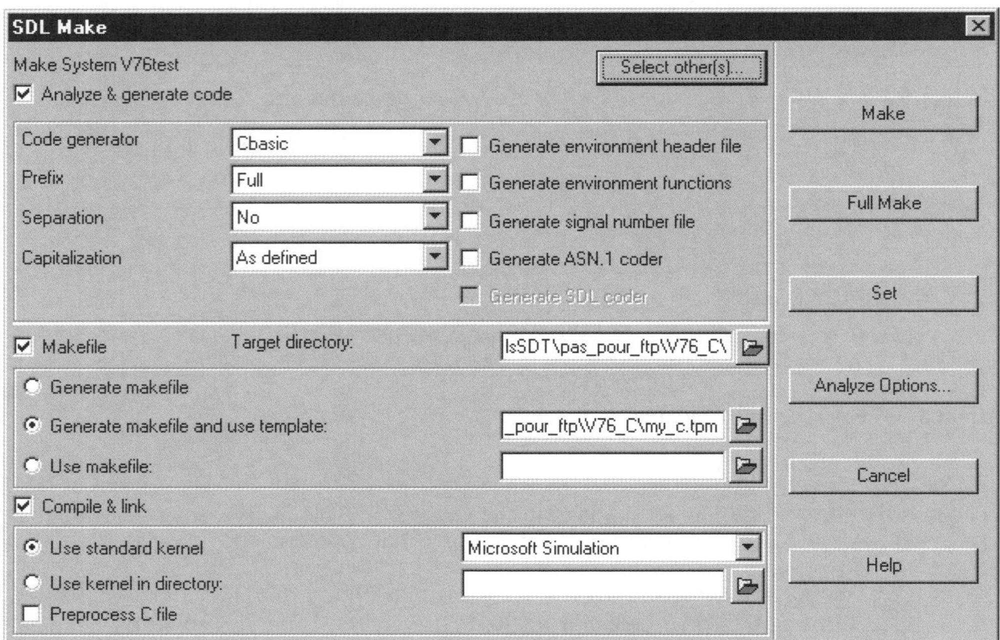

Figure 8.7 The SDL Make window set for simulation

O. The package *ctypes* has been added automatically to the Organizer. You can move it to the *Used Files* part, as shown in Figure 8.8.

P. In the Organizer, press the *Save* button.

Q. In the Organizer, press the *Simulate* button. Execute the command script *cnx1.com*, send signal *L_DataReq(0, 25)* to process *DLC* in block *DLCa*, press the button *Trace: SDL*, simulate using the *Symbol* button until the call to procedure *CRCok*, and check that the *zero* answer is executed, as the parameter passed is positive.

8.1.3 Simulating ASN.1 data types

The Simulator and the Validator accept SDL models whose types are based on external ASN.1 modules, as described in [Z105_2]. ASN.1 is more powerful than the SDL data types, for example, it allows the CHOICE construct (similar to a C union). In addition, several protocol standards use ASN.1 to describe data.

8.1.4 Adding buttons to the Simulator

8.1.4.1 The three Simulator definition files

As shown in Figure 8.9, the buttons and menus present in the Simulator are defined in the file *def.btns*. The content of the *Command* Window is defined in *def.cmds*, and the variables to display in a watch are defined in *def.vars*.

The names and location of each definition file can be changed in the Preferences Manager (from the Organizer). The Simulator loads the first of each file, searching in the following order: the current directory, the user's home directory and the installation directory.

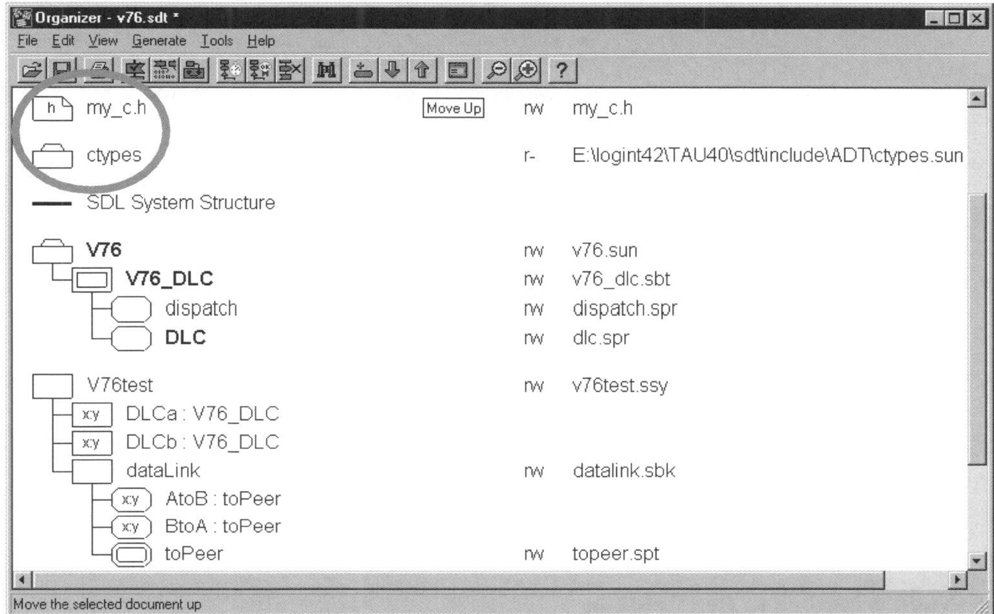

Figure 8.8 The Organizer after adding *ctypes*

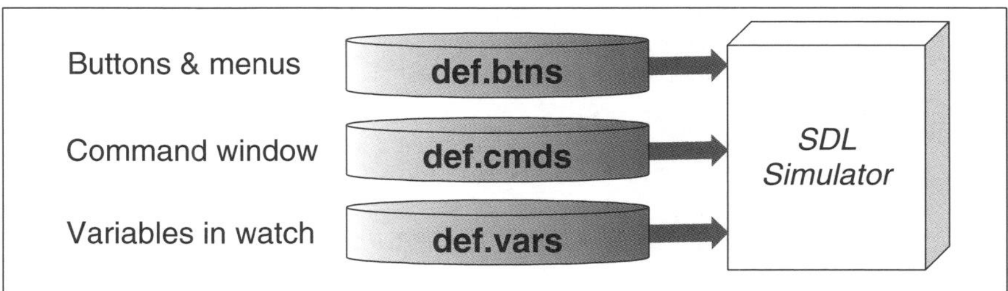

Figure 8.9 The three Simulator definition files

8.1.4.2 Adding four buttons to the Simulator

We are going to add two groups *Service User A* and *Service User B* to the Simulator. Then we will add four buttons to these groups, as shown in Figure 8.10.

A. In the Simulator, *select Buttons > Add Group*, enter *Service User A* and press *Apply*, enter *Service User B* and press *OK*.

B. Press the *button* Group in *Service User A* and select *Add*.

C. Type *EstabReq(0)* in the field *Label*, and *Output-To L_EstabReq(0) <<Block DLCa>> dispatch* in the field *Definition*. Press *Apply*.

D. Enter *DataReq(0,5)* in the field *Label*, and *Output-To L_DataReq (0, 5) <<Block DLCa>> dispatch* in the field *Definition*. Press *Apply*.

E. Enter *Release(0)* in the field *Label*, and *Output-To L_ReleaseReq(0) <<Block DLCa>> dispatch* in the field *Definition*. Press *OK*.

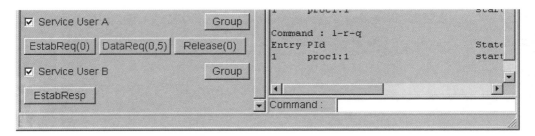

Figure 8.10 Four buttons added to the Simulator

F. Press the button *Group* in *Service User B* and select *Add*.

G. Type *EstabResp* in the field *Label*, and *Output-To L_EstabResp <<Block DLCb>> dispatch* in the field *Definition*. Press *OK*.

H. Save the buttons definitions: select *Buttons > Save As* and enter *def.btns*. At the next launch of the Simulator, you will automatically get the newly created buttons.

I. You can now test the new buttons to transmit signals to the SDL model.

8.1.5 Adding buttons to the Validator

The same principles as those for the Simulator apply to add buttons to the Validator. Note that the buttons defined in Section 8.1.4.2 cannot be used in the Validator, as the command *Output-To* does not exist. The file names are different: *def.btns* becomes *val_def.btns*, *def.cmds* becomes *val_def.cmds* and *def.vars* becomes *val_def.vars*.

8.1.6 Setting breakpoints in the Simulator

Breakpoints can be specified in the Simulator, to stop the simulation when a certain SDL symbol is reached, or on a transition, on a signal output, or on the modification of a variable. One or more Simulator commands can be executed automatically when a breakpoint is reached.

To set a breakpoint on the input of signal *V76frame* in process type *toPeer*:

A. Start the Simulator on the V.76 model.

B. In the Simulator, select *Breakpoint > Connect sdle*. A new *Breakpoints* menu appears in the SDL Editor.

C. In the Editor, open the process type *toPeer*, select the input of signal *V76frame* and do *Breakpoints > Set Breakpoint ...*: in the *Prompt* window, enter *ex-pid*. A red stop symbol appears near the input, as shown in Figure 8.11.

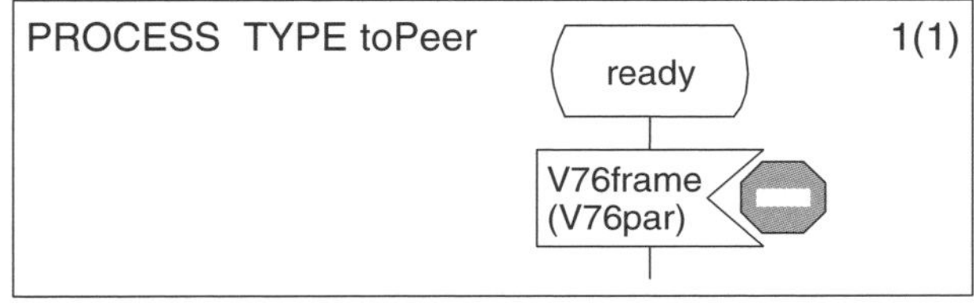

Figure 8.11 A breakpoint on an input symbol

D. In the Simulator, press on *Send To*, select signal *L_SetParmReq*, choose *<<Block DLCa>> dispatch*, and press on *Go*: when the breakpoint is reached, the execution stops, and the specified command *ex-pid* (*Examine-Pid*) is executed:

```
Breakpoint matched at #SDTREF(SDL, topeer.spt(1),
  140(35,35),1,1)
Instance of process type : toPeer

Parent    : null
Offspring : null
Sender    : <<Block DLCa>> dispatch:1
PId         AtoB:1                        in state ready
```

8.1.7 Running several communicating Simulators

Several Simulators can be started, and they can exchange SDL signals. For example, our V.76 SDL model can be split, and each instance of block type *V76_DLC* can run in one Simulator and communicate with its peer instance running in the other Simulator.

A. Create a new directory, and copy all the files (except the MSCs) of the V.76 example into it.

B. Load *v76.sdt* in the Organizer.

C. If you added an observer process to the model as specified in Chapter 5, go back to the version without observer process: in the Organizer, select *Edit* > *Connect*, choose *To an existing file*, press the folder-shaped icon and connect to the file *v76test.ssy*.

D. Open the system *V76test* and remove blocks *DLCb* and *dataLink* as indicated in Figure 8.12.

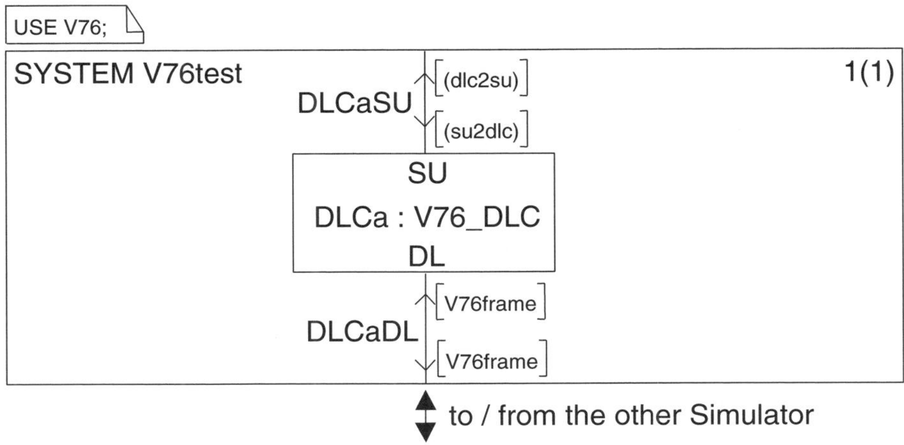

Figure 8.12 System *V76test* split in two

E. In the Organizer, press the *Save* button, select the system *V76test* and select *Generate* > *Make*. Check that the options are correct, especially: *Microsoft* (or other) *Simulation* and *Generate Makefile* must be selected.

F. Press *Full Make*.

G. In the Organizer, press the *Simulate* button. The first Simulator starts.

H. In the Organizer, select *Tools > SDL > Simulator UI*. The second Simulator starts: select *File > Open* and choose *V76test_smc.exe* (or the equivalent Unix suffix).

I. In each Simulator, select *General > Start SDL Env*: the two Simulators, represented in Figure 8.13, can now exchange SDL signals (through the PostMaster).

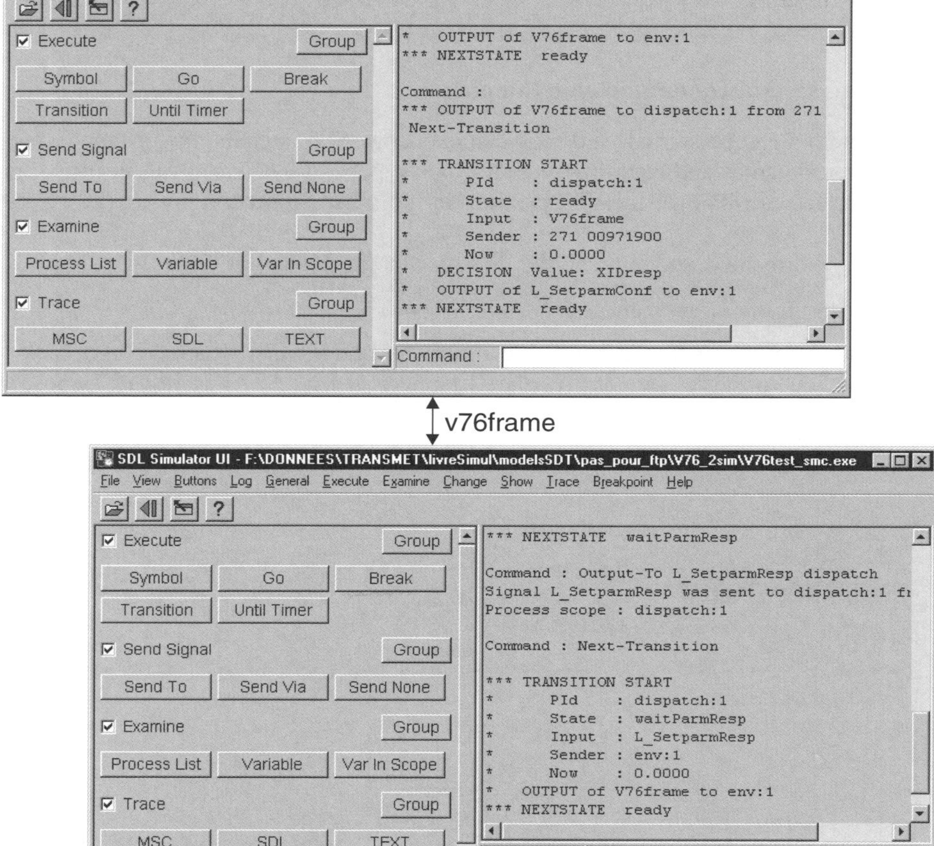

Figure 8.13 The two Simulators after the XID exchange

J. Simulate an XID exchange: in the first Simulator, transmit an *L_SetParmReq* to process *dispatch* (using the button *Send To*), execute the transitions: the signal *v76frame* is transmitted to the second Simulator. Execute the transitions to input it; after output of *L_SetParmInd*, transmit an *L_SetParmResp* to process *dispatch* (always in the second Simulator), execute the input transition: the signal *v76frame* is transmitted back to the first Simulator. In the first Simulator, you see the reception of *v76frame*. Execute the transitions to input it: an *L_SetParmConf* is transmitted. The XID exchange is finished.

8.1.8 Real-time simulation

During simulation, the Simulator uses discrete time; after starting an SDL timer with a value of five minutes, you are not forced to wait five minutes before its timeout: you can trigger the timeout immediately. Note that the time unit is tool-dependent, here it is seconds (on Windows).

If necessary, for example, to run the Simulator connected to a real system, the Simulator can run in real time: if you set a timer with a value of five minutes, it will actually timeout after five minutes.To use real-time simulation, you must change the kernel used in the Organizer Make window, as shown in Figure 8.14.

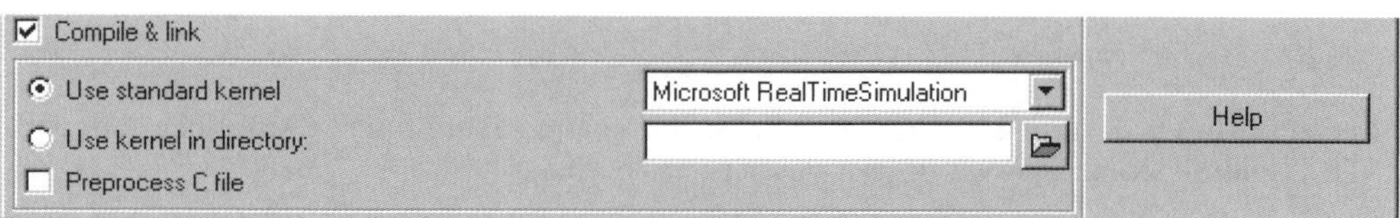

Figure 8.14 Selecting real-time simulation in the Make window

Then, if you set a ten-second SDL timer, you will see that the timeout occurs ten seconds after pressing the Go Simulator button.

8.1.9 List of Validator options

Most of the important Validator options can be changed using Validator menus, but a few require to type a textual command. To avoid using menus to change settings each time the Validator is invoked, it is handy to put the corresponding textual commands into the Validator startup file, *valinit.com*.

The following sections are presented in the same order as the result of the command *show-options*.

8.1.9.1 Bit-state options

- *Hash table size: 1000000 bytes*: number of bytes in the hash table (1 byte can store 8 states). Must be (in states or bits) 50 or 100 times the number of global system states to minimize the collision risk.
- *Search depth: 100*: defines the maximum exploration depth; when reached, the Validator backtracks to explore other transitions, the graph is truncated.
- *Iteration Step: 0*: do not use.

8.1.9.2 Exhaustive search option

- *Search depth: 100*: defines the maximum exploration depth; when reached, the Validator backtracks to explore other transitions, the graph is truncated.

8.1.9.3 Random walk options

- *Search depth: 100*: defines the maximum simulation depth; when reached, the random simulation stops, returns to the initial state and begins a new random simulation.
- *Repetitions: 100*: number of times the random simulation is restarted after reaching the search depth. With 100, and a search depth of 100, the maximum number of transitions executed will be 10000.

8.1.9.4 Tree search options

- *Search depth: 100*: defines the maximum simulation depth.

8.1.9.5 Power walk options

As this algorithm is designed for TTCN test case automatic generation for Autolink, to maximize the SDL symbols coverage, we do not describe its options.

8.1.9.6 Autolink options

As Autolink is the tool used for TTCN test case automatic generation, we do not describe its options.

8.1.9.7 Event priorities

This is used to change the priority between events during simulation. The default is maximum priority (1) for internal events (inputs and outputs etc.) and channel output (if channel queues are used), and lower priority (2) for input from ENV (input of external signals), timeout events and spontaneous transition (input NONE). The menu command *Options1 > Advanced* assigns the same priorities to all events, leading to richer behaviors but larger number of global states.

```
Event priorities:              1 2 3 4 5
Internal events              : 1 0 0 0 0
Input from ENV               : 0 1 0 0 0
Timeout events               : 0 1 0 0 0
Channel output               : 1 0 0 0 0
Spontaneous transition       : 0 1 0 0 0
```

8.1.9.8 State space generation options

- *Max input port length: 3*: maximum number of signals present in each process input queue.
- *Max symbols in transition: 1000*: maximum number of symbols in a transition, useful to detect infinite loops within a transition.
- *Scheduling: First*: the simulation uses a ready queue to execute first the oldest transition. If changed to *all*, a more recent transition can be executed first, generally leading to richer behaviors but larger number of global states.

- *Transition: SDL*: do not use.
- *Max no of instances: 100*: do not use (seems equivalent to specifying the maximum number of instances in the process reference in the SDL model).
- *Spont. transition progress: On*: means that a spontaneous transition (input none) is considered as progress during nonprogress loop detection.
- *Timer progress: On*: means that a timer expiration is considered as progress during non-progress loop detection.
- *Timer check level: 1*: when verifying an MSC, 0 means that checking of timer events is not performed, 1 means that checking of timer events is performed, but a timer event can be missing in the MSC, and 2 means that checking of timer events is performed and that timer events in the SDL model and in the MSC must match exactly.
- *MSC condition check: Off*: conditions in MSC are ignored.
- *Symbol time: Zero*: if the SDL model is not blocked, a set timer cannot timeout. If set to *undefined*, as soon as a timer is set, it can be timed-out.
- *Max state size*: 100000: do not change.

8.1.9.9 Report actions

The menu command *Options1 > Report* can be used to change the default actions after the detection of an event triggering a report. The default is *Prune – Log one* for each event.

For example, by default, after reaching a state where the maximum number of signals in an input queue is exceeded, the simulation does not continue beyond this state (*prune* means cut), and backtracks to explore another state, if any. The other values are *abort*, meaning that the simulation will stop, or *continue*, meaning that the simulation will continue beyond this state and explore its successors.

Log one means that even if the event occurs more than once, only one report will be generated. The other values are *log all*, where each occurrence of the event will generate a report, and *log off*, where no report is generated.

Here is the list of reports sorted alphabetically, and their meaning:

- *Assertion*: the predefined procedure *Report* has been called (which calls the function *xAssertError*), generally from an observer process. See the observer process example.
- *ChannelOutput*: error in channel output, rarely used.
- *Choice*: error in CHOICE.
- *Create*: error in create.
- *Deadlock*: deadlock detected.
- *Decision*: error in a decision.
- *ImplSigCons*: implicit signal consumption, no input or save specified for a received signal.
- *Import*: error in import.
- *Index*: array index out of range.

- *Loop*: loop in the states graph.
- *MaxQueueLength*: the maximum number of signals allowed in a process input queue has been exceeded.
- *MaxTransLen*: infinite loop within a transition.
- *MSCVerification*: the MSC has been verified (the simulated behavior matches the MSC).
- *MSCViolation*: the MSC has been violated.
- *Observer*: an observer process has not been able to execute a transition.
- *Operator*: error in an SDL predefined operator.
- *Optional*: error in an ASN.1 optional field.
- *Output*: error in an output.
- *PowerWalk*: reports used for MSC and TTCN test cases generation (Autolink).
- *RefError*: concerns SDL models using pointers (an extension provided by Tau).
- *Subrange*: range overflow in a syntype value.
- *TreeWalk*: reports used for MSC and TTCN test cases generation (Autolink).
- *UnionTag*: error in an ASN.1 union.
- *UserRule*: a user-defined rule is satisfied.
- *UserSpecified*: not found in the Validator documentation.
- *View*: error in view-revealed.

8.1.9.10 MSC verification options

- *Search depth: 1000*: the maximum exploration depth (number of transitions) executed to verify an MSC. When the depth is reached, the Validator backtracks and explores other transitions, if any.
- *Search mode: Violation*: after trying this option, its effect has not been found.
- *Algorithm: BitState*: by default, a bit-state exploration is performed to try to verify the MSC. The other alternative is *TreeSearch*.
- *Ignore parameters: Off*: if set to *On*, MSC parameters are ignored.

8.1.9.11 Other options

- *Report viewer autopopup: On*: the Report Viewer will automatically appear after the end of the Validation.
- *MSC trace autopopup: On*: the MSC Editor will automatically appear.
- *MSC trace states: On*: the state of SDL processes is written in the MSC trace.

- *MSC trace actions: Off*: the actions (tasks etc.) executed are written in the MSC trace.
- *MSC trace channels: Off*: the env instance is not split into one instance for each channel connected to env in the MSC trace.

8.2 OBJECTGEODE

8.2.1 Writing in the Simulator trace

To write a message in the Simulator trace window, the ObjectGeode-specific *write* and *writeln* procedures can be called, from a procedure call symbol (if you use a task symbol, you will get an error). Executing the example shown in Figure 8.15 produces the trace:

```
*** n = 0 ***
```

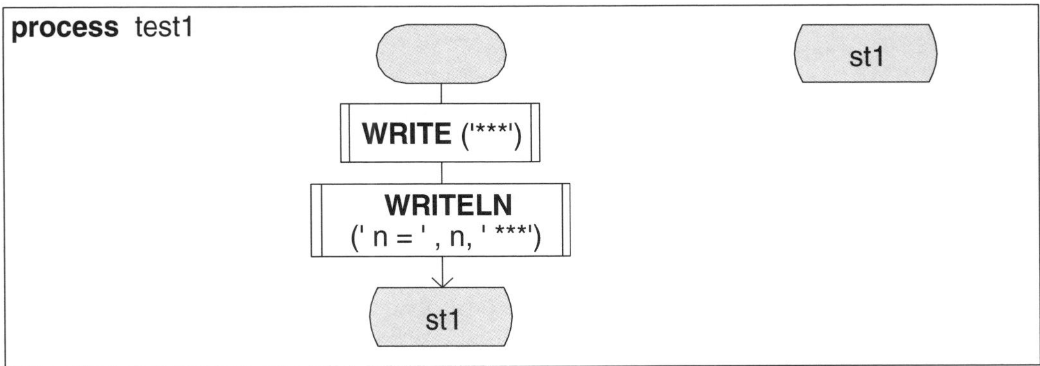

Figure 8.15 Calling *write* and *writeln*

8.2.2 Calling external C code

8.2.2.1 Introduction

You may want to reuse existing C code: for example, in the V.76 SDL model, instead of writing a complex CRC (a kind of checksum) computation in the procedure *CRCok*, you could call an existing C function.

The ObjectGeode Simulator provides several ways to call external C (or C++) code: SDL operators and SDL procedures can be implemented as C functions. The same interfacing mechanism is provided in the ObjectGeode C application generator.

8.2.2.2 Example of SDL procedure implemented as a C function

We will modify our SDL V.76 model to replace the SDL procedure *CRCok* by the C function *crcok*.

A. In a new empty directory, make a copy of any version of *v76.pr* (copy also *v76.startup* and the files it loads), and load it into the SDL Editor.

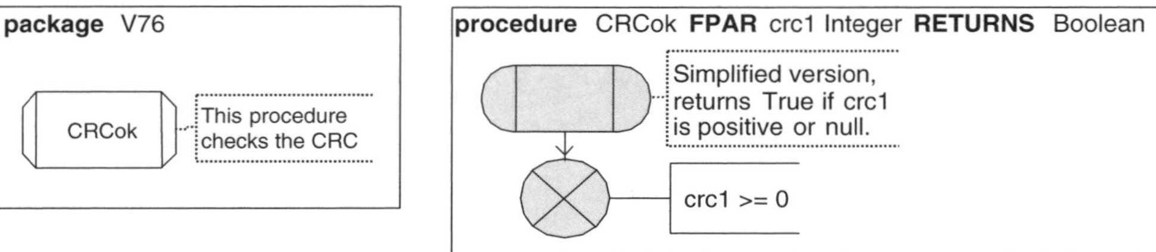

Figure 8.16 The procedure *CRCok* in SDL

B. In the package *V76*, delete the procedure *CRCok* shown in Figure 8.16.

C. In the package *V76*, create a text symbol containing the external declaration of procedure *CRCok*, as shown in Figure 8.17.

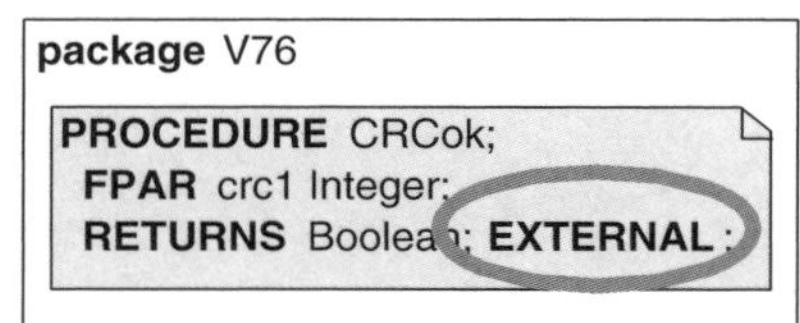

Figure 8.17 The procedure *CRCok* declared external

D. Select *Tools > SDL & MSC Simulator*, and check that the working directory is correct.

E. In the ObjectGeode Launcher, remove any file other than *v76.pr*, and press the *Build* button: as the keyword *EXTERNAL* has been used in the procedure declaration, the Simulator has generated the file *v76.h*, shown in Figure 8.18, containing the declaration of *crcok*.

```
#ifndef CODE_INCLUDE_INCLUDED
#include "code_include.h"
#endif
#ifndef GEODESM2
This_file_can_only_be_used_with_ObjectGEODE_Simulator
#endif
#ifndef GEODESM_EXPORT
#define GEODESM_EXPORT
#include "hpredef.h"
typedef char SDL_CHARSTRING[GX_STRMAX];
struct _SDL_CHARSTRING_struct { SDL_CHARSTRING a; };
extern _SDL_CHARSTRING _SDL_CHARSTRING_empty;
#define SDL_CHARSTRING_empty (_SDL_CHARSTRING_empty.a)

extern SDL_BOOLEAN crcok (SDL_INTEGER crc1);
#endif
```

Figure 8.18 The generated file *v76.h*

F. Copy the file *v76.h*, rename it *my_c.c*, and modify it to obtain the content shown in Figure 8.19. The function *crcok* is defined, performing the same behavior as the SDL version. Here you could paste an actual CRC computation.

```
#include "v76.h"

/* Test of the CRC, simplified version: if
   parameter crc1 is negative, the CRC is
   incorrect else it is correct: */
SDL_BOOLEAN crcok (SDL_INTEGER crc1)
{
if (crc1<0) return SDL_FALSE;
else return SDL_TRUE;
}
```

Figure 8.19 The file *my_c.c*

G. Create the file *v76.lub* (list of user binaries) as shown in Figure 8.20.

```
my_c.obj
```

(a)

```
geodecc_sm my_c.c
```

(b)

Figure 8.20 The files (a) *v76.lub* and (b) *geodesm_ubld.bat*

H. Create the file *geodesm_ubld.bat* (*geodesm_ubld.cmd* in Unix) as shown in Figure 8.20.

I. In the ObjectGeode Launcher, press again the *Build* button: the Simulator compiles the C code generated from the SDL model, the file *my_c.c*, and generates the executable *v76.sim* (the file *hpredef.h* is also generated).

J. Press the *Execute* button: the Simulator executes the SDL model plus the C function *crcok*.

8.2.3 Simulating ASN.1 data types

The Simulator accepts SDL models containing ASN.1 data types definitions, as described in [Z105_1], with a few minor restrictions. ASN.1 is more powerful than the SDL data types, for example, it allows the CHOICE construct (similar to a C union). In addition, several protocol standards use ASN.1 to describe data.

8.2.4 Adding buttons to the Simulator

To simplify repetitive commands, buttons can be added to the Simulator interface. The default button definitions are in the file *geodesim.but* located in the installation directory of Object-Geode, as shown in Figure 8.21.

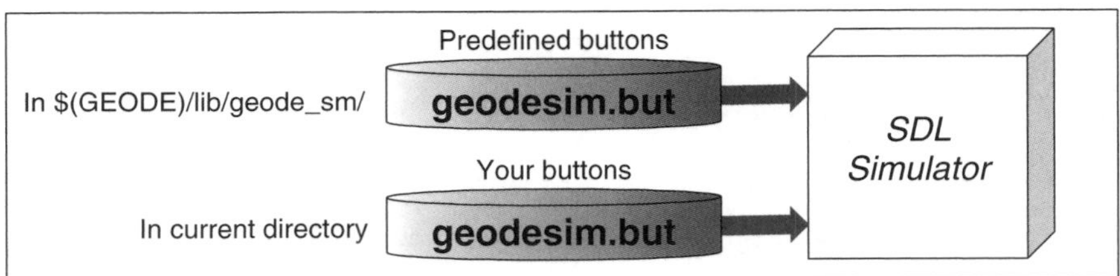

Figure 8.21 The Simulator button definition files

If a file named *geodesim.but* is found in the current directory, the Simulator loads button definitions from it, and will not read the file *geodesim.but* in the installation directory.

If you put the file depicted in Figure 8.22 into your current directory, the Simulator will have three more buttons, as shown in Figure 8.23.

```
! To get also the Simulator standard buttons:
include "$(GEODE)/lib/geode_sm/geodesim.but"

Panel main
{
"_____________" label
"V.76 test" label

! Inits the Simulation and plays start.scn:
"Re-init V.76" cmd init; source start.scn

! Inits the Simulation and plays cnx1.scn:
"Connect DLC 0" cmd init; source cnx1.scn; print state

! Displays the name of the first signal in the queues:
"Queues head" cmd echo "Input queues head:"; print pr(1)!queue(1)!name \
    for all pr in process if length(pr(1)!queue)/=0

" " label
}
```

Figure 8.22 The file *geodesim.but*

Figure 8.23 Three buttons added to the Simulator

The first statement includes the file containing the standard button definitions, then three buttons specific to the V.76 model are added to the main Simulator panel. The first button reinitializes the model to Step 4, ready to begin a simulation, automatically executing the process start transitions. The second button places the model in a state where DLC number 0 is established. The last button displays the name of the first signal present in the queue of each process instance.

It is also easy to create a button opening a new panel with specific buttons, such as the panel opened by the button *Verify*.

8.2.5 Simulation scheduling like in Tau SDL Simulator and Validator

When an SDL model contains many process instances, the list of firable transitions in the Simulator can be sometimes long. This is compliant with the execution semantics of SDL.

The ObjectGeode Simulator (like the Tau SDL Validator when using the command *Define-Scheduling All*) does not use a ready queue like the Tau SDL Simulator, to propose only the oldest transitions, but proposes all the ready transitions at the same time.

To simplify the choice by reducing the number of transitions, a GOAL observer delivered with ObjectGeode, named *scheduling*, can be compiled with any SDL model.

To illustrate this, we have created the SDL model *test1*, represented in Figures 8.24 and 8.25. This model contains one process *TX* transmitting *sig1* to process *RX_1* and then *sig2* to process *RX_2*.

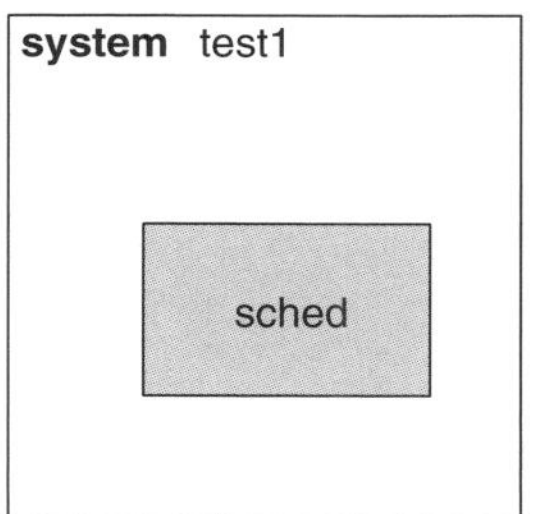

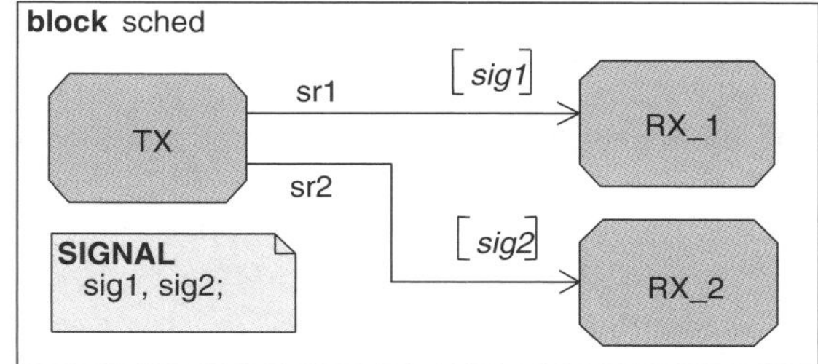

Figure 8.24 The system *test1*

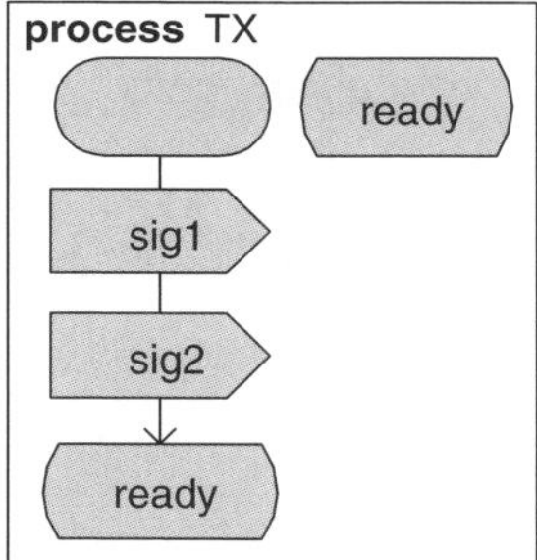

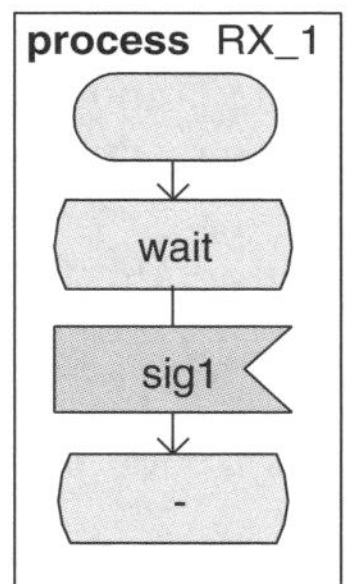

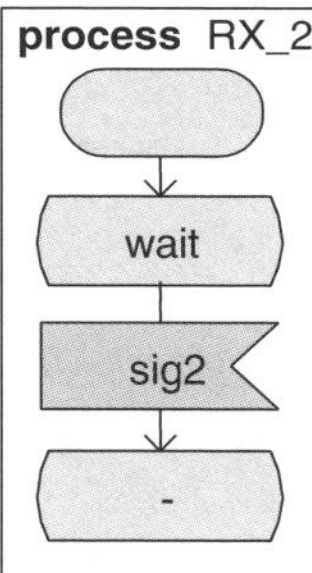

Figure 8.25 The three processes in block *sched*

By default, after executing process *TX*, the Simulator proposes two transitions:

- input of *sig1* by *RX_1* and
- input of *sig2* by *RX_2*

To use the *scheduling* observer:

A. Create the model *test1* with the SDL Editor (or load any model of your choice).

B. In the SDL Editor, select *File > Load*, change the *File type* to **.obs* and select *Og_sdl\examples\geode_sm\scheduling\scheduling.obs* in ObjectGeode installation directory.

C. Select *Tools > SDL & MSC Simulator*, and check that the working directory is correct.

D. In the ObjectGeode Launcher, press *Build* and then press *Execute*.

E. Execute the three start transitions (double-click on them). There are two firable transitions:

```
rx_1(1)  : from_wait_input_sig1
rx_2(1)  : from_wait_input_sig2
```

F. In the Simulator, type the textual command: *filter scheduling!filter*. There is now only one firable transition, corresponding to the first process that is ready (because it received a signal before process *RX_2*):

```
rx_1(1)  : from_wait_input_sig1
```

G. Execute the transition. As expected, the second transition appears:

```
rx_2(1)  : from_wait_input_sig2
```

8.2.6 List of Simulator settings

8.2.6.1 Define commands

Most of the important Simulator settings can be changed using Simulator menus, but others require to type a define command. To avoid using menus to change settings each time the Simulator is invoked, it is handy to put the corresponding define commands into a Simulator startup file.

The following list is the result of the command *define*, sorted alphabetically. A few defines listed below are absent from the Simulator documentation.

- *define alpha_order_trans 'true'*: if true, the firable transitions are sorted alphabetically.
- *define build_graph 'false'*: do not use.
- *define call_depth_limit '100'*: limits to 100 the number of recursive procedure calls.
- *define client_external_encoding 'true'*: do not use.
- *define client_external_env 'true'*: do not use.
- *define client_external_processes 'false'*: do not use.
- *define compose_unit '0'*: if not 0, a second level of global states compression is performed.
- *define compress_unit '0'*: if not 0, a second level of global states compression is performed.
- *define coverage_go 'false'*: do not use.
- *define coverage_list 'false'*: if true, the Simulator displays in front of each firable transition the number of times it has been simulated.
- *define depth_limit '0'*: defines the maximum depth during exhaustive simulation. If 0, no limit.
- *define depth_limit_stop 'false'*: if false, the exhaustive simulation explores another branch of the states graph when depth_limit is reached, otherwise the exhaustive simulation stops.
- *define edges_dump ' '*: if a *name* is specified between the single quotes, the transitions of the states graph are written into the file *name*. For training only.

- *define expand_limit '20'*: when calling an undefined SDL operator, if the return type has less than 20 values, then one firable transition will be proposed for each value, otherwise a window will prompt the user for the value.
- *define flush_log_file 'true'*: do not use.
- *define forced_interactive 'false'*: do not use.
- *define graph_file ' '*: do not use.
- *define hash_fill_limit '10'*: specifies the first (default) level of global states compression (no effect on supertrace). The second level is set by *compose_unit* and *compress_unit.*
- *define hash_size '1000'*: specifies the first (default) level of global states compression (no effect on supertrace).
- *define HOME 'C:\WINNT\Profiles\Administrateur\ObjectGEODE'*: the location of the current home directory.
- *define loose_time 'false'*: see Chapter 4.
- *define main_hash_size '100000'*: specifies the first (default) level of global states compression (no effect on supertrace).
- *define map_1param_signal_to_sequence 'false'*: reserved for TTCN test case generation.
- *define marglim '1000'*: used for performance simulation.
- *define max_lines_watch '1000'*: defines the maximum lines displayed in a watch window.
- *define MODEL 'ping'*: current SDL file name.
- *define msc_always_dynamic 'true'*: if true, the MSC trace is updated during automatic simulation (go, redo etc.), otherwise it is updated after the end of the execution.
- *define msc_fly 'true'*: if true, an MSC trace is created during the simulation.
- *define msc_global 'true'*: if true, time in the generated MSC trace is global, otherwise local to each instance.
- *define msc_xspace '140'*: horizontal space between two instances in the generated MSC trace; it seems that this option is no longer active, especially on the Windows version.
- *define msc_yspace '20'*: same as above for the vertical space between two signals.
- *define mscinst_by_event 'true'*: if true, the MSC instances are placed in their order of chronological appearance, otherwise the instances are placed according to the *msc for* command (i.e. if you type *msc for proc1, proc2*, if true you get *proc1* drawn on the left and *proc2* on the right, even if *proc2* is first to receive a signal). Very handy to avoid reordering instances after MSC generation.
- *define print_filter_condition_errors 'false'*: do not use.
- *define print_hook ' '*: if a name is entered, the user-defined printing operators (useful for opaque Abstract Data Types, especially if implemented in C) matching this name are executed. For example, if you define the type:

```
NEWTYPE opaque1
   OPERATORS
```

```
         print1: opaque1 -> Boolean;
      OPERATOR print1;
         FPAR p1 opaque1; RETURNS res Boolean;
            START;
            WRITELN('*** Hello world!');
            RETURN True;
      ENDOPERATOR;
   ENDNEWTYPE;
```

If you declare a variable *x* of type *opaque1* in the SDL model, when you enter the command *print x*, you get no result because the Simulator does not know what to print (because the NEWTYPE *opaque1* is neither a struct, nor a literals list, nor an array etc.). If you enter *define print_hook 'pr*'*, then typing *print x* activates the operator *print1* and **** Hello world!* is displayed.

- *define print_stop_condition_errors 'false'*: do not use.
- *define range_check 'false'*: if true, the range overflow is checked when the Simulator evaluates a command containing ranges.
- *define real_prec '6'*: number of digits after the decimal point for real numbers.
- *define reasonable_feed 'true'*: see Chapter 4.
- *define run_forever 'false'*: do not use.
- *define scc_sink_limit '2'*: maximum number of livelock scenario files generated during an exhaustive simulation, here 2.
- *define show_optionals 'false'*: prevents the Simulator from displaying the OPTIONAL fields (generally from ASN.1) that are not present in the value of the sequence.
- *define significance_level '0.05'*: used for performance simulation.
- *define states_dump ' '*: if a *name* is specified between the single quotes, the states of the states graph are written into the file *name*. For training only.
- *define states_limit '0'*: maximum number of explored (unique) global states. When the limit is reached, the exhaustive simulation stops. 0 means no limit. In *supertrace* mode, it specifies the number of bits in the hash table, which must be 50 or 100 times the number of global system states to minimize the collision risk.
- *define stop_cut 'true'*: if true, the states after a state where a stop condition is satisfied are not explored (cut is equivalent to prune in Tau Validator).
- *define tc_engine 'default'*: reserved for TTCN test case generation.
- *define time_horizon '10000.0'*: used for performance simulation.
- *define timescale '1'*: contains the value specified for the timescale option when launching the Simulator. Cannot be changed.
- *define tp_coverage '0%'*: reserved for TTCN test case generation.

- *define tp_coverage_limit '100'*: reserved for TTCN test case generation.
- *define tp_dir '.'*: reserved for TTCN test case generation.
- *define tp_interpretation 'complete'*: reserved for TTCN test case generation.
- *define tp_msc_gen 'false'*: reserved for TTCN test case generation.
- *define tp_obs_step 'true'*: reserved for TTCN test case generation.
- *define trace_stmt 'true'*: when true, the SDL statements are traced in PR (textual) form in the Simulator window.
- *define trans_events_limit '1000'*: used to detect infinite loops in the SDL model. Here, after 1000 events, the transition is considered infinite and an error is reported.
- *define trap_multiple_receiver 'true'*: when true, detects an error if a signal (or a remote procedure call) is transmitted and several process instances can receive it through the same path (channels and routes). When false, no error is raised, as specified in [SDL92]. If the signal can be received through different paths, no error is raised, as specified in [SDL92].
- *define trap_no_receiver 'true'*: when true, detects an error if a signal (or a remote procedure call) is transmitted and no receiver exists (for example, the process instance has stopped). When false, no error is raised and the signal is discarded, as specified in [SDL92].
- *define trap_unexpected_signal 'true'*: when true, detects an error if a signal is transmitted and no input exists for such signal in the current state of the receiver. When false, no error is raised and the signal is discarded, as specified in [SDL92].
- *define ts_controllable 'true'*: reserved for TTCN test case generation.
- *define ts_default_testcase 'DEF_0'*: reserved for TTCN test case generation.
- *define ts_language 'text'*: reserved for TTCN test case generation.
- *define ts_name 'ping'*: reserved for TTCN test case generation.
- *define ts_purpose_comment 'from state %s of %p, receive %i, send (%o) and go to state %f'*: reserved for TTCN test case generation.
- *define ts_test_groups 'false'*: reserved for TTCN test case generation.
- *define ts_test_steps 'false'*: reserved for TTCN test case generation.
- *define verify_stats 'true'*: when true, the number of states for each process and each input queue is displayed at the end of an exhaustive simulation. A must to detect which queues contain too many signals and must be limited using the filter command.
- *define watch_expand_depth '3'*: number of levels displayed expanded in a watch.
- *define windows 'true'*: true if Windows is used, otherwise false (Unix).

8.2.6.2 Other settings

A few Simulator settings are not define commands. They are displayed textually by typing the command *verify options*. As the define commands, they can be put into a Simulator startup file.

- *deadlock limit 2*: maximum number of deadlock scenario files generated during an exhaustive simulation, here 2.
- *error limit all 2*: same as previous for errors detected by MSC or GOAL observers.
- *exception limit 2*: same as previous for exceptions.
- *stop limit 2*: same as previous for stop conditions.
- *success limit all 2*: same as previous for success detected by MSC or GOAL observers.

Bibliography

Web sites

www.etsi.fr, ETSI: European Telecommunications Standards Institute.

www.itu.int, ITU: International Telecommunications Union.

www.sdl-forum.org, SDL Forum Society: information about SDL tools, training, events such as the SDL-Forum, SDL news electronic mailing list, etc.

perso.wanadoo.fr/doldi/sdl, the site of the author of this book.

Books

[Belina91] F. Belina, D. Hogrefe, A. Sarma, *SDL with Applications from Protocol Specification*, ISBN 0-13-785890-6, Prentice Hall International Ltd, 1991.

[Doldi01] L. Doldi, *SDL Illustrated, Visually Design Executable Models*, ISBN 2-9516600-0-6, TMSO, 2001.

[Holz91] G. J. Holzmann, *Design and Validation of Computers Protocols*, ISBN 0-13-539834-7, Prentice Hall Software Series, 1991.

[Reed94] R. Reed, A. Olsen, O. Færgemand, B. Møller-Pedersen, J. R. W. Smith, *Systems Engineering Using SDL-92*, ISBN 0-444-89872-7, Elsevier, 1994.

[Sari93] B. Sarikaya, *Principles of Protocol Engineering and Conformance Testing*, ISBN 0-13-012642-X, Simon & Schuster International, 1993.

[Telen00] Telenor, *Languages for Telecommunication Applications*, ISSN 0085-7130, No. 4-2000, Telektronikk Volume 96.

[Thiel01] A. M. Thiel, *Systems Engineering with SDL – Developing Performance-Critical Communication Systems*, ISBN 0-471-49875-0, John Wiley, 2001.

ITU recommendations

[MSC96] Z.120 (1996), Message Sequence Chart (MSC).

Validation of Communications Systems with SDL: The Art of SDL Simulation and Reachability Analysis.
Laurent Doldi ISBN: 0-470-85286-0

[SDL92] Z.100 (1993), Specification and Description Language (SDL), Version SDL-92.

[SDL00] Z.100 (1999), Specification and Description Language (SDL), Version SDL-2000.

[Meth97] Supplement 1 to Z.100 (05/97), SDL + Methodology.

[Z105_1] Z.105 (1995), SDL Combined with ASN.1.

[Z105_2] Z.105 (1999), SDL Combined with ASN.1 Modules (18 pages).

[Z106] Z.106 (1996), Common Interchange Format for SDL.

[Z107] Z.107 (1999), SDL with Embedded ASN.1 (3 pages).

[Z109] Z.109 (1999), SDL Combined with UML (41 pages).

[Z110] Z.110 (1996), Criteria for the use of Formal Description Techniques by ITU-T.

SDL Forum proceedings

[For87] SDL'87: State of the art and future trends, *Proceedings of the Third SDL Forum*, North Holland, Amsterdam, 1987.

[For89] SDL'89: The language at work, *Proceedings of the Fourth SDL Forum*, North Holland, Amsterdam, 1989.

[For91] SDL'91: Evolving methods, *Proceedings of the Fifth SDL Forum*, North Holland, Amsterdam, 1991.

[For93] SDL'93: Using objects, *Proceedings of the Sixth SDL Forum*, North Holland, Amsterdam, 1993.

[For95] SDL'95, *Proceedings of the Seventh SDL Forum*, North Holland, Amsterdam, 1995.

[For97] SDL'97: Time for testing, *Proceedings of the Eighth SDL Forum in Paris*, Elsevier, 1997.

[For99] SDL'99: The next millenium, *Proceedings of the Ninth SDL Forum in Montreal*, Elsevier, 1999.

[For01] SDL2001: Meeting UML, *Proceedings of the Tenth SDL Forum in Copenhagen*, ISBN 3-540-42281-1, Springer LNCS, 2001.

Papers

[Alga91] B. Algayres, L. Doldi, H. Garavel, Y. Lejeune, C. Rodriguez "VESAR: a pragmatic approach to formal specification and verification", *Computer Networks and ISDN Systems*, Special Issue on Tools for FDTs, Vol. 25, No. 7, North Holland, February 1993.

[Alga93] B. Algayres, Y. Lejeune, F. Hugonnet, F. Hantz, "The AVALON project: A VALidatiON environment for SDL/MSC descriptions", *SDL'93 Forum*, 1993.

[Com94] P. Combes, S. Pickin, B. Renard, F. Olsen, "MSCs to express service requirements as properties of an SDL model: application to service interaction detection", *SDL'95 Forum*, Oslo, 1995.

[Doldi92] L. Doldi (Verilog), P. Gauthier (DGAC/STNA), "VEDA 2: Power to the protocol designers", *FORTE 92, 5th International Conference on Formal Description Techniques*, 1992.

[Doldi95] L. Doldi (Verilog), F. Goudenove (Aerospatiale – Airbus), "Use of SDL to specify Airbus future air navigation systems", *SDL'95 Forum*, Oslo, 1995.

[Doldi96] L. Doldi, V. Encontre (Verilog), J.-C. Fernandez, T. Jeron (INRIA), S. Le Bricquir, N. Texier (Cap Sesa), M. Phalippou (CNET), "Assessment of automatic generation methods of conformance test suites in an industrial context", *IWTCS'96, 9th International Workshop on Testing of Communicating. Systems*, Darmstadt, 1996.

[Jard88] C. Jard, R. Groz, J.-F. Monin, "Development of VEDA: a prototyping tool for distributed algorithms", *IEEE Transactions on Software Engineering*, March 1988.

[Jard89] C. Jard, J.-M. Jezequel, "A multi-processor Estelle-to-C compiler to experiment distributed algorithms on parallel machines", Protocol Specification, Testing and Verification, IX, *Proc. IFIP WG6.1 9th International Symposium*, June 1989.

[Jeron91] T. Jeron, *"Contribution a la validation des protocoles: test d'infinitude et verification a la volee"*, These de Doctorat de l'Universite de Rennes, France, April 1991.

[Holz94] G. J. Holzmann, "Proving the value of formal methods", *7th International Conference on Formal Description Techniques*, Berne, Switzerland, 1994.

Index

Validation of Communications Systems with SDL: The Art of SDL Simulation and Reachability Analysis.
Laurent Doldi ISBN: 0-470-85286-0